ANNALS OF THE NEW YORK ACADEMY OF SCIENCES
Volume 365

THE THREE MILE ISLAND NUCLEAR ACCIDENT: LESSONS AND IMPLICATIONS

Edited by Thomas H. Moss and David L. Sills

The New York Academy of Sciences
New York, New York
1981

The cover photograph of the Three Mile Island nuclear plant was provided by Metropolitan Edison Company.

Library of Congress Cataloging in Publication Data

Main entry under title:

The Three Mile Island nuclear accident.

(Annals of the New York Academy of Sciences; v. 365)
Result of a conference held on April 8–10, 1980 by the New York Academy of Sciences.
1. Three Mile Island Nuclear Power Plant (Pa.) —Congresses. 2. Atomic power-plants—Pennsylvania—Accidents—Congresses. I. Moss, Thomas H. II. Sills, David L. III. New York Academy of Sciences. IV. Series.
Q11.N5 vol. 365 [TK1344.P4]   500s   81-38323
ISBN 0-89766-115-X   [363.1'79]   AACR2
ISBN 0-89766-116-8 (pbk.)

SP
Printed in the United States of America
ISBN 0-89766-115-X (Cloth)
ISBN 0-89766-116-8 (Paper)

ANNALS OF THE NEW YORK ACADEMY OF SCIENCES
VOLUME 365
April 24, 1981

THE THREE MILE ISLAND NUCLEAR ACCIDENT: LESSONS AND IMPLICATIONS*

Editors and Conference Organizers
THOMAS H. MOSS AND DAVID L. SILLS

◆

CONTENTS

*This volume is the result of a conference entitled The Three Mile Island Nuclear Accident: Lessons and Implications, held on April 8–10, 1980 by The New York Academy of Sciences.

Part IV. Societal Reaction to the Event: Public Reactions

Part V. Societal Reaction to the Event: Institutional Reactions

Part VI. Major Lessons and Issues

Financial assistance was received from:
- ELECTRIC POWER RESEARCH INSTITUTE
- THE FORD FOUNDATION
- NATIONAL SCIENCE FOUNDATION
- OFFICE OF NUCLEAR POLICY, U.S. DEPARTMENT OF ENERGY
- SANDIA LABORATORIES
- U.S. NUCLEAR REGULATORY COMMISSION

PREFACE

Thomas H. Moss

Subcommittee on Science, Research, and Technology*
U.S. House of Representatives
Washington, D.C. 20515

David L. Sills

Social Science Research Council
New York, New York 10016

The accident at the nuclear power plant at Three Mile Island (TMI), which began on March 28, 1979, has not gone unnoticed by students of technology and society. In addition to the reports of the president's commission and the Nuclear Regulatory Commission, there have been numerous official and unofficial reports, articles, and books.[1,2] A book coedited by one of us contains an extensive bibliography of social science reports.[3] The November 1979 issue of *Spectrum*, the journal of the Institute of Electrical and Electronics Engineers, is devoted entirely to Three Mile Island and represents only the beginning of an extended discussion in the technical literature. "Pennsylvania is everywhere!" became a slogan of the anti-nuclear-power student demonstrators in Europe; it is a slogan that also describes the desks and shelves of all students of modern technology.

This book, a report on an April 1980 conference on TMI, contains more than an account of the accident and its immediate consequences for the plant personnel, for the neighboring communities, for the nuclear power industry, and for the government agencies responsible for public safety. In the first place, it contains "one-year later" accounts by many of the major participants in the event, and thus constitutes a valuable form of personal testimony. In the second place, it seeks to generalize to the complex, population-threatening accidents that can—and indeed do— occur in many segments of our technology-based society. It thus also is about accidents in general.

Under the stress of population-threatening accidents such as that at Three Mile Island, two key questions are whether the relevant responsible institutions are able to respond promptly and wisely, and whether they are able to communicate effectively with each other and with the public. Will we be able to rely on the media to keep the public informed of the technical, political, and social issues raised by such accidents? Can we be sure that the media themselves are provided with accurate and timely information and are prepared to use the information effectively? Will the public be prepared to deal with the uncertainties, probabilities, and technical language used in describing the accident and its after-

---

*Dr. Moss is science advisor to this committee, not a member of Congress.

math? Will there be realizable plans to deploy communications facilities, medical personnel, and technical specialists? Will the numerous governmental and nongovernmental institutions responsible for these deployments be able to mesh their individual approaches into coherent and integrated activity? The Kemeny commission concluded that the accident at TMI was caused by "people-related problems and not equipment problems,"[1] and the Rogovin report stated that "the principal deficiencies in commercial reactor safety today are not hardware problems, they are management problems."[2] In social science language, individual confusion over role behavior contributed to a massive social systems failure.

An analysis of the accident and its social impact offers some tentative answers to these questions. One important observation concerns the impact of various forms of stress upon the anticipated behavior patterns of participants. The behavioral impact of stress began with the actions of the control room operators, extended to the behavior of officials at press briefings, contributed to the reaction of the public to the information provided, and ultimately shaped the actions of officials at the highest levels of government. Clearly, the impact of future technology-related accidents must be anticipated in the context of behavior under stress— behavior that cannot be envisioned in periods of calm.

Questions concerning communication, individual behavior, and organizational effectiveness constitute some of the major issues raised by the accident. In all instances, it seems clear that mere speculation about future behavior, effective institutional arrangements, and means to ensure optimal decisions is inadequate. Since deliberate experimentation on these matters is difficult or impossible, it is important to seize upon unplanned experiments as they occur. The analysis of an unplanned, or "natural," experiment is what this book is all about.

The book illustrates what can be learned from an interdisciplinary approach to technological issues. The editors, who cochaired the conference, are a physicist and a sociologist. Our collaborators include (in addition to other physicists and sociologists) utility executives, nuclear engineers, government officials, journalists, and psychologists. We are grateful to them all for lending their talents to this attempt to learn useful lessons from the disruption and the near catastrophe of TMI, and we are grateful to the officers and staff of The New York Academy of Sciences for sponsoring both the conference and this publication.

REFERENCES

1. KEMENY, J. G., et al. 1979. Report of the President's Commission on the Accident at Three Mile Island. U.S. Government Printing Office. Washington, D.C.
2. ROGOVIN, M. 1980. Three Mile Island: A Report to the Commission and to the Public. **1.** U.S. Nuclear Regulatory Commission. Washington, D.C.
3. SILLS, D. L., C. P. WOLF & V. B. SHELANSKI, Eds. 1981. Accident at Three Mile Island: The Human Dimensions. Westview Press. Boulder, Colo. (In press.)

THREE MILE ISLAND IN PERSPECTIVE: KEYNOTE ADDRESS

Alvin M. Weinberg

Institute for Energy Analysis
Oak Ridge Associated Universities
Oak Ridge, Tennessee 37830

INTRODUCTION

Reactions to Three Mile Island often reflect prior biases toward nuclear energy. For those who hold nuclear fission to be an important long-term source of energy, Three Mile Island proved that a reactor can be treated badly without anyone being hurt. For those who consider nuclear energy to be an abomination that must be extirpated, Three Mile Island proved that reactors are unsafe—the sooner we have done with nuclear fission, the better. I would suggest, therefore, that each speaker at this symposium announce his position on nuclear energy before he speaks, much as bridge players announce their bidding conventions before they play.

I myself am a critical advocate of nuclear energy. I have spent most of my life developing nuclear reactors. I recognize that the nuclear breeder is the only large and inexhaustible source of energy, aside from the sun, that is known to be technically feasible; and it is probably much cheaper than is production of electricity using solar energy. To give up on nuclear energy as some advocate, or even to decide now that nuclear fission is merely a temporary bridge to a fission-free future based on the sun and fusion, this I cannot accept. I recognize that a serious accident could cause widespread radioactive contamination. My response is to reduce greatly the probability of any serious accident through technical and institutional fixes, not to destroy nuclear fission. In short, I propose that we construct an acceptable nuclear future. I see this symposium as contributing to the design of an acceptable nuclear energy system.

The Institute for Energy Analysis (IEA) has, under Department of Energy sponsorship, conducted two workshops in Gatlinburg, Tennessee, to discuss the design of an acceptable nuclear future.

At the first conference, held in December 1976, those opposed to nuclear energy laid down what they considered to be *necessary* conditions for the survival of nuclear energy: (1) deal with the military wastes at Hanford, Savannah River, and Idaho Falls; (2) stop reprocessing; and (3) defer or abandon the breeder. At least the last two conditions had, until Three Mile Island, been the cornerstones of the Carter administration nuclear policy; but as Three Mile Island has demonstrated, these conditions—perceived as necessary by nuclear opponents—have not proved *sufficient* for continued success of nuclear energy.

1

0077-8923/81/0365-0001 $01.75/2 © 1981, NYAS

Three Mile Island has demonstrated that the central problem of nuclear energy is not waste disposal (which is largely a political, not a technical, question) or even the connection between proliferation and nuclear power (a connection I believe to be exaggerated). It is, rather, reactor safety. I expressed this at the 1976 Gatlinburg conference in the following:

> A massive meltdown . . . even though it has essentially no off-site consequences, is still a very major thing . . . as far as the public's perception is concerned. It could well bankrupt a utility. It raises . . . questions about what we do after a meltdown . . . [we have therefore been] casting about for ways of reducing the probability of meltdown . . . and of mitigating the consequences even further.

Increasing reactor safety was therefore the focus of the second Gatlinburg Conference on an Acceptable Nuclear Future, held last December just after the Kemeny commission issued its report. At this meeting, ideas for improving the nuclear system were discussed. Much of what I have to say is based on the deliberations at the second Gatlinburg conference and on the studies we have done at our institute on the design of an acceptable nuclear energy system.

REQUIREMENTS FOR AN ACCEPTABLE NUCLEAR SYSTEM

In the wake of Three Mile Island, many have suggested what changes in the nuclear enterprise they believe to be necessary. Unfortunately, no one can prove that the proposed changes are sufficient to ensure the survival or even growth of nuclear energy. I shall begin by listing what my colleagues at IEA and I consider to be necessary elements of an acceptable energy system. I shall then compare these theoretical criteria with the actual remedies that already have been put in place since Three Mile Island.

We propose four requirements for an acceptable nuclear energy system.

Technical Fixes

These are the most obvious remedies. They include, for example, ways to monitor the coolant so that the operator knows whether steam is blanketing the core, or direct measurements of the water level in the core. Technical improvements of this sort will reduce the likelihood of a repetition of Three Mile Island. Many of them are already being put in place; more will undoubtedly be backfitted as the technical lessons of Three Mile Island are learned.

Beyond such incremental fixes is a broader issue. If nuclear energy is viewed as a very long-term energy source, should we not reconsider whether there are other reactor systems that depend less on engineered safety features and more on inherent characteristics of the reactor to prevent accidents than do the light water reactor and the liquid metal fast breeder reactor? We in the United States sometimes forget that reactor types other than the light water reactor are used commercially elsewhere in the world: the D_2O-moderated Canadian CANDU, the gas-cooled graphite in France and England, and the Russian steam-cooled graphite. In addition, various thorium breeders and near breeders have been proposed and partly developed, most notably the molten salt reactor and the high temperature gas-cooled reactor. In the early days of nuclear energy, new reactor types were proposed, and even studied seriously, with remarkable frequency. Today the enterprise has been so preoccupied with immediate survival that it seems to be out of the question to propose, let alone develop, completely new reactors. I would suggest that the present moratorium might be used to advantage to study what reactor types might be best over the very long run and to develop the most promising.

Professionalization of the Nuclear Cadre

Three Mile Island has demonstrated that nuclear energy requires excellence at every level. It is not only the operators themselves, it is the executives of the utility who must understand the responsibility they undertake in going nuclear, and it is the people who design reactors. Three Mile Island was as much a design failure as it was an operator failure; the designers no less than the operators and their managers could be faulted (although had communications in the industry been better, the incident, being a repetition in part of a previous incident, could have been avoided). Achieving a higher level of professionalism in every segment of the enterprise is a first order of business. We must incorporate such excellence into every organization responsible for nuclear operations.

Technical fixes and professionalization of the cadre, through better training and better pay, are measures that make sense in the long run and can be put into effect practically immediately. If adopted, they could help cure nuclear energy of its current disaffection. But for the long run and possibly sooner, the next two measures seem important to us.

Restructuring of the Nuclear Utility Industry

Is the present structure of the utility industry adequate for nuclear energy? The responsibilities involved in owning and operating a nuclear power plant are much more far-reaching than are those entailed in operating a conventional power plant. Should not the operation of the nuclear enterprise therefore be in the hands of powerful consortia or other entities that do nothing but own and operate nuclear plants? The responsibilities of the president of such a consortium would not be diluted by operation of nonnuclear facilities; the whole outfit would talk, think, act nuclear. Such consortia have been in existence; for example, the Yankee Atomic Power Company. But because the U.S. utilities are so fragmented, it would be difficult to create nuclear consortia. There are 17 utilities that operate or will soon operate but a single nuclear reactor. For them, nuclear plants are, I suppose, orphanlike, certainly more so than for utilities that operate many reactors. I consider this arrangement undesirable in the long run. What I propose is consolidation of the nuclear operating entities into relatively few, powerful organizations.

The Kemeny commission, without quite saying so, implied that the existing structure of the utility industry may be inadequate for the task of running the nuclear business; and Governor Babbitt, a member of the commission, urged consolidation in his minority report. But the range of possible changes is very great—all the way from a single national nuclear electric utility to the present highly fragmented utilities, strengthened, to be sure, by the new institutes proposed by the utilities. In between might be consolidation, perhaps through several single-reactor utilities joining to create a single nuclear corporation comparable in technical, financial, and organizational strength to the very large nuclear utilities (like Duke Power or Commonwealth Edison).

It is instructive to examine the structure of utilities in other countries. In both England and France, there is but a single national electric utility, the Central Electric Generating Board and Electricité de France; in Canada, each province has a single utility. Japan, with one-fifth the generating capacity of the United States, has one-twentieth as many utilities. In all these countries, the nuclear enterprises seem to be conducted more coherently than in the United States. In fairness, it must be pointed out that in Germany and Switzerland, utilities are less consolidated, yet Switzerland's reactors have the best operating record of any in the world. I do not believe we fully understand the reasons for this difference.

Confined Siting

An overriding criterion for a long-term nuclear system is a rational siting policy. The Institute for Energy Analysis has argued that the ultimate nuclear system in the United States ought to consist of about 100 large, permanent sites rather than 300 to 500 small sites. The advantages of confined siting seem compelling to me: internal lines of transport, stronger operating entities on a large site rather than a small one, better organizational memory, improved construction practices, tighter security and better control of fissile material, a limit on the land at risk in the event of an accident, and a population living close to the plant that is sophisticated and experienced in nuclear matters. What we propose is to return the nuclear enterprise to the original siting pattern represented by

Hanford, Savannah River, and Oak Ridge. These are large, isolated enclaves that are notably capable of coping with emergencies. Many of the people in the nearby towns work at the nuclear plants. They and their neighbors understand radiation and are able to place its risks in perspective.

We propose conferring on these sites an imputation of permanence, perhaps like national parks or even like the nuclear test sites, such as Johnson Island. Once permanence is accepted, decommissioning of old reactors or even handling of low-level wastes becomes easier: old reactors and low-level wastes are held on site until their radioactivity has so decayed that they need little special handling.

Clustering reactors into a few sites has some disadvantages; for example, an accident in one reactor conceivably could incapacitate the whole site. Balanced against this is my sense that the probability that an accident, once started, will progress to the point of posing a public hazard is significantly lower in a big, powerful site than among the same number of reactors on isolated sites. Because of the many intangible advantages in organization and support at big sites, it may be that the probability of creating an actual public hazard *per site* is smaller on large sites than on small sites. This, of course, is only a conjecture that would be hard to prove. Nevertheless, it is clear that once the accident at Three Mile Island started, experts mobilized from everywhere were instrumental in containing the damage. On a big site, such experts would be on hand all the time. We tend to forget that the vast majority of accidents develop slowly; Three Mile Island demonstrated that timely intervention (particularly in handling the off-gas system) can contain the damage. It is this attribute that leads me to expect large sites with their powerful on-site cadres to be safer than small ones.

Can we convert the current dispersed siting pattern into a more consolidated one? Certainly not in the short run, since we already have sited all the reactors that are actively planned. But if our purpose is to design an acceptable long-term nuclear energy system, then it is important to prevent proliferation of nuclear sites. Calvin Burwell of the IEA staff has suggested that we confine new reactors essentially to those of the existing sites that are judged to be adequately remote. Of the 100-odd sites in the United States, all but about 12 have fewer than 300 people per square mile within 50 or more miles and all but 11 are farther than 20 miles from a city of 100,000 or more. Burwell has shown how to accommodate a 615-GWe nuclear system, say in the year 2025, on 100 sites of which 80 are existing sites. The largest site contains nine reactors; the average site holds six reactors. I would propose that the 20-odd sites that are less well situated be decommissioned, but only after the reactors on these sites have become obsolete—say, in 25 or 30 years. The risk imposed thereby seems to me to be acceptably small.

In effect, what we propose is almost a moratorium on new nuclear sites, not a moratorium on new nuclear reactors. Whether this policy would affect public perception is hard to say. It is revealing that in a

Roper poll held six months after Three Mile Island, 80% of the U.S. public supported nuclear energy if reactors were placed 50 miles from where they live. Insofar as most existing sites are already rather remote, an existing-sites policy would be responsive to the public's desire for remoteness, yet would take advantage of the generally favorable attitude toward nuclear power held by people who have lived close to existing sites for a long time.

The United States, perhaps because its utilities are so heavily fragmented, has moved toward siting reactors in clusters much more slowly than have many foreign countries. Of the 500 reactors in the world either operating, under construction, or planned, more than 170—representing one-third of all nuclear power (one-half of nuclear power outside the United States)—are on sites with 4 or more plants. There are even now three sites with 8 reactors planned for them; and it is not unlikely that very large complexes will become the rule. Of these clustered sites, only four will be in the United States. The issue is whether we ought to adopt a policy of actively encouraging confined siting or whether we should merely let nature take its course. In the latter case, the average capacity per site would probably slowly increase because new sites would be larger than older ones, but such a laissez-faire siting policy would, I suspect, not affect the acceptance or rejection of nuclear energy.

THE UTILITIES' RESPONSE

Three Mile Island has made clear to all in the nuclear business that an accident anywhere is an accident everywhere. A repetition of Three Mile Island could close down the entire nuclear enterprise. This point was brought home to us at Gatlinburg by the utility executives there. The response of the utilities has been correspondingly far-reaching; the details of this response we will hear from Chauncey Starr.

Three new institutions have been created:

1. The Institute of Nuclear Power Operations (INPO), an agency that examines and accredits the nuclear operations of utilities belonging to INPO. INPO should raise both the professional level of those responsible for nuclear operations and the standards of operation by the utilities;

2. The Nuclear Safety Analysis Center (NSAC), which will analyze incidents throughout the nuclear enterprise and will alert operators to the significance of incidents in other reactors. In effect, it will help provide the organizational memory so necessary for safe nuclear operation. The sequence leading to Three Mile Island had occurred several times before in other reactors, yet operators at Three Mile Island were not aware of the significance of these incidents. But Three Mile Island already has had an effect in this regard: at Crystal River recently, when the system went "water solid," the high pressure coolant injection pumps were not turned off and as a result the core was kept cool;

3. A nuclear insurance pool that will pay participating utilities up to $152 million for replacement electricity in the event of an accident.

In addition, the utilities have made and will continue to make technical improvements that will reduce the probability of new Three Mile Island incidents. Clearly, they have already moved to meet two of the criteria I have set forth: technical fixes and professionalization of the cadre. As for restructuring the nuclear industry, many of the utilities have moved to set up their nuclear organizations as rather separate entities within the utility instead of being submerged by the rest of the generating division. Only the Tennessee Valley Authority has adopted a confined-siting policy; TVA has announced that all of its nuclear generation will be confined to the 7 existing sites rather than to the 13 or so sites expected earlier. Fuel elements and low-level wastes will be stored on the existing sites also.

THE GOVERNMENT'S RESPONSE

The main thrust of the Kemeny commission report was the need for change. Though the Kemeny commission was careful not to generalize about the deficiencies of the nuclear industry, it faulted the operators and, to a degree, the designers of Three Mile Island. Mostly, its criticism was directed at the Nuclear Regulatory Commission: NRC had not ordered its priorities properly. Day-to-day licensing questions left it little time to decide on much broader issues, such as whether licensing of a utility ensured its capacity to operate safely. The commission did not call for a moratorium, though several moratorium motions failed by only one vote.

The Nuclear Regulatory Commission is responding in a variety of ways to these criticisms, as will be discussed by Mr. Budnitz of NRC. In addition, the Rogovin commission, set up by NRC after Three Mile Island, has recommended among other things that a utility-wide consortium be given responsibility for operating nuclear plants for utilities that are not now fully qualified, and that new reactors be sited more remotely.

ARE THESE RESPONSES SUFFICIENT?

The central question that remains is whether the measures now being taken in response to Three Mile Island will be sufficient to preserve nuclear energy, and if they are, whether they are sufficient to allow nuclear energy to play an increasing role in the future. One can only admire what the private utilities have done: INPO, NSAC, and the insurance pool. But I would suggest that they consider seriously the next step, the establishment of nuclear operating consortia (as suggested by the Rogovin commission) that pool the very best talent for management and ownership of nuclear energy. As matters stand now, 100 new reactors are scheduled to come on-line during the next 10 years. There will be 17 new utilities that have never operated a reactor; and altogether, in 1990, there will be 47 utilities that operate but 1 or 2 reactors. The Nuclear Regulatory Commission in 1990 will have to monitor the operation of over 60 separate nuclear utilities. I would suggest that consolidation into many fewer entities—possibly by allowing large nuclear utilities to take over responsibility for operating smaller "orphan" sites—might lead even in the short run to a more manageable as well as safer enterprise overall.

But it would be unfortunate if our preoccupation with immediate survival of the nuclear enterprise in the wake of Three Mile Island caused us to ignore the longer term matter: How should a very large, permanent nuclear energy system be designed? As I have explained, I consider the siting question to be central to this. I am convinced that confined siting based on existing sites is the way to go. Moreover, a moratorium on nuclear sites, not on nuclear reactors, with a planned site-replacement and decommissioning at end-of-life of poorly located sites, could help the survival of nuclear energy. I would urge that the Nuclear Regulatory Commission, in its current attempts to formulate a new siting policy, take seriously the proposal to use the existing sites as the nuclei for a large and long-term nuclear energy system.

Is Nuclear Energy Necessary?

All that I have said reflects the bias that I announced originally; I advocate nuclear energy and am therefore anxious to help put in place the improvements needed to make nuclear energy acceptable. I agree with President Carter who, in his response to the Kemeny commission report, reaffirmed the necessity of nuclear energy, even though I disagree with his seemingly grudging support of this option.

Our judgment as to what measures are sufficient as well as necessary to ensure a future for nuclear energy depends on our assessment of the need for nuclear energy. We cannot assure *complete* safety, since it is impossible to prove that something that does not violate a law of nature (a serious reactor accident) is impossible. (I do point out that Three Mile Island demonstrated that catastrophic failure of the containment was impossible in the one case where the experiment has been performed.) How much we are prepared to spend toward achieving an acceptable level of safety depends on how badly we need nuclear power and on how the risks of nuclear power compare with the risks of alternative ways of producing energy.

That a shutdown of existing reactors would be catastrophic I believe is self-evident. It is not only the energy that we would lose, it is the $100 billion investment whose write-off would cause a violent shock to our financial institutions. Even a moratorium that allowed no new reactors beyond those in the pipeline would be extremely awkward because of the heavy pressure such a moratorium would place on coal. Whether new reactors are built in the short run ought to be determined by economics; the nuclear enterprise ought to be fixed so that doctrinaire, as contrasted with economic, objections to it would have no standing.

At issue is the role of electricity in the future. I am aware of a currently fashionable distaste for electricity as something we ought to do with as little as possible. Centralized electricity is the epitome of the coercive, wicked, centralized society. But for every decentralized utopia envisioned by those who oppose electricity, one can construct a centralized utopia in which electricity plays a key role. It becomes again a matter of cost and convenience. If the electric automobile can be developed, and if heat pumps (or even that bane of the electric opponents, resistive heating) become widespread, electricity might grow from its present 30% use of U.S. primary energy to, say, 40-45%. In any event, the day of large, central power plants will certainly not have passed. Instead, by the mid-21st century, we might have an energy system in which, say, 60 quads or 1,000 1-GWe power plants are operating. To provide all this with coal would require a fivefold increase in our production of coal. This I consider to be all but impossible.

I cannot prove the long-term necessity of nuclear energy. I can only argue that in view of the enormous uncertainties—Will the future be electric? What about the CO_2-greenhouse effect? What about the feasibility of digging and transporting 4 or 5×10^9 tons of coal per year? How expensive will solar energy be?—it would be extremely imprudent to close down the nuclear enterprise. Thus, Three Mile Island could well be the salvation of nuclear energy: it has pinpointed the real, rather than imaginary, weaknesses in our existing nuclear system, and thus has pointed to the directions in which we must go to construct an acceptable nuclear energy system.

DISCUSSION

T. H. Moss* (*Subcommittee on Science, Research, and Technology, U.S. House of Representatives, Washington, D.C.*): Thank you very much, Dr. Weinberg. I hope all the speakers will look into the particular systems that they are addressing as deeply as you have into the nuclear system. I don't quite agree with Dr. Weinberg that everyone need state their position on nuclear power. There will be speakers who probably don't care about nuclear power one way or the other, that is, speakers who are dealing with information systems, with emergency preparedness systems, with public attitudes, and with economic situations that only peripherally touch nuclear power. But in every case, I think there's an

*Dr. Moss is science advisor to this committee, not a member of Congress.

opportunity to draw lessons. I hope that all look into the systems that they are interested in as deeply as Dr. Weinberg has looked into the one about which he admittedly feels rather strongly.

C. P. WOLF (*Polytechnic Institute of New York, Brooklyn, N.Y.*): My question to Dr. Weinberg is, Aren't you being dogmatic about the economics driving the rate of nuclear development?

A. M. WEINBERG: What I said was that at present, there are many considerations other than economics that are driving the nuclear system, or driving it to the ground, if you like. The thrust of my remarks was that I would hope that we can fix nuclear energy, that we can design a nuclear energy system that is sufficiently acceptable so that considerations other than economics will no longer play a role in the determination of whether we have nuclear energy or how much nuclear energy we have.

I have the feeling that unless we succeed in making nuclear energy acceptable in this very basic sense, we will not have nuclear energy. I often draw the analogy between the job of the Nuclear Regulatory Commission and the job of the Federal Aviation Administration. The problem with the Nuclear Regulatory Commission in licensing reactors is that we do not have a full consensus as to whether we should have nuclear energy. Imagine the job that the Federal Aviation Administration would have in licensing the DC-10 if there were a large minority (which perhaps someday might become a majority) of people who believe—for reasons having nothing to do with economics—that aviation is an abomination that must be annihilated. The job of the FAA would be impossible.

The main job of the nuclear enterprise—the main job of the country, if you like, because I think nuclear energy is important—is somehow to design a nuclear system that will be fully acceptable and will command an enormous consensus in the same way that aviation has an enormous consensus.

UNIDENTIFIED SPEAKER: What is your opinion of nuclear parks and also of organizing in those nuclear parks some research and development activities that would add to the safety factor mentioned before?

A. M. WEINBERG: About 15 years ago, when the Tennessee Valley Authority started its nuclear program, which was going to involve about a dozen reactors, I suggested that the Tennessee Valley Authority put all of its reactors on the Oak Ridge site. The reason I gave was that if anything ever happened, then right there you would have 5,000 people who knew their way around. Of course, one of the main reasons that I am so insistent on this question of keeping reactors together is that when they're together, you have lots of good people who know their business. You really didn't have that at Three Mile Island.

A. R. TIEMANN (*General Electric Corp., Schenectady, N.Y.*): I'm a private consultant in energy and environmental issues. The suggestion by Dr. Weinberg that massification and centralization of the nuclear industry would be a solution runs a little bit counter to public opinion, which suggests that people would like to see things less centralized. How will that help nuclear power?

A. M. WEINBERG: I suppose that to some degree, the outcome of the presidential election will be a measure of the extent to which the assertion you've just made is, in fact, correct. I think that for certain segments of the public, you're quite right: the fear of centralization—as you put it, massification—is a very real one. In other parts of the public, I think the issue of reliability of electrical supply is a more important consideration.

As I said in my closing remarks, I am aware of the fact that centralized electricity is viewed by many as being somehow the epitome of the centralized, wicked, coercive society. But I would assert that for every decentralized utopia, one can construct an equally plausible centralized utopia.

You must recognize, and I'm sure you do, that the issue of centralization versus decentralization is almost the fundamental issue in the whole history of political thought. It is very interesting, in my mind, that we find in the nuclear issue the centralization-decentralization issue coming again to the surface. My feeling about it is that however this generation decides the issue, weaknesses in the one posture will be recognized by the next generation. If this generation opts for decentralization, then the generation after that will decide that perhaps we ought to be more centralized.

D. BENNETT (Sandia Laboratories, Albuquerque, N.M.): Concerning your confined siting, have there been any studies on the thermocapacities for existing sites that would be confined and on the economics of load distribution? In other words, how do you give the power to the users with confined siting?

A. M. WEINBERG: Let me say that with respect to load distribution, there is a misunderstanding. Many people don't realize that the transmission corridor requirements for confined siting are much less stringent or much less difficult to achieve than if you continue with dispersed siting. You already have the transmission corridors, and it's very easy or relatively easy to strengthen those transmission corridors. Our findings were that the confined-siting system, 100 sites for 615 gigawatts, in fact involved less in the way of additional transmission than would a dispersed-siting system.

As far as the thermal load is concerned, that was looked at in the Nuclear Regulatory Commission study on nuclear parks. Our impression is that the thermal load is probably not all that big a problem up to something like 10 gigawatts, possibly to as much as 15 gigawatts.

H. KIHN (H. Kihn Associates, Lawrenceville, N.J.): I must say I was very impressed with your presentation. I just wonder, in view of the point that you made that the population doesn't really understand the worth of nuclear energy, why the industry hasn't pointed out that many states and many communities depend on nuclear. In New Jersey, 25% of our electricity comes from nuclear energy. In Chicago, a goodly portion of the energy comes from nuclear energy. I don't believe the population really understands that. Now why hasn't this been brought out?

A. M. WEINBERG: All I can say is that it has been talked about many, many times—100 times, 1,000 times in the past few years. You cast pebbles on water, and maybe something happens to them.

With respect to the confined-siting policy, a practical reason why I think that confined siting is a good idea as far as the public is concerned is because of the experience that I have had. I've lived next to 10 reactors all my life and have brought up a family, and so on. We in Oak Ridge love nuclear energy. We understand nuclear energy. I'm not talking merely about Alvin Weinberg. I'm talking about my neighbors. We understand radiation. The *Oak Ridger,* which is the local newspaper, every week has a little column that reports the radiation readings for the week. This is the kind of thing that I would hope would develop in the communities close to reactors. The people in Oak Ridge do understand the difference between a milliroentgen and a million roentgens, something I don't think the average man on the street does understand.

What I deeply believe about nuclear energy is that it cannot survive until and unless the public as a whole can view the issue of low-level radiation in the same way that it views other low-level industrial insults. It is my belief that in communities close to the reactors, at least properly operating reactors, people do acquire that understanding.

SAFETY CONSIDERATIONS IN THE DESIGN AND OPERATION OF LIGHT WATER NUCLEAR POWER PLANTS

John R. Lamarsh

Department of Nuclear Engineering
Polytechnic Institute of New York
Brooklyn, New York 11201

There are essentially only two different types of nuclear power plants being built and operated in the United States today: one type uses a boiling water reactor, or BWR; the other uses a pressurized water reactor, or PWR. The Three Mile Island plant is in the latter category, a PWR plant.

The basic principles of the operation of both types of nuclear plant are quite straightforward. First, energy is released in the reactor fuel as the result of nuclear fission. This energy is then absorbed from the fuel by a passing coolant, which is ordinary water in the case of both the BWR and the PWR. With the boiling water reactor, as the name implies, the heat absorbed from the fuel causes the water to boil within the reactor itself. With the pressurized water reactor, on the other hand, the pressure of the water is kept so high that it does not boil within the reactor. Instead, after passing through the reactor core, the heated water goes to a heat exchanger called the steam generator, where the energy from the primary reactor cooling water causes water to boil in a secondary cooling-water loop. In short, in a BWR plant, the reactor itself produces steam; in a PWR plant, the steam is produced in the steam generator. In either case, the steam is used to turn a turbogenerator to produce electricity, the output of the plant. While none of these processes involve either new or especially erudite technology, nuclear plants must be designed and operated with some care due to the fact that they contain large amounts of radioactive materials.

Most of the radioactivity in a nuclear power plant originates in the fission process itself. In this reaction, a neutron strikes a nucleus (in either the BWR or the PWR, this is usually the nucleus of the uranium isotope uranium-235), the neutron is absorbed by the U-235 to form uranium-236, and then this nucleus, the U-236, splits into two parts with the release of a considerable amount of energy. The two pieces of the fissioning nucleus that remain after the fission is completed are called fission products. Unfortunately, the majority of these fission products are radioactive, and spontaneously decay with the emission of a variety of β-rays and γ-rays.

Most of this fission product radioactivity dies away rapidly, on the order of seconds to hours. Several fission products have half-lives—a measure of how rapidly they disappear—on the order of days. A

0077-8923/81/0365-0013 $01.75/2 © 1981, NYAS

couple—strontium-90 and cesium-137, in particular—have half-lives of about 30 years; these tend to be the most troublesome in the disposal of radioactive wastes. Finally, a few fission products, such as iodine-129, have half-lives that are so long that they are essentially stable.

There is one group of fission products that is of special concern from the standpoint of reactor safety. These are the fission products that are gases under normal circumstances, namely, the isotopes of krypton, xenon, and iodine. These gases are the first fission products to be released in a reactor accident in which the fuel sustains damage. The krypton and xenon are noble gases, and they do not combine with or remain in constituent parts of the body. As a consequence, their biological effect is relatively mild. The iodines, by contrast, are chemically active, and when ingested or inhaled, they tend to travel to the thyroid gland, where one isotope, I-131, continues to deliver a radiation dose for several weeks. By the same token, of course, the release of radioactive iodine from a nuclear power plant can be controlled readily by various chemical and physical means.

It is the major objective of nuclear plant design to keep the fission products confined at all times—during the normal operation of the reactor and under accident conditions. They must be prevented from coming into contact with either plant personnel or the surrounding public. There are a few other sources of radioactivity in a nuclear plant: the so-called activation products, materials that have become radioactive by the action of the neutrons in the reactor; and the transuranic nuclides, mostly isotopes of plutonium, which are produced by the absorption of neutrons by U-238 in the fuel. But by and large, it is the fission products whose release to the public is the major source of concern in nuclear plant design.

Multiple Barrier Design

To prevent the escape of these fission products, nuclear plants are designed with a sequence of obstacles, or barriers, between the fission products and the public at large. There are seven such barriers associated with all BWR and PWR nuclear plants as they are designed in the United States.

The first fission product barrier is the nuclear fuel itself. This is in the form of uranium dioxide pellets—little cylinders about ½ inch in diameter and 1 inch in length. Uranium dioxide is a hard ceramic material with a high melting point, about 2,900°C (5,200°F). Practically all of the fission products remain trapped within the uranium dioxide near the point where they are formed in fission. Only a very small fraction of the fission product gases, usually less than one percent, diffuses out of the fuel at normal operating temperatures.

These fuel pellets are placed in long, hollow tubes that are sealed at both ends. Fission products formed near the surface of the fuel and the fission product gases that diffuse out of the fuel are collected in these fuel tubes and, in this way, are prevented from passing into the adjacent water coolant. The fuel tubes, or fuel cladding, as it is also called, are the second safety barrier in nuclear plant design.

The third barrier arises from the fact that all BWRs and PWRs use closed cooling loops to remove the heat produced in the reactor core. There was a time when reactors were cooled by simply passing air or water through the reactor core and discharging it directly to the environment. With those reactors, any fission products that may have leaked from the fuel into the coolant exited the reactor with the coolant. All of today's power reactors use closed cooling systems, and whatever radioactivity is picked up by the coolant from the fuel remains within the system. In addition, the coolant water is cleansed and purified on a continuing basis, and the radioactive residues are retained for decay and disposal.

In all reactors being built or operated today in the United States, the reactor core is located within a heavily built pressure vessel. The craft of designing, fabricating, and caring for reactor vessels is a very special branch of technology by itself. Needless to say, the reactor vessel, which is constructed of stainless steel six to eight inches thick, represents another important barrier inhibiting the release of fission products from the reactor.

The last barrier at the plant itself is the containment structure that houses the reactor and its cooling system. This is a heavily reinforced concrete building, on the order of three or four feet thick, with an interior steel liner. Any fission products—in particular, the fission product gases—that escape from the reactor vessel or the cooling system are released into and confined within the containment building. In PWR plants, the containment building contains an overhead spray system to reduce the pressure of the steam in the building in case of a break in a high pressure coolant pipe. The sprays also tend to wash out or remove radioactive iodine from the containment atmosphere. Containment buildings also house elaborate filter systems to reduce the fission product concentration within the building.

These last two barriers, the closed cooling system and the containment structure, are tremendously effective in holding fission products. At the Three Mile Island accident, for example, a total of only between 13 and 17 curies (Ci) of radioactive iodine were released from the plant during the course of the accident. However, about 7.5 million Ci were retained within the cooling system, and 10.6 million Ci were held within the containment building.

If the Three Mile Island accident showed anything, it was the wisdom of containment philosophy, which originated in the United States in the early 1950s. Until very recently, the Soviet Union did not place containment structures around their reactors. I understand they are beginning to do so now.

The sixth barrier to the exposure of the public to radioactivity is the manner in which nuclear plants are sited. There is much to be said about reactor siting. The most important is that nuclear plants are restricted to areas with relatively low population densities. Whether they should be more remotely sited than they are now is the subject of debate at the present time. Clearly, if remote siting is required to regain public confidence in nuclear power, then remote siting will simply have to be adopted as national policy if we are to retain the nuclear option.

The seventh and last of the barriers separating nuclear power plant radioactivity from the public is evacuation. If all else fails, the other barriers have been breached, and dangerous amounts of radioactivity may be released, then the threatened population simply has to be moved. Much has been made of the evident lack of planning for evacuation in the vicinity of Three Mile Island. The fact is, however, that the evacuation of large numbers of people is ordinarily not an especially difficult problem, and evacuations occur all the time. Last autumn, a town west of Toronto was evacuated of about 250,000 people because of the derailment of chemical tank cars; a few years ago, 25,000 people were evacuated near Cicero, Illinois, for the same reason; last December, 8,000 people were removed from the outskirts of Detroit because of fire in a gasoline storage facility; and only last week, 9,000 people were evacuated in Somerville, Massachusetts, because of a minor rail crash involving chemical tank cars. And all of these evacuations were accomplished without the benefit of elaborate, federally approved evacuation plans.

Reactor Accidents

The foregoing sequence of seven barriers is designed to prevent radioactivity from coming in contact with the public, both while the plant is operating normally and in the case of accidents. In normal operation, nuclear plants emit only trivial amounts of radiation. Nuclear plants are extremely tight. Incidentally, the tightest kind of plant appears to be the sodium-cooled fast breeder reactor that the Carter administration so strongly opposed.

Now what about nuclear accidents? There are any number of things that can happen at a nuclear plant that can lead to the release of radiation. Most of these releases are of no consequence, however. But significant releases can occur if a substantial portion of the uranium dioxide fuel reaches the melting point, in short, if there is what is popularly known as a meltdown, and the multiple barriers fail in one way or another. The most important thing to recognize about a meltdown is that it would not be the end of the world. The most probable consequence of a meltdown is essentially no release of any radiation whatsoever.

Melting of fuel can occur in a number of ways, the most likely being some sort of loss-of-coolant accident, that is, a circumstance in which the reactor fuel is not properly cooled. It is necessary to cool the reactor core, i.e., the region containing the fuel, both when the reactor is in operation and for some time after it is shut down, owing to the continuing decay of the fission products. The amount of energy released by the fission products is not negligible. About 7% of the total energy released in the core of an operating reactor originates in fission product decay, after the reactor has been in operation for a month or so. Thus, in a nominal 1,000-MWe* nuclear plant that operates at a thermal efficiency of about 33%, so that the reactor is actually operating at a power level of 3,000 MW, approximately 210 MW (7% of 3,000) come from fission product decay. Following the shutdown of a reactor, the fission product decay heat drops off rapidly at first and then more slowly. After one minute, it drops from 210 MW to 120 MW; after one hour, it is down to 30 MW. But at the end of a full day after shutdown, 15 MW are still being produced by the fission products. If this heat is not removed, the fuel, or a portion of it, will melt.

When a normally operating reactor is shut down, the residual fission product heat is removed by special components of the reactor cooling system. However, if the reactor cooling system should fail while the reactor is operating, it is necessary to supply an alternative source of cooling. This is the function of the emergency core cooling system—the ECCS, for short—a required component in every nuclear power plant used in the United States today.

Actually, the problem that engineers must solve in designing an ECCS is not especially world-shaking, and indeed, the ECCS is neither a particularly complicated nor a sophisticated system compared with other areas of modern technology. All that is required is to assure that cooling water can be pumped continually through the reactor core at all times— when the reactor pressure is high and when it is low; when there is a break in a large cooling pipe or in a small cooling pipe; when a valve in the cooling system sticks open, as it did at Three Mile Island, letting the coolant escape into the containment building; when electric power on the utility grid is not available to operate pumps and valves; and so on. Providing a continuing flow of coolant to the core under these circumstances is a straightforward mechanical engineering problem that can be solved, and has been solved, by the installation of a multiplicity of independently operated high pressure pumps, low pressure pumps, multiple sources of reserve cooling water, redundant emergency electrical power sources, and so forth. Although there have been only a couple of loss-of-coolant incidents to date, emergency core cooling systems appear to operate properly. Recent tests in the Loss of Fluid Test Facility at the Idaho National Engineering Laboratory, which was designed and built specifically to test emergency core cooling systems, indicate that the ECCS works as designed. Of course, if some one shuts off the ECCS, as they did at Three Mile Island, then regardless of how well the system is engineered, it is not going to work at all and some melting of the fuel may result.

*e stands for electrical.

Summary

U.S. nuclear power plants are designed with a sequence of barriers to prevent the escape of fission products. These include the fuel itself, its cladding, a closed cooling system, and a massive containment structure. In addition, nuclear plants are sited away from major population centers so that the affected population can be evacuated if the need arises. While it is highly unlikely, a release of fission products could occur if the core were to melt due to an unintentional lack of cooling. To avoid such occurrences, all nuclear plants are equipped with emergency core cooling systems. The fundamental soundness of this design philosophy has been demonstrated by the, to date, untarnished record of minimal population exposure from all nuclear power plant operations.

DISCUSSION

P. STRUDLER (*National Institute of Occupational Safety and Health, Rockville, Md.*): Coming back to the question of clustering reactors, isn't the danger that if there is a major accident, with environmental venting and possible contamination and evacuation of the area, you are going to have to shut down not one but four or five reactors, with a much greater loss of electrical power?

J. R. LAMARSH: Here I'm speaking really for Dr. Weinberg, who's the authority in this area. That is one of the disadvantages, obviously, that is ordinarily mentioned in connection with a nuclear park–confined siting concept. I believe the answer is that, first of all, the probability of an accident of such a nature as Three Mile Island actually occurring would be reduced if the reactors were, in fact, clustered. Also, I can't imagine that the entire electrical power for regions of the country—the entire electrical supply—would be dependent upon any one of these reactor sites. I myself think that would be very foolish. I don't know how you feel about that, Dr. Weinberg.

A. M. WEINBERG (*Institute for Energy Analysis, Oak Ridge, Tenn.*): That's actually the answer that I would give. I would point out again, however, that the clustering of reactors is not a theoretical concept. In Canada, Pickering has eight reactors right next to each other; Bruce has eight reactors right next to each other.

P. STRUDLER: Are they all power reactors?

A. M. WEINBERG: Oh, yes. The Canadians, in examining the issue, have concluded that the gain that you get in reducing the probability of something happening by clustering more than offsets whatever loss you get if there are multiple failures. But, of course, it's an arguable point.

P. STRUDLER: One final comment on your figures about evacuation. Pardon me, but I would not use those in front of large public gatherings, if I were you, because there's a great deal of difference in perceived danger by the public for various dangers. Also, following your reasoning, since they've seen all the films about the evacuation of London during the blitz, the people in Middletown should know exactly what to do and should have left immediately for Hershey, Pennsylvania, which was not the case. So I don't think that's a good argument to use—that repeated evacuations in the country make it more feasible to evacuate people from nuclear accidents. People don't perceive it as something that they will do every day.

J. R. LAMARSH: People may not perceive it as something that they do every day, but I have had occasion to discuss this issue with people who have been involved with evacuations. For some years, I was a consultant to a town, and we looked into the issue of how difficult it would be to evacuate the full 1,200 people in the town. I discussed the whole matter of evacuation with civil defense authorities, for example, and they pointed out to me that the general public has no conception of how easy it is to evacuate people, particularly two or three thousand people. Evacuations involving very large numbers of people do, in fact, occur routinely in the United States and throughout the world.

F. ROSENTHAL (*New York University Medical Center, New York, N.Y.*): I also have a question about evacuation. I wasn't sure if your remarks about the ease of evacuation were meant to apply to a major urban area, such as New York City.

J. R. LAMARSH: No, I don't think that a city the size of New York City could be evacuated in a matter of hours. I think that's relatively clear. On the other hand, presumably there would be no reason to evacuate New York City as long as reactors are sufficiently remotely sited.

F. ROSENTHAL: I was wondering if you could comment on a recent letter to *Chemical and Engineering News* with respect to the Zircaloy cladding that's used in fuel elements. The letter stated that this particular design was unsafe because of the instability of this material to chemical reactions at high temperature.

J. R. LAMARSH: Well, it certainly is true that at very high temperatures, Zircaloy does react with water to produce hydrogen. I don't know that this proves that it's a particularly undesirable substance to use. Zircaloy is used, of course, because the primary component of that alloy is zirconium, and zirconium has a very small absorption cross section for neutrons. So from a neutronic standpoint, it is a very desirable material to use.

I don't know that any reactor has ever run into any particular difficulty as a result of zirconium and the production of hydrogen. I realize that the production of hydrogen led to certain difficulties in the Nuclear Regulatory Commission during the Three Mile Island accident, but I don't believe it ever led to any particular difficulties with the reactor.

METHODS OF HAZARD ANALYSIS AND NUCLEAR SAFETY ENGINEERING

Norman C. Rasmussen

Department of Nuclear Engineering
Massachusetts Institute of Technology
Cambridge, Massachusetts 02139

INTRODUCTION

The accident at Three Mile Island has focused considerable attention on the methods of reliability analysis that can provide estimates of the risks associated with hazardous activities. It has become fairly common practice to use the term "hazard" to refer to the potential for producing some undesired consequence, while the term "risk" is used to refer to some function of the undesired consequence and its probability of occurrence. The most commonly used function is the simple product of probability and consequence magnitude. Reliability methods are those techniques for identifying system failure modes and estimating their probability of occurrence. The purpose of this paper is to review how these methods have been applied to nuclear power plants and, when combined with calculations of consequences, have been used to make risk estimates of nuclear power plants. Because of the author's close association with the Reactor Safety Study (WASH 1400), the discussion will focus mainly on the approach used in that study.[1] The paper will also examine the strengths and shortcomings of the study in predicting an accident like Three Mile Island.

THE REACTOR SAFETY STUDY APPROACH

The goal of this study was to estimate the risks to the public from accidents in nuclear power stations of the type we build and operate in the United States today. It is clear that these risks are associated almost entirely with the potential for a release of a large amount of radioactivity. The vast majority of the radioactivity in the plant is produced in the fuel and remains tightly bound in the fuel matrix under normal conditions. Thus the major area of concern is events that seriously overheat or melt the fuel and thus permit a fraction of the otherwise trapped radioactivity to be released. It is generally recognized that events that seriously overheat or melt the fuel dominate the accident risk. Thus the risk analysis is reduced to identifying and determining the probability of those chains of events that can cause serious fuel overheating and, of course, to calculating the consequences of these events.

The Reactor Safety Study (RSS) was divided into the seven major tasks shown in FIGURE 1. Task 1 was to identify the accident sequences,

20

0077-8923/81/0365-0020 $01.75/2 © 1981, NYAS

and task 2 was to determine their probability. In task 3, the amount of radioactivity released in each sequence was determined. Task 4 calculated how the radioactivity was dispersed by the prevailing weather conditions, and task 5 calculated the number of people exposed to various doses of radiation. Task 5 also calculated, in dollars, the damage to property outside of the plant. In task 7, all the individual accident sequences were combined to determine the overall risk. To give some basis for comparison to other currently accepted risks, other potentially large risks were examined; this is identified as task 6.

THE EVENT TREE METHOD

Nuclear power plants are very complicated machines containing literally thousands and thousands of individual parts, any of which can fail. Thus a logical method is needed to subdivide the problem into general classes of conditions that might cause serious fuel failure. In

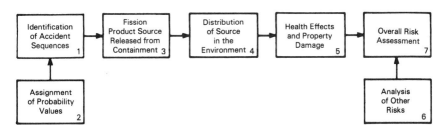

FIGURE 1. Seven basic tasks in the Reactor Safety Study.

order to identify those failures that could produce serious overheating or melting of the fuel, the study group noted that such overheating could occur only for conditions that produced a serious heat imbalance in the plant, that is, conditions that undercooled the fuel by failing to remove the heat generated, either because the heating rate was in excess of the cooling system capacity, or because the cooling system was not removing heat at the rate for which it was designed. These were identified as the overpower case and the undercooling case. Analysis led us to conclude that although certain conditions that could lead to the chain reaction rate in excess of the normal power level were possible, their likelihood was relatively small as compared to failures that could lead to improper cooling of the fuel. Thus we found in the study that the major contribution to the risk of release of radioactivity was caused by failure of the cooling system to perform properly in one of two ways. The first condition, called loss-of-coolant accident (LOCA), is those failures that lead to the loss of the primary coolant itself. This can come about by ruptures in the primary system or by safety valves that remain open when they should not. The second mode of failure is failures in the heat removal system in which the coolant is present but where various functions required to move the heat from the fuel to its ultimate heat sink fail.

To identify the accident sequences in an organized, logical way, the study used a technique called event trees for the initial analysis. FIGURE 2 shows a simplified version of an event tree to illustrate the method. The event tree starts with an initiating event, in this case a pipe break, which is a form of LOCA. The analyst must then, through his knowledge of the

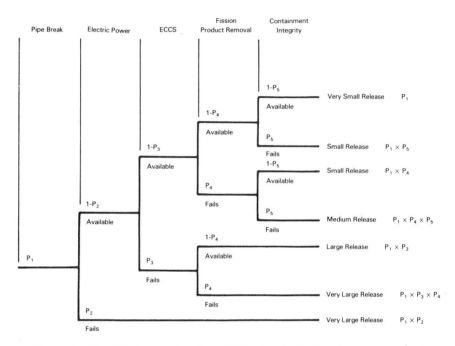

FIGURE 2. Simplified event tree for a LOCA in a typical nuclear power plant.

system, identify various plant functions that will affect the outcome of this initiating event. In this case, the functions identified are the availability of electric power; the availability of the emergency core cooling system (ECCS); the fission product removal system, which is a system that washes radioactive aerosols out of the containment; and finally the integrity of the containment itself. In FIGURE 2, at each branch point, the path that goes up represents successful function or availability of the system that provides this function. When the path goes down at the branch point, it represents failure of this system to perform its design function. Thus we see a variety of possible paths through the tree, each of which represents a possible state of the plant following the initial event pipe break. Not all possible branches are shown because, for example, if electric power fails, the ECCS cannot operate, so there is no choice shown on that line. In addition, the fission product removal system cannot operate; and if these two systems fail, the core will melt and rupture the containment so that containment integrity is not available either. Thus we see that the bottom line on this tree has one step, failure of electric power, and leads directly to a very large release of radioactivity.

The P's on the figure are meant to indicate the probability of that particular path at any particular branch point. Thus any one path's total probability can be obtained by multiplying the probabilities at the various branches together. The final column on the chart shows this. It should be noted that, in general, these probabilities—the values of P—are quite small, so that a good approximation is to assume that $1 - P$ is essentially 1. That approximation has been made in determining the probability shown in the final column. Clearly, there may be dependencies between the various probabilities. That is, the value of the probability of ECCS function failure, P_3, may depend in some way upon the conditions created by the pipe break itself, P_1. Thus, in the analysis, one must be careful to assess any dependencies that may exist between the probabilities. This analysis is often called the common cause or common mode failure analysis of the system. Thus, if done with care, the event tree allows the analyst to define all the possible final states that might result from the initial event, in this case a pipe break, and indicates what the probability of the various outcomes might be if a method is available to determine the values of the P's in the chart. In the study, six different event trees were developed to cover the spectrum of initiating events that could lead to core melt.

THE FAULT TREE METHOD

To determine the values of the probability, another logic method was used. This method is called fault tree analysis and is indicated in a very simplified way in FIGURE 3. In the figure, we have shown a simplified fault for the failure of the electric power system to provide power to the emergency safety features (ESF) of the plant. You will note that the logic is somewhat the reverse of the event tree logic. In the fault tree, we start with some undesired outcome and ask how it might have happened. In the event tree, we start with some initial event and ask what are all the possible outcomes. Referring to FIGURE 3, we see that loss of power to the emergency safety features can be caused by loss of DC power or loss of AC power. This is so because there is a direct current system that activates the controllers to the various electrical equipment and there is an alternating current system that provides the power to operate the equipment. Thus loss of either means loss of ability for those systems to

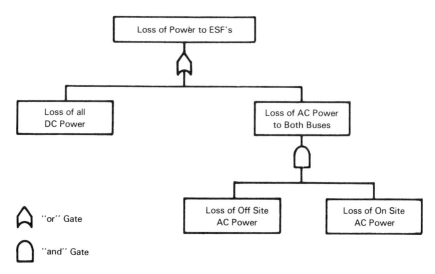

FIGURE 3. Simplified fault tree on electric power.

function properly, and on the chart it is indicated by the symbol called an "or" gate, which simply means that the top event will occur if either one or the other of the two lower events happens.

To illustrate another type of logic that occurs in the tree, we further analyze loss of AC power. AC power is provided by two different sources, each of which is capable of operating the emergency safety features. These are the off-site power, that is, the connection of the utility grid, and the on-site power, the station diesel generator. To lose all AC power, both sources must be lost, so they are related to the box loss of AC power through a symbol called an "and" gate, which indicates that both must occur for the event above to happen. The analyst knows that the probability of the top event in this case is the product of the probabilities of the two lower events, whereas in the case of the "or" gate, the probability of the event above is—to a first approximation—the sum of the probabilities of the two lower events. If the tree is developed to lower and lower levels, that is, to smaller and smaller parts of the plant, one finally obtains a series of failures—failure of the operator to throw the switch, failure of the switch to work properly, the solenoid sticks, the valve sticks, the pump fails, and so on. From experience with equipment of this type in a variety of operations under similar conditions, we can obtain an estimate of how likely these events are to occur. With the probabilities of these bottom events, called primary events, and the relationships that indicate how their probabilities are related to each other, one can propagate the probabilities upward through the tree and determine the probability of the top event.

Fault trees based on these principles were drawn for some 20 different plant systems. These were systems identified by the event trees as affecting the outcome of various initiating events. Thus the fault trees were used to obtain the probabilities, and the probabilities were combined, as indicated, by the event trees to determine the probability of various failure states of the plant. Careful analysis of these conditions was required to determine whether serious fuel overheating would result or not. The application of fault trees and event trees thus provided a method for carrying out tasks 1 and 2 indicated in FIGURE 1.

CONSEQUENCE ANALYSIS

Task 3 was carried out by an analysis of the conditions created by the failures identified in task 1. These analyses allowed the group to determine whether fuel melting had occurred or not. And, given that fuel melting had occurred, further analysis was done to estimate how much radioactivity would be released. The release of radioactivity could come about by various kinds of failures of the containment itself. The group assumed that if the core melted, the containment would surely fail. However, there were different ways in which the containment might fail,

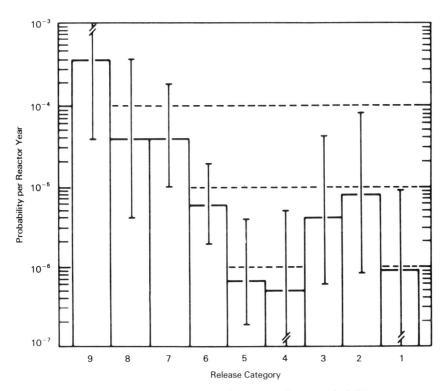

FIGURE 4. Histogram of PWR radioactive release probabilities.

TABLE 1

SUMMARY OF ACCIDENTS INVOLVING CORE

Release Category	Probability (yr^{-1})	Time of Release (hr)	Duration of Release (hr)	Warning Time for Evacuation (hr)	Elevation of Release (m)	Fraction of Core Inventory Releases*							
						Xe-Kr	Org-I	I-Br	Cs-Rb	Te	Ba-Sr	Ru†	La‡
PWR 1	7×10^{-7}	1.5	0.5	1.5	25	0.8	6×10^{-3}	0.6	0.4	0.4	0.05	0.4	3×10^{-3}
PWR 2	5×10^{-6}	2.5	0.5	1.5	0	0.9	7×10^{-3}	0.7	0.5	0.3	0.06	0.02	4×10^{-3}
PWR 3	5×10^{-6}	2.0	1.0	1.5	0	0.8	6×10^{-3}	0.2	0.2	0.3	0.02	0.02	3×10^{-3}
PWR 4	5×10^{-7}	2.5	3.0	1.5	0	0.5	2×10^{-3}	0.09	0.04	0.03	5×10^{-3}	3×10^{-3}	4×10^{-4}
PWR 5	1×10^{-6}	2.5	4.0	1.5	0	0.2	2×10^{-3}	0.03	9×10^{-3}	5×10^{-3}	1×10^{-3}	6×10^{-4}	7×10^{-5}
PWR 6	1×10^{-5}	12.0	10.0	1.5	0	0.2	2×10^{-3}	8×10^{-4}	7×10^{-4}	1×10^{-3}	9×10^{-5}	7×10^{-5}	1×10^{-5}
PWR 7	6×10^{-5}	10.0	10.0	N/A	0	5×10^{-3}	2×10^{-5}	2×10^{-5}	1×10^{-5}	2×10^{-5}	1×10^{-6}	1×10^{-6}	2×10^{-7}
PWR 8	4×10^{-5}	0.5	0.5	N/A	0	2×10^{-3}	5×10^{-6}	1×10^{-4}	5×10^{-4}	1×10^{-6}	1×10^{-8}	0	0
PWR 9	4×10^{-4}	0.5	0.5	N/A	0	3×10^{-6}	7×10^{-9}	1×10^{-7}	6×10^{-7}	1×10^{-9}	1×10^{-11}	0	0
BWR 1	9×10^{-7}	3.0	2.0	2.5	25	1.0	7×10^{-3}	0.50	0.40	0.70	0.05	0.5	5×10^{-3}
BWR 2	2×10^{-6}	3.0	0.5	2.5	0	1.0	7×10^{-3}	0.60	0.30	0.10	0.04	0.07	2×10^{-3}
BWR 3	1×10^{-5}	28.0	5.0	2.5	0	1.0	7×10^{-3}	0.08	0.05	0.20	0.03	0.06	3×10^{-3}
BWR 4	3×10^{-5}	9.0	0.5	2.5	0	1.0	7×10^{-3}	0.10	0.07	0.07	9×10^{-3}	6×10^{-3}	9×10^{-4}
BWR 5	1×10^{-5}	5.0	2.0	2.5	0	0.6	3×10^{-3}	0.05	0.02	0.05	2×10^{-3}	3×10^{-3}	6×10^{-4}
BWR 6	1×10^{-4}	30.0	5.0	N/A	0	4×10^{-4}	3×10^{-8}	6×10^{-12}	4×10^{-11}	8×10^{-14}	8×10^{-16}	0	0

*A discussion of the isotopes used in the study is found in Appendix 6 of Reference 1. Background on the isotope groups and release mechanisms is found in Appendix 7 of Reference 1.

†Includes Mo, Rh, Tc.

‡Includes Nd, Y, Ce, Pr, Pm, Np, Pu, Zr.

and these different ways could release different amounts of radioactivity. The result of step 3 was to generate a histogram of the magnitude of release of radioactivity versus the likelihood of release of radioactivity.

The final result of this analysis was a histogram of the type shown in FIGURE 4. This figure was the result obtained for a pressurized water reactor (PWR) in the study [of course, a similar histogram was developed for a boiling water reactor (BWR)]. The ordinate gives the probability per year of a release. The abscissa gives the release magnitude as one of nine different categories. Category nine represents quite small releases; category one is the largest release. In TABLE 1, the fraction of the core inventory of various fission products released in each category is tabulated.

In the analysis, all identified accidents are assigned to one of the nine categories according to the calculated release of radioactivity. The probabilities of all accident sequences assigned to that category are then summed to get the final histogram shown in FIGURE 4.

The next step was to develop a model for calculating the consequences of the release of radioactivity. This computer code calculated five different health effects, plus the economic loss in dollars due to property damage, cleanup costs, and cost of relocating people. The five health effects considered were early fatalities (fatalities within one year); early injuries (nonfatal injuries requiring medical care); cancer fatalities (cancer deaths that might be expected in a 10- to 40-year period after the accident); thyroid injury (latent nonfatal effects of the thyroid gland that would require medical care); and genetic effects.

The consequence code, called CRAC, used a Gaussian plume model to predict how radioactivity was dispersed under prevailing weather conditions. The input data required were the release histogram, demographic data from the Census Bureau out to 500 miles from the site, and a year's worth of weather data from the site. Using a Monte Carlo method, the code calculated the magnitude of a large number of possible accident consequences and the probability of these various consequences. The reader interested in more details about the model is referred to Appendix 6 of the *Reactor Safety Study*.[1]

RESULTS

The results of the analysis were presented in three different ways often used for expressing risk. The societal risk was defined as the average annual impact of a 100-reactor industry with the type of reactors analyzed in the report. The individual risk was the average probability that any particular individual would suffer a given health effect as a result of a nuclear plant accident. The results for these two types of risks are given in TABLE 2.[1]

The third method of presentation was complementary, cumulative probability distributions that expressed the probability of an accident of

TABLE 2

APPROXIMATE AVERAGE SOCIETAL AND INDIVIDUAL RISK PROBABILITIES PER YEAR FROM POTENTIAL NUCLEAR PLANT ACCIDENTS*

Consequence	Societal	Individual
Early fatalities†	3×10^{-3}	2×10^{-10}
Early illness†	2×10^{-1}	1×10^{-8}
Latent cancer fatalities‡	7×10^{-2}/yr	3×10^{-10}/yr
Thyroid nodules‡	7×10^{-1}/yr	3×10^{-9}/yr
Genetic effects§	1×10^{-2}/yr	7×10^{-11}/yr
Property damage ($)	2×10^{6}	—

*Based on 100 reactors at 68 current sites.[1]

†The individual risk value is based on the 15 million people living in the general vicinity of the first 100 nuclear power plants.

‡This value is the rate of occurrence per year for about a 30-year period following a potential accident. The individual rate is based on the total U.S. population.

§This value is the rate of occurrence per year for the first generation born after a potential accident; subsequent generations would experience effects at a lower rate. The individual rate is based on the total U.S. population.

any given size or larger. Such plots were prepared for each of the consequences. FIGURE 5 gives such a plot for the consequence of early fatalities. The results of the curves for the three latent health effects are summarized in TABLE 3. The numbers in the table are the expected annual rate of incidence of these three effects over a 30-year period, starting about 10 years after the accident. For comparison, the normal incidence rate experienced by the exposed population is also given. From TABLE 2, it can be seen that the latent cancer fatalities are on average about 700 times greater than the early fatalities. This is because in a large fraction of accidents, exposure levels are so low that no early fatalities are expected, but the relatively low doses to a fairly large population do lead to some expected latent cancer fatalities. In the largest accident identified, at a probability of one in 10,000,000 per 100 plants per year, the ratio of latent fatalities to early fatalities is 45,000/3300 = 13.6.

THE REACTOR SAFETY STUDY AND THREE MILE ISLAND

A reasonable question to ask is, How well did this approach predict an event of the type that occurred at Three Mile Island (TMI)? There are in fact several ways to approach this question. First one might ask how likely was a radioactive release of the magnitude that occurred at TMI according to the RSS analysis. One could also ask how well RSS analysis identified the specific chain of events that occurred at TMI. Let us consider each of these questions.

The actual radioactive release at TMI was reported to be about 10 Ci of radioiodine and about 10^{6} Ci of noble gases. According to the RSS, in a category 9 release, 10^{-7} of the iodine and 3×10^{-6} of the noble gases will be released. The original iodine inventory is about 10^8 Ci, so on that basis it was a category 9 release. In terms of the noble gas ^{133}Xe, the inventory is about 2×10^{-8} Ci and the release nearer to that predicted for category 8. However, the iodine is orders of magnitude more serious in terms of public health effects, and a category 8 release of iodine would be 1,000 times larger than actually occurred at TMI. For these reasons, it seems reasonable to conclude that in terms of potential public health effects, this release most nearly resembles a category 9 release. From FIGURE 4, we see that such a release is predicted to have a nuclear probability of 4×10^{-4} (1 in 2,500) with a range from 1 in 250 to 1 in 25,000. There had been about 400 plant years of PWR experience at the time of TMI. The observation of one such release in 400 plant years is certainly quite consistent with the RSS prediction. Clearly this event would also be

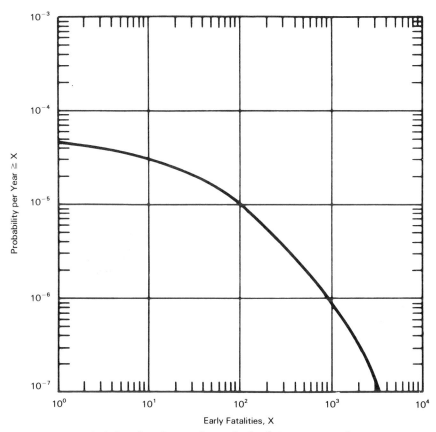

FIGURE 5. Probability distribution for early fatalities per year for 100 reactors. Approximate uncertainties are estimated to be represented by factors of 1/4 and 4 on consequence magnitudes and by factors of 1/5 and 5 on probabilities.

TABLE 3

CONSEQUENCES OF REACTOR ACCIDENTS FOR VARIOUS PROBABILITIES FOR 100 REACTORS

	Consequences		
Chance per Year	Latent Cancer Fatalities† (per year)	Thyroid Nodules† (per year)	Genetic Effects‡ (per year)
One in 200*	<1.0	<1.0	<1.0
One in 10,000	170	1400	25
One in 100,000	460	3500	60
One in 1,000,000	860	6000	110
One in 10,000,000	1500	8000	170
Normal incidence	17,000	8000	8000

*This is the predicted chance per year of core melt for 100 reactors.[1]

†This rate would occur approximately in the 10- to 40-year period after a potential accident.

‡This rate would apply to the first generation born after the accident. Subsequent generations would experience effects at decreasing rates.

consistent with a higher value of the release probability. Using a similar analysis, the President's Commission on Three Mile Island concluded:

The 1974 WASH 1400 Reactor Safety Study (the Rasmussen Report) analyzed events, equipment failures, and human errors that could happen during reactor accidents, including those associated with the TMI accident. However, NRC [Nuclear Regulatory Commission] has not made systematic use of WASH 1400, a major study commissioned by the Atomic Energy Commission (AEC) in its design review analyses. WASH 1400 showed that small-break LOCA's similar in size to the accident at TMI were much more likely to occur than the design basis large-break LOCA's, and can lead to the same consequences. Further, the probability of occurrence of an accident like that at Three Mile Island was high enough, based on WASH 1400, that since there had been more than 400 reactor years of nuclear power plant operation in the United States, such an accident should have been expected during that period.[2]

Now let us examine how well the RSS predicted a chain of events similar to TMI. The initiating event at TMI was a loss of feedwater to the steam generator. This was followed by a pressure rise in the primary system, which opened the safety relief valve (SR). The auxiliary feedwater system (AFWS) was turned on, but it failed to work for the first eight minutes. The reactor protection system (RPS) was automatically initiated to shut down the plant, and it functioned properly. After the pressure surge, the safety relief valve should have closed; it failed to do so, leading to a small LOCA.

Events such as this should appear in the transient event tree for the PWR. FIGURE 6 from the RSS shows this tree. The initial event in the figure, designated T, was caused by the loss of feedwater. Event K, the reactor protection system, functioned properly; thus one follows the line upward. The secondary system relief valve (SSR) worked, but the power conversion system (PCS) failed because of loss of feedwater, so function M failed; thus we go down at this branch point. The auxiliary feedwater failed, so function L is a failure. The safety relief valve opened, so function P is a success. The safety relief valve failed to reseat, so function Q is a failure. The chemical volume control system (CVCS) was not needed, nor was the residual heat removal system (RHRS), so they need not be considered. Basically what happened initially was event TMLQ.

The event TMLQ was predicted to lead to core melt, except that a footnote suggests that this event leads to a small LOCA, and under some conditions core melt may not occur. Clearly, the RSS analysis identified the status of the plant for the first few minutes of the accident as a serious condition that might lead to core melt. However, within eight minutes the AFWS was recovered, so event L became a success; and later the open

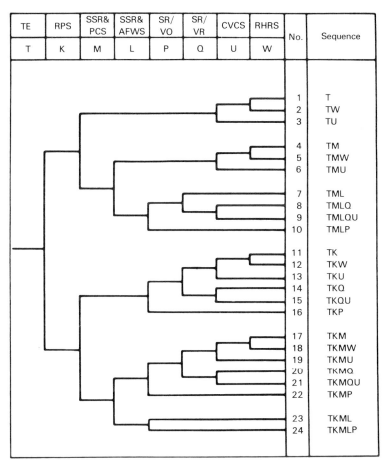

FIGURE 6. PWR transient event tree.

relief valve was isolated so that Q became a success. In the end, core melt was averted, but not until very serious overheating of the fuel had occurred. Thus the analysis did identify in a general sense the possibility of an accident of the type that occurred and, further, that it was potentially a very serious accident. The analysis, of course, did not predict in precise detail the actual events.

It is also of interest to review what frequency the analysis gave for the events TMLQ. For the event TM, a transient initiated by loss of main feedwater, the frequency of occurrence was given as ~ 3 per year based on experience. The loss of auxiliary feedwater for more than half an hour was estimated at $\sim 4 \times 10^{-5} \pm$ a factor of 3 per demand, and the failure of the valve to reseat was $10^{-2} \pm$ a factor of 10 per demand. Thus TMLQ's frequency would be $3 \times 4 \times 10^{-5} \times 10^{-2} = \sim 1 \times 10^{-6}$ if we assume AFWS failed, or 3×10^{-2} if we assume it did not. Since the failure of AFWS for eight minutes did not affect the course of the accident, it was not a failure in the sense of event K. The real problem of TMI was the shutting off by the operators of the high pressure injection system that replaced water lost through the stuck-open valve. The probability of failing to provide make up for lost inventory was given as 2×10^{-3} per demand, with an error spread of $\pm$ a factor of 4. Thus the RSS would predict a probability for TM-Q-U (U is failure to make up inventory) of $(3)(10^{-2})(2 \times 10^{-3}) = 6 \times 10^{-5}$, or about 1 in 16,000 with an uncertainty of about a factor of 10. Although the probability may in fact be correct for the Westinghouse plant that was analyzed in the RSS, I believe most would agree it is too low for the Babcock & Wilcox plant and the state of operator training that existed at TMI. It seems quite possible that a careful analysis of a Babcock & Wilcox-type plant that integrated the Davis-Besse experience might well have shown this prior to TMI. One cannot help but note one of the prophetic findings of the Lewis committee, issued some six months before TMI occurred:

> The achievements of WASH-1400 in identifying the relative importance of various accident classes have been inadequately reflected in NRC's policies. For example, WASH-1400 concluded that transients, small LOCA, and human errors are important contributors to overall risk, yet their study is not adequately reflected in the priorities of either the research or regulatory groups.[3]

Despite the successes of the RSS analysis, noted above, there was one aspect of the TMI accident that the study failed to recognize. This was the fact that a large amount of hydrogen might collect in the primary cooling system and could possibly interfere with the proper cooling of the core. The study took account of the fact that overheating of the cladding would produce a large volume of hydrogen and that it might burn or explode in the containment and threaten containment integrity. It concluded that such hydrogen burning would not fail a containment of the type that existed at TMI. However, the study assumed that if the fuel got hot enough to produce large volumes of hydrogen, then the core would melt, and melt through the vessel, and the hydrogen would not be trapped in the primary system. At the time, this was felt to be a conservative assumption. We now know that even after very serious cladding degradation, it is possible to cool the fuel and prevent core melt; so the conservative assumption led the study to an incorrect conclusion. This, once again, points out how careful one must be in the use of what are thought to be conservative assumptions. In failing to recognize this possibility, the study failed to point out that a relatively easy fix, such as a remotely operated valve on top of the vessel, could easily cope with such a problem.

A second shortcoming was that the study implicitly concluded that a category 9 release—because it would have a minimal health impact— was a trivial accident. Even though TMI will clearly have a minimal health impact, no one can conclude it was a trivial accident in terms of its psychological impact on the public or its impact on the nuclear industry. Thus I conclude that these small accidents in terms of health impacts must be considered in more detail in future analyses of overall reactor safety.

CONCLUSIONS

It seems clear that reflection on the TMI accident has shown that a careful risk analysis of a plant can in fact be an extremely valuable tool in understanding risks so that resources can be better allocated in reducing them. However, such analysis techniques have not yet developed to the point where they are ready to fully replace the older historic approaches to the problems of risk reduction. Thus they should, in my opinion, be added to the tools of a good safety analysis, but clearly they are far from the point where they should replace good engineering judgment. I thoroughly support the current trend of applying these methods more widely in the nuclear field. However, I would add a word of caution, for if they are not applied with care and good judgment, the results may be very misleading. Their improper use could lead to a sense of complacency that is not warranted. This is particularly true at this time, when the demand for such analyses exceeds the resources of well-trained personnel for carrying them out.

REFERENCES

1. NRC. 1975. Reactor Safety Study: An Assessment of Accident Risk in U.S. Commercial Nuclear Power Plants. Report No. WASH-1400 (NUREG-75/014). U.S. Nuclear Regulatory Commission. Washington, D.C.
2. KEMENY, J. G., et al. 1979. Report of the President's Commission on the Accident at Three Mile Island. U.S. Government Printing Office. Washington, D.C.
3. NRC. 1978. Risk Assessment Review Group Report to the U.S. Nuclear Commission. Report No. NUREG/CR-0400. U.S. Nuclear Regulatory Commission. Washington, D.C.

Discussion

A. M. Weinberg (*Institute for Energy Analysis, Oak Ridge, Tenn.*): Dr. Rasmussen, could you give us an estimate of how much you think the probability has to be reduced by.

N. C. Rasmussen: Well, we've done a lot of studying of the perception of risk. As you know, the nuclear industry calculates the risk and then tries to convince the public that it's low enough. We've just finished a thesis where we did the opposite. We tried to study what risk the public would accept if their behavior was consistent with their past behavior. We concluded that nuclear has to be about 10 times safer because it's a new technology, it should be about a factor of 30 safer because it can have catastrophic accidents, and it should be about a factor of 30 safer than coal because its consequences are delayed and not immediate. Overall, by this estimate, it should be about a factor of 9000 safer than coal. If our estimates are correct, that's just about what it is—the risks of early fatalities from nuclear accidents are about 9000 times lower than the delayed risks from coal burning.

Unidentified Speaker: In evaluating the radiation release impacts of the Three Mile Island accident, have you included future radiation releases that may occur incidental to the cleanup?

N. C. Rasmussen: I haven't calculated Three Mile Island under any conditions. You realize that all these calculations were for hypothetical accidents that we studied. This long after the accident, however, the potential for any further significant dose to the public is very small.

Unidentified Speaker: What is the situation if you are multiplying the probability of a small accident with a corresponding fatality and the large accident probability with the fatalities? What does it look like?

N. C. Rasmussen: That's why we draw curves to show the probability versus magnitude. An accident that has a probability of 10% times 100 fatalities gives you 10 fatalities a year. An accident that has a probability of 1 in a million and exposes a 10-million population also has 10 per year. That's why we made the other kinds of plots, which are magnitude versus probability, so you can see whether large events are possible. Those curves show that the largest event was a little over 3000 early fatalities and 45,000 cancers over 30 years. Those kinds of curves reveal that type of information; so you can think of it that way as well. But there's no way that I know to put both in the same equation.

G. Kuenstler (*In These Times, Chicago, Ill.*): How do operators work at the board? Do they work with permutations of these variables in front of them? It would seem that you've outlined possibly a thousand variables. How do they work? You spoke of the innovative operator.

N. C. Rasmussen: There are always at least two people in the control room, an operator and a supervisor, and they work as they do with any complicated machine. They've been trained to understand how the machine behaves in normal operation, and hopefully they've been trained on a lot of unusual situations that might occur, and there are typical, trained responses: if the water level gets too low, do this; if the water level gets too high, do that.

They are also supposed to understand the fundamental theory of what's happening physically in the machine so that if a set of events that they haven't been specifically trained to identify occurs, they will understand what's going on, what is moving from where to where, what heat is moving from where to where and, by these things, try to identify what the cause or the condition is.

If you've ever seen a picture, the control room is a big place. It's got a lot of dials and a lot of handles and a lot of information to present to two men. At Three Mile Island, I'm told—and I guess Mr. Jaffe can say more—hundreds and maybe a thousand or more bells rang and lights lit within a period of minutes. That's too confusing a situation for anybody to comprehend fully, and they didn't. They looked at some things that they thought were important and did what they had been told to do when they saw the water level go to the top of the pressurizer, which turns out to be the wrong thing for them to be paying attention to at that time. Unfortunately, they weren't trained adequately enough to realize the problem, so they did the wrong thing by shutting off the water. I don't think we dare blame them individually, as much as we should blame the system that didn't train them in all of the implications.

But what operators do is look at all that information and try to integrate it, based on their training and experience, to come to some conclusion about the condition of the machine. I don't know how to explain it to you any better than that. The airplane pilot does it; the ship navigator does it. If you ever see the control panel of any one of the installations in an oil refinery, your first impression is that it's much too much for anyone to comprehend. If you've ever looked in the pilot's cockpit of a 747, your first reaction is, God, the guy can't keep track of all that. But there are systematic ways of doing it, and there are annunciator lights that help them focus attention on the right thing. I just think we haven't made as much progress in the nuclear business as modern technology allows in analyzing and presenting that information. I would think that that is one thing we learned.

H. LEWIS (*University of California, Santa Barbara, Calif.*): It has been suggested that it is not too late to take the top 50 sequences in your famous report and go through them in great detail and ask whether the instrumentation is in place so that we would know that they were happening if they did happen and the operators would know how to recognize them if they happened. Do you think it's a dumb idea? If you don't think so, what is the right number instead of 50?

N. C. RASMUSSEN: It's at least 25. If you just take the page that summarizes our results, there are at least 25 sequences, and it doesn't take that long to go through and do it. I'm convinced that this is the most rational, logical way to sort out what's going on. I personally think your suggestion is 100% correct. If you have this information already analyzed, why not do it? It doesn't take months to do it. Two or three or four good people could go through a system in weeks, I would think, and identify what would happen in those sequences. I think it makes good sense, but I've had trouble convincing the Nuclear Regulatory Commission of the validity of this approach, and they haven't been quick to use it.

E. LERCH (*Interfaith Center on Corporate Responsibility, New York, N.Y.*): My question has to do with the Nuclear Regulatory Commission. Do you think the NRC ought to do an evaluation of operating reactors in terms of the probability of class 9 accidents?

N. C. RASMUSSEN: My personal opinion is that part of the risk of reactor operations comes from accidents well beyond their class 8 accidents, and so a careful and a thorough assessment of reactor risk must include all these accidents. That does not necessarily mean that I believe that elaborate new systems to cope with class 9 accidents are required. But I think that in their responsibility for looking at the risk to the public, they must evaluate and make some judgment on the likelihood of a variety of accidents worse than the design basis and whether or not their contribution to risk is small enough so that no further design features are needed to cope with them.

TECHNICAL ASPECTS AND CHRONOLOGY OF THE THREE MILE ISLAND ACCIDENT

Leonard Jaffe

Technical Task Force
The President's Commission on the
Accident at Three Mile Island

and

National Aeronautics and Space Administration
Washington, D.C. 20546

I indeed feel very privileged to be able to address you today on the subject of the Three Mile Island accident. I felt it both an honor and a terrible obligation to be a part of the staff of the president's commission investigating the accident. It was an awesome responsibility.

The Three Mile Island (TMI) accident may be to the nuclear power industry what the Apollo fire was to this country's manned space flight program. That fire, which killed three astronauts, made the space industry aware of the necessity to renew its commitment to provide the management, technical, and quality assurance talent that would pay great attention to the smallest of details and every bit of experience to insure the success of the Manned Lunar Landing Program and the safety of future astronauts.

It is important, at this point, for me to do what the president's commission did very early in its report.[1] It indicated very clearly what it did not do. It did not examine the entire industry. It did not deal with the very important questions of nuclear waste disposal, for example. It looked solely at the TMI-2 accident and factors that related to it.* In this same spirit of setting the record straight, I want to tell you that I am not a nuclear engineer. I am an electrical engineer by training, and my professional career has been devoted to aeronautics and space. For the last 22 years, I have been involved in the development of the applications of space. I have directed the development of large, technically risky projects in which the public exhibited keen interest. I have had some considerable experience in the investigation of failures. Perhaps that is why I was asked to become involved.

With the help of a small, but excellent, group of real nuclear experts from the Navy, the Department of Energy, the National Energy Laboratories, and the National Aeronautics and Space Administration and many consultants from industry and the universities, we looked at the events leading to the accident, the accident itself, and the response of individuals and organizations to the accident. This involved six months of a most

---

*TMI-2 refers to one of two plants at Three Mile Island, the one in which the accident occurred.

0077-8923/81/0365-0037 $01.75/2 © 1981, NYAS

intensive effort, and resulted in hundreds of thousands of pages of records and testimony that had to be digested. It is not possible nor would it be appropriate to cover that effort in detail. I will try to tell you broadly what did happen and then give you my impression of possible contributors to the events that led to TMI-2.

At 4:00 A.M. on March 28, 1979, TMI-2 was operating normally at nearly 100% rated power. At this power level, approximately 2,700 megawatts of thermal energy are being generated by the reactor, which heats the primary coolant to 600°F at a pressure of about 2,200 psig. This coolant is pumped through steam generators where it gives up its energy to create steam in the secondary loop, which drives a conventional steam turbine to generate about 1,000 megawatts of electrical energy. TMI-2 first went critical exactly one year to the day before the accident. It had been declared operational at the end of December 1978.

It all began just after 4:00 A.M. when the main feedwater pumps that circulate water in the secondary loop tripped, interrupting the removal of heat from the primary system. Upon sensing this loss of feedwater flow, auxiliary feedwater pumps were started, which should have reestablished a supply to the steam generators. However, a pair of block valves in the discharge lines of these emergency pumps, which were supposed to be open at all times during normal operations, were in fact closed, resulting in the steam generators essentially boiling dry.

The primary system temperature and pressure began to rise almost instantly with the loss of heat removal, causing the pressure relief valve on the top of the pressurizer to open. In about 15 seconds, the pressure in the primary loop had dropped to the point where the relief valve should have closed, but it failed in an open condition, creating a leak in the system. The pressure continued to drop to the point where emergency high pressure injection pumps were automatically turned on to replace the lost liquid. The closed block valves were discovered and opened at about 8 minutes after the onset of the event. This would have been soon enough to prevent damage to the plant, but the open relief valve was not discovered and blocked off for 2 hours and 20 minutes.

Because of the open relief valve and the increasing water temperature and the high pressure injection of emergency coolant, the water level in the pressurizer rose, causing the operators to believe that they had too much water in the primary system when in fact there was too little. The reduced pressure permitted steam pockets to form, contributing to the increase in pressurizer water level. As a result, the operators took several actions that made matters worse rather than better. In an effort to control the level of liquid in the pressurizer and to prevent the system from becoming solidly filled with water, they throttled high pressure injection and let water out of the system via the "let-down line." Shortly thereafter, they turned off the emergency high pressure injection system, which had functioned properly up to that point. This caused further increases in the amount of gas (steam) in the system to the point where the main circulation pumps began to cavitate, and then they were turned off to prevent damage. It was at this point that the two phases of the coolant in the primary system separated and the upper portion of the core became uncovered. The temperatures there rose to the point where the core cladding material, Zircalloy, could react with the steam. This reaction produces zirconium oxide and hydrogen, a noncondensable gas, which accumulated at the high points in the system. This made it difficult to reestablish coolant circulation.

Before the cladding reacted with the steam, fuel rod internal pressure buildup had caused some of the cladding to rupture, releasing radioactive fission products to the coolant, which flowed out through the open relief valve into a drain tank, which in turn eventually overflowed onto the floor of the containment building. When the level of water in the sump area rose to a specified level, a sump pump automatically started and pumped some of the radioactive coolant into storage tanks in the auxiliary building; these tanks also overflowed onto the floor of that building.

At about 1:30 in the afternoon of March 28, there was an explosion, or ignition, of a pocket of hydrogen that had accumulated in the containment building, but this apparently resulted in little damage, if any.

Circulation was reestablished about 16 hours into the accident, but the damage to the core had taken place. Although some say the accident was essentially over at that time, consternation continued through the weekend to Monday, April 3, partly due to misunderstandings and misinformation. On Friday morning, March 30, an evacuation recommendation was made by the Nuclear Regulatory Commission (NRC) to the governor of Pennsylvania in response to a radioactive gaseous release from the plant. The entire weekend was one that frightened the community with discussions of the possibility of a hydrogen explosion within the primary system and of dire consequences.

Let me go back to the beginning and look at some of these events with a view toward shedding light on some possible root causes of the accident.

The commission reported that the most probable initiator of the events of that March 28 was the shutdown of the condensate polishing system. It was this that probably interrupted the flow of secondary feedwater. The condensate polisher is a device that is used to remove impurities from the condensed water before it's returned to the steam generator. It's an ion exchange device that catches the impurities in a bed of resin that must periodically be replaced. Such a maintenance procedure was in fact being executed at 4:00 that morning on one of eight parallel units of the system. This routine maintenance procedure called for the injection of high pressure air to break up the bed so that the resin could be transported out of the system with the fluid prior to replacement with fresh resin.

Although the failure had not been duplicated at the time of the commission report, it was felt that the most probable cause for the shutdown was water getting into the instrument air lines that controlled the entire system. The controls were designed or modified so that the entire system might shut down instead of only a portion of it doing so.

Why did this happen? Problems with the condensate polisher had occurred before at TMI-2, with at least one operator having noted that something should be done to avoid a serious problem. The system, as it existed (and it may have been modified from the original design), was apparently not forgiving of a shutdown. The other sister plant on the site, TMI-1, is not completely identical to TMI-2. TMI-1 routinely bypasses 50% of the flow around the condensate polisher, so there is no possibility of a complete loss of all feedwater. TMI-1 also incorporates an automatic bypass valve that would open upon shutdown of the polisher. TMI-2 had only a manually operable bypass valve, which was normally closed. Taking all this into account and recognizing that the maintenance procedures might result in shutdown, procedures could have been considered that would have required that the bypass valve be opened during these maintenance procedures. Was engineering consideration of the original design or of design modifications adequate? Where were the quality and assurance people who could have observed maintenance procedures and reviewed not only the adherence to procedures but the adequacy of the procedures themselves? Our impression was that the quality and assurance staffing was such that only a small percentage of procedures could be monitored. Secondly, the condensate polisher was not labeled a "safety related" system and, therefore, was not as high on the priority list of things to look at. This term "safety related" comes up time and time again.

The next system that failed to provide its function was "safety related"—the auxiliary feedwater system. It came on automatically on loss of feedwater flow as it should, but two blocking valves were closed instead of open as prescribed and thus prevented circulation until discovered closed eight minutes later. Why were the valves closed? We don't know! The operator's control panel indicators showed them to be closed. One of the indicators was covered by a "tag" hanging from a control above it. A routine surveillance procedure had been performed on this system two days before the event. That surveillance checkoff record certified that the valves had been left open at the conclusion of the test. At least six operator shift changes had occurred between that surveillance procedure and 4:00 A.M. on March 28. No one had observed the valves in the wrong position. There is no routine checklist requiring operators to check all valve alignments on shift changes.

The above raises several questions about (1) the adequacy of the surveillance procedures and quality assurance monitoring; (2) the adequacy of control room procedures and discipline; and (3) the adequacy of control panel indicators, displays, and tagging practices.

The surveillance procedure called for certification of procedural completion by individuals who did not actually witness the restoration of correct valve lineup. Operational procedures in the control room are not independently monitored by the quality assurance people. It is left to the operating and engineering staffs to review their own performance.

Valves have been found misaligned at TMI before. The TMI-2 control room does not routinely employ computer aids to assist in determining correct alignment of systems for the conditions prevailing or desired. This should be a simple system to implement. Even the color coding employed on the control panel is not conducive to easy realization of out-of-normal conditions. In the space business, we try to adhere to color-coding practices on control panels in which normal conditions or correct operating status is indicated by green lights. In general, red lights are few. They indicate trouble. A glance at the board quickly tells you where the trouble spots are. What is correct or normal in a nuclear power plant changes with the condition or function required at the time. The same is true for the various phases of a space mission. This requires computational aids to check status of systems and alignments for each operational situation, and with adequate telemetry, it can be done.

The next and perhaps the most important *equipment* failure in terms of its ultimate consequences was the failure of the electromatic pilot-operated relief valve (PORV) to close after properly opening to relieve pressure in the primary coolant system. Note that I said *equipment* failure. The control panel indicator showed the valve closed, but this indicator really monitored the electrical input to the valve and not the mechanical valve position. For 2 hours and 20 minutes, this condition went undetected and uncorrected even though there were many other indicators of loss of coolant through that valve. The actions taken by the operators (and concurred in by engineering and supervisory people on the scene) during this period clearly indicated a lack of understanding of the system under the conditions of this size loss-of-coolant accident. This occurred in spite of the following:

1. Pilot-operated relief valves have experienced failures.

2. Almost precisely the same situation (including operator misinterpretation and inappropriate initial response) had occurred in other plants of the same basic design.

3. Numerous indications of loss of coolant from the primary system.

4. Previous warnings that pressurizer level was not a reliable indicator of coolant level in the core.

5. Previous warnings (in the Rasmussen report) that an accident of the TMI-2 type had a higher probability of occurrence than did the more severe design basis accidents.

Again the question is, Why? There are numerous contributors.

1. Inadequate monitoring and follow-up of equipment difficulties. We could find no comprehensive industry-wide record kept of equip-

ment problems with a view toward using this information to eliminate recurring equipment problems.

2. Operating and maintenance procedures were inadequate or deficient in light of recurring problems and recognized and documented potential for operator misinterpretation of plant condition.

3. Inadequate consideration and follow-up of prior experiences and concerns.

4. Inadequate training of operators in system fundamentals.

5. Inadequate training of operators in the response of the specific system to problems.

6. Inadequate simulator experience. Prior to the accident, the Babcock & Wilcox simulator used to train TMI operators could not simulate boiling water in the system other than in the pressurizer.

7. Lack of analytic aids to operators in the control room to assess the problem or to monitor the required alignment of the system.

8. Poor location and indication of some useful measurements.

9. Inadequate range on some instrumentation to cope with the excursions experienced during the transient.

10. A plethora of alarms.

11. Inadequate engineering capability to back up operations.

There are others.

But let's continue the scenario, for even after the PORV was discovered open and then closed at 2 hours and 20 minutes into the accident, the operators did not recognize that they had lost coolant, and actions taken turned out to be inconsistent with the fundamental requisites of keeping the reactor fuel covered and insuring heat removal from the system. The automatic emergency systems were designed to do this and did what they were supposed to do. They were interfered with. They were prevented from doing their job during the critical period of the accident. This was the major cause of TMI-2 damage.

Other plants have had similar experiences. Davis-Besse had an almost identical situation in which the effects of a stuck-open PORV were misinterpreted by operators and emergency high pressure injection was throttled in response to water level increases in the pressurizer. But Davis-Besse was operating at a small fraction of rated power at the time, and they discovered the open PORV in 12 minutes instead of 2 hours and 20 minutes. This was reported in a licensee event report (LER). Why wasn't that LER given more attention? Possibly because the event was terminated without serious consequence, or because the LER did not acknowledge an operational error, or because the attention that LERs generally get can be questioned.

Emergency control of a nuclear power plant should be concerned with basically two functions: keeping the core covered and getting rid of the decay heat. An emergency control station should be considered in which only those controls and indicators necessary to these functions are located. This control station could be overriding and be such that return of the plant to "safe" conditions is its sole concern and its control should override all other instruction. This control station could be provided with appropriate assistance by computer analysis of plant status and required action to return to safe conditions against which the safety operator could check his judgment and the results of actions taken.

A control room should be designed to handle emergencies. It must be designed so that the operator can handle the unexpected, not only the norm. When events are occurring according to the book, it's relatively easy to know where to look for confirmation of actions taken. In TMI-2 the unexpected occurred, and the operators did not interpret properly. Ultimately, hundreds of alarms were on at the same time; there were so many that some operators complained of the additional confusion.

But before I go on with possible engineering fixes, let's look at some more of the accident. There are human interfaces with an accident other than those associated with controls and indicators on control panels.

At about 1:30 P.M., a hydrogen explosion occurred in the reactor building. A 28-psi pressure pulse was experienced in the containment building. This was not generally recognized as a hydrogen explosion for some time, even though it has long been recognized that hydrogen can be generated and must be attended to in the event of high temperatures in the core.

Later, the accumulation of incompressible gases in parts of the primary system made it difficult to reestablish coolant circulation. Methods for ridding the system of the hydrogen bubble that developed in the primary system had to be devised during the accident.

Frantic considerations and evaluations took place on the ensuing weekend, on March 30 and 31 and April 1, regarding the potential effect of a hydrogen-oxygen explosion in the primary system, when it should have been known that oxygen could not accumulate due to a normal overburden of hydrogen in the coolant.

Questions about what would happen if the core melted could not be answered. A distinction was not made between some melting of the core and a complete "meltdown."

This information should have been available, i.e., in the files, in well-considered studies executed in a thorough and unrushed atmosphere. Information generated on a crash basis is generally subject to question. It usually is encumbered with sufficient uncertainty that one is driven to assume the worst-case scenario in order to be safe. In my view, one should give consideration to worst-case scenarios when contemplating the design of a system. But when you are in an emergency—when you have to take action to prevent catastrophe—you must understand the situation as it most probably is. To base action and decisions solely on worst cases and hastily generated studies may remove realistic solutions from your list of options and can cause unnecessary problems, as did the scare caused by the potential of an H_2-O_2 explosion in the primary system over that critical weekend.

We apparently also were not prepared to deal with an emergency that lasted for any length of time. This was probably the first time that the NRC had time to deploy forces, reorganize its response, and take part in a firsthand way in the resolution of a nuclear plant transient. We were unprepared. Communication channels were inadequate, changed, or otherwise ignored. This resulted in a certain amount of confusion and—at least in the case of the March 31 release of radioactivity from the plant and the resulting advice to the governor that an evacuation was in order—in a serious error.

Communications channels that did exist were not used to fully understand the circumstances of the release. Again, *ad hoc* provisions and arrangements can result in such confusion. Accidents and responses to accidents should be planned and well rehearsed.

But here I am again, engineering fixes to a particular set of problems. I'm afraid this is what the NRC has been doing for years. They have instituted "fixes" or patches to take care of particular problems experienced, rather than making fundamental and institutional changes that will insure that the system will become self-correcting and will take appropriate action based on experience, thorough analysis, and rigorous follow-up with corrective measures.

Some of the fundamental concepts really work. Defense in-depth is one. In spite of the problems and errors of TMI-2, no one was killed. The plant was returned to a safe, stable condition. There are, however, some design aspects that may compromise some of the defense in-depth measures. Defense perimeters may be breached, either for operational convenience or to provide a measure of protection against the large design basis accidents that have low probability of occurrence. The use of PORV may be a case in point. Any break of the primary coolant system boundary should be clearly recognized as something that should not generally occur and therefore, if it occurs, as representing an abnormal or failure situation. In TMI-2, the opening of the PORV on March 28 was not immediately viewed with alarm. It was an expected response to a transient.

Another possible breach of a defense boundary involves double failures. Primary coolant is circulated outside the containment building even though this water may be contaminated (as it was in TMI-2) when core damage occurs. Since the let-down/makeup system is used to control the composition and quantity of coolant in the primary system even during accident conditions, consideration should be given to establishing the containment isolation boundary so that gaseous leakage from the let-down/makeup system would not result in direct releases to the atmosphere.

In trying to look for contributing factors to the accident at TMI-2, we took a look at the utility–industrial supplier–regulatory agency relationships that have developed.

The utility selects a contractor to supply a plant for a price. He expects the contractor to meet the licensing requirements of the NRC, and the supplier agrees to meet those on record at the time the contract is signed. If the NRC determines that changes are required, they are made at additional cost to the utility. Changes that appear worthwhile to the supplier but that are not required by the NRC probably do not get too much attention because they are deemed nonessential. Thus the NRC, in effect, is assuming the responsibility for saying what is adequately safe and what is not. Under these circumstances, the utility may feel secure with that assurance and perhaps even absolved from having to make these difficult technical and financial decisions. This in turn can minimize the requirement for the kind of technical staffing and expertise that the utility might need to have to make these decisions.

Secondly, the ordinary building period for a nuclear power plant is 12 years. The entire microelectronic revolution took place in less time. If you don't commit to the very latest in technology at the outset, you will surely end up with outdated designs. Several studies were executed in the 1973–74 time period that pointed up some inadequacies of existing control room designs, but apparently these have not motivated extensive changes as yet. The mystique of standardization, which says that it is desirable to duplicate previous designs, may still be dominating the industry.

Dr. Harold Lewis, in his recent article "Safety of Fission Reactors," said that the president's commission report was "notable for its lack of specific technical recommendations for the enhancement of reactor safety."[2] We could have recommended specific fixes so that the scenario of TMI-2 per se would not occur again. This is what the NRC tends to do—recommend specific fixes. But how does this prevent a yet unimagined sequence of events from becoming an accident?

If there is one thing that I have learned through the TMI-2 investigation, it is this: Nuclear power plants are very large, very complex systems that cannot be completely accurately modeled. Dangerous transients cannot be incurred deliberately so that the actual plant response to all events can be experienced and tested. The total amount of experience with individual components is still relatively low. It is with this in mind that I feel that the industry must seize upon every transient, every excursion, every abnormality, every operator mistake, and every component failure and learn from it. The industry must pull itself up by its bootstraps. Current plant performance statistics must not be accepted as "good enough" because they may not be good enough for the future, and one accident is one too many.

In the space business, if we have a failure—let's say of a component, a transistor perhaps—we try not to merely replace the faulty component, we try to investigate thoroughly the reason for the component failure. Was it a generic problem? a design problem? a materials problem? a quality control problem? or a one-of-a-kind problem? Why didn't the system tolerate this kind of failure? Can it be made to tolerate this kind of failure? Were our procedures adequate to cope with the situation produced by the failure? We try to pursue these questions vigorously and rigorously before we close out the story of a failure.

In a similar manner, the nuclear industry must be made to vigorously and rigorously examine every implication of a transient or failure experienced by anyone in the business, and this is a specific technical recommendation of the president's commission. It recommended that the following questions be asked—and answered on the occasion of occurrences not previously experienced:

1. Did the system design codes accurately predict the transient?
2. Did the design accommodate it?
3. Was the operational control room capable and convenient to the handling of the problem?
4. Did the operational procedures properly accommodate it?
5. Do the training and the training simulator address this accurately and adequately?
6. Is the component that failed a continuing problem?
7. Are the quality control and quality assurance adequate?
8. Are the organizations appropriate and capable of coping with results?
9. Do we understand the input of the occurrence on the probability of future accidents?

This kind of rigorous investigation and attention should be given to all failures in all systems regardless of whether they are currently defined as "safety related" or not. Many failures or problems could be initiated by non-safety-related systems. If the result can cause confusion and operator error, the definition of what is safety related or not is a moot point.

To do what I have suggested requires two things: (1) a regulatory requirement to insure and demonstrate the adequacy of the abnormal event closeout effort in all of the above areas; and (2) an adequate running recording of all critical plant conditions to insure that transients can be accurately reproduced for evaluation purposes. In other words, an on-line telemetry (flight) recorder at all times. In TMI-2, we were very fortunate to have a recorder called a reactimeter recording many useful parameters. Without it, the postmortem analysis would have been much more difficult indeed, if not impossible. There are currently no regulatory requirements for a recorder and instrumentation adequate for this function.

Only if the kinds of records that I am suggesting are kept and rigorously used to critically examine our prior understanding and ability to handle the situation can we be confident of preparedness. Only if we plan for failure and accidents will we be able to react properly to minimize the effects. Only if we continuously use the statistics of experience and the tools of failure mode analysis to determine where the weak points are and judiciously eliminate them will we be able to constantly reduce the risk of nuclear reactors. It can be done.

The WASH 1400 study reduced the probability of occurrence of a problem at the Peach Bottom plant because it used that plant as a case to study.[3] A failure analysis was made—a fault tree/event tree of the specific plant was analyzed—that did reveal the weakness. Such analyses should be living exercises on each plant. Weak links should be well known and tolerated only if carefully considered in documentation for the record.

It is only if we introduce this kind of rigor into the system and provide a utility-industry-regulator relationship that places the responsibility clearly on the utility for the safe operation of the nuclear power plant—which he ordered built and which he operates and he maintains—that we can be reasonably assured of continued improvement in the safety of these facilities.

In this light, it is interesting to note that Dr. Lewis in his article seems to deplore the fact that the NRC does not assume a greater role for assuring safety akin to that of the National Transportation Safety Board in the aircraft industry. At the same time, he acknowledges that there exists "an Advisory Committee on Reactor Safeguards [on which Dr. Lewis is a member] that can perform this function, and it does so on an ad hoc basis. There is certainly enough information in the operational experience of other reactors to have alerted us to the possibility of an accident of the kind that happened at TMI, but for one reason or another we seem not to have been alerted."[2] Isn't it interesting that almost a year after TMI-2, the Advisory Committee on Reactor Safeguards has not determined why they weren't alerted? Perhaps if they examined that question, they could better recommend ways in which to assure the avoidance of another TMI-2.

The cost of TMI-2 will not be known for a number of years. Our lowest estimate was approximately $1 billion. This is in excess of $10 million per existing nuclear power plant in this country. This does not take into account costs paid for more expensive energy incurred because of the hiatus in the development of new nuclear facilities. I don't know whether nuclear power can be made acceptably safe in the ultimate picture. I do know that we can ill afford not to make the expenditures required to rigorously pursue every experience and every piece of relevant information to drive the probability of an accident continuously toward zero. Each utility should be able to show every year that this year's plant performance and problem record is better than last year's.

REFERENCES

1. KEMENY, J. G., et al. 1979. Report of the President's Commission on the Accident at Three Mile Island. U.S. Government Printing Office. Washington, D.C.
2. LEWIS, H. 1980. Safety of fission reactors. Sci. Am. **242**(3): 53.
3. NRC. 1975. Reactor Safety Study: An Assessment of Accident Risks in U.S. Commercial Nuclear Power Plants. Report No. WASH-1400 (NUREG-75/014). U.S. Nuclear Regulatory Commission. Washington, D.C. (The Rasmussen report.)

BACKGROUND OF THE THREE MILE ISLAND
NUCLEAR ACCIDENT, I: GENERAL DISCUSSION

Moderator: Thomas H. Moss

Subcommittee on Science, Research, and Technology*
U.S. House of Representatives
Washington, D.C. 20515

J. R. LAMARSH (*Polytechnic Institute of New York, Brooklyn, N.Y.*):
I'd like to ask a question about the Advisory Committee on Reactor
Safeguards. I've always had the feeling that ACRS was not a terribly
effective organization over the years. For example, it didn't get its hands
on the fact that they were using candles to determine the pressure inside
the containment structure at Brown's Ferry. Is ACRS becoming more
effective now?

H. LEWIS (*University of California, Santa Barbara, Calif.*): Historical-
ly, the ACRS has not put as much attention, in my personal view, into the
cosmic issues of reactor safety as it should. There's been a very big
licensing load, and the law still requires that ACRS write a letter on
every license application. Fifteen people working part-time just can't do
that and—you know, it's the standard government disease—cannot have
the time to sit back and say, Hey, are we doing a good job? In a certain
sense, the *de facto* moratorium may be good because more attention is
being paid now to the so-called cosmic issues. So I'm an optimist, but we
all know the definition of an optimist.

A. M. WEINBERG (*Institute for Energy Analysis, Oak Ridge, Tenn.*): I'd
like to comment on that. I would only point out that broad criticisms of
that sort about an institution that's been in existence for 32 years are
unfair. ACRS was first started—it was then called the Committee on
Reactor Safety—in 1948. It was founded by Edward Teller. One must not
forget the overriding fact that it originally was the only show in town with
respect to reactor safety.

The fact that over these 32 years, lots has happened is, of course,
correct; but perhaps you can set some of these things right.

N. C. RASMUSSEN (*Massachusetts Institute of Technology, Cambridge,
Mass.*): Let me just say one thing. I think, Dr. Lamarsh, you might have a
point on failures in reactor safety; but if you look over the years, the
ACRS has had more impact on the way the staff reacted and dealt with
reactor safety than has any other organization. They've been very
effective in conveying their philosophy to the staff. I think they've had
more impact than you give them credit for. Their job is not to look at how
the pressure is measured in Brown's Ferry but to see that the overall
system is running well and general principles are adhered to. I think
they've had a big impact on the Nuclear Regulatory Commission [NRC]

*Dr. Moss is science advisor to this committee, not a member of Congress.

48

0077-8923/81/0365-0048 $01.75/2 © 1981, NYAS

staff, because I know how the staff members quake when they go up to ACRS; they do their homework a lot better when they have to go before that board and explain their position on an issue. It's been very effective in that. What I'm saying is that without them, we'd be in a lot worse shape.

J. R. LAMARSH: The point of my question was not to derogate ACRS completely. I think we're obviously better off with ACRS than without it. I certainly would not contend that ACRS should be done away with. However, it does seem to me that its functions could be strengthened somewhat, and I hope that you will strengthen them so that they will be more aware of the safety problems in actual operating plants.

H. LEWIS: I think it's moving in that direction, actually.

T. H. MOSS: We'll check with you, Dr. Lewis, in a year or two to see whether you've perfected the system.

A. M. WEINBERG: I'd like to put a question to Dr. Lewis. You remember at the very end of your comments, you applauded what utilities were doing, but you didn't applaud very vigorously.† The point that I would disagree with you on is the immense pressure that the utilities are now under, for reasons that are very bread and buttery, i.e., no utility president is going to bet his existence on a device if he doesn't think that the device is going to be really safe. I think that pressure cannot be underestimated; it is enormous pressure that is put on them.

It's too early to say whether the actual response by the utilities industry is sufficient. My own instinct in the matter is that it may be sufficient. I think more ought to be done, but I don't think one should leave the impression that there isn't enormous pressure now for them to move on the matter.

UNIDENTIFIED SPEAKER: I have two minor technical questions on the accident for Mr. Jaffe. One, in what part of the system was the hydrogen explosion at 1:30 in the afternoon? Two, what was the maximum temperature measured in the fuel coil? I believe those were both in your report.

L. JAFFE (*Technical Task Force, President's Commission on the Accident at Three Mile Island*): The explosion occurred in the containment building, not in the primary system. We don't know precisely where. There was a pressure pulse noted inside the containment building at that particular time.

The maximum temperatures really are not known well. Unfortunately, although there were thermocouples in the system capable of measuring the temperatures, the temperature monitors did not have a range adequate for this kind of temperature excursion. We believe that the temperatures in some portions of the core reached very close to 5000°F.

M. EISENBUD (*New York University Medical Center, New York, N.Y.*): I'd like to address this question to Mr. Jaffe. I want to focus on the period that began some hours after the terrible, tragic errors were made

†Dr. Lewis spoke at the conference, but was unable to submit a paper for publication.

by the operators. In your talk, you began to develop the network of experts in various places who were intercommunicating. During that period, which I guess began on the second day, there were some terrible errors made—terrible errors of technical judgment. My question has to do with your opinion or the opinion of the commission as to what extent these errors, the faulty judgment, may have been due to the fatigue brought on by the development of demands, discoordinate demands, for information?

L. JAFFE: The effect of the demands was very great. They had to make judgments on a very part-time scale. As a case in point, the need to evacuate was a very serious question, a very urgent question, in the eyes of the commissioner, certainly. And there was a need to provide that information to Governor Thornburgh, who was pressing very hard for that information. They were pressed into making a decision.

UNIDENTIFIED SPEAKER: I have a question for Dr. Rasmussen. On the question of evaluating the probability of inappropriate operator response, I can see how we can use past experience to evaluate the probability of something like flipping the wrong switch or turning on the wrong valve. But when we get into very complex situations or accident situations, it seems to me that we don't really have that many data points.

N. C. RASMUSSEN: Well, you're wrong about that. We looked at the data from airplanes—the accidents they've had—and from submarine operations and a variety of sources. What we found was that under normal conditions for a typical operation with no anxiety factor, an operator did something he was trained to do with about 1 failure in 1,000 with an uncertainty of plus or minus 10. When the bells are ringing and the lights are flashing, he goes way back to 1 in 10, which is 90% and gets an A grade at MIT or any other school. But his failure rate really degrades under tense situations; we have enough data to show that, and we put an uncertainty factor of 10 on that and said that his failure rate lay between 1 and 10^{-2} in emergency conditions. Of course, some people freeze and do nothing, and they surely fail; so we don't have high degrees of accuracy. These numbers are not good to even one significant figure, as you see, but they give us a range of the uncertainty that we can use in our answer.

UNIDENTIFIED SPEAKER: My question is for Dr. Lewis. Do you think it's important for the general public to be better informed on the nature of radiation? Do you feel that the scientific community has a responsibility to assist in this process?

H. LEWIS: Of course. It is the responsibility of anyone who is informed in a society about something that affects that whole society to do two things: to comport himself in his professional work as well as he can in support of the society, using the special knowledge he has; and to transmit as much information as he can to society. I'm a professor by trade, and that means that I profess, I suppose. But my job is both the generation and the dissemination of knowledge; and I think all of us who are in this business feel that way.

UNIDENTIFIED SPEAKER: Do you think it's a failure in the school system, perhaps? Why is it that people don't know enough about radiation?

N. C. RASMUSSEN: The answer is pretty simple: people aren't very interested in it. There are books in the library that go back for 40 years on the effects of radiation, but nobody is very interested in it. It was only something that affected radiologists in the early days. Now that it's become understood and there is the bomb fallout business and so on, people have become more interested. There are more books, and we're learning more. But you know, the old story is that we do a lot of teaching, it's just that we don't get much learning done in some of these schools.

J. R. LAMARSH: I'll just make one very brief comment on that. I'm also a professor, and for some years I've been trying, in various capacities, to get groups of people to understand radiation, particularly people in the general public. I did this in particular in connection with a town that I mentioned earlier. After a period of years and years of explaining radiation over and over and over again, I came away with the conclusion that essentially nobody understood regardless of how I tried. So, Dr. Lewis may not have made the statement that the public will never understand radiation, but I happen to be much more pessimistic about that. I think that a certain fraction of the public will, in fact, understand radiation, but I certainly don't think everybody will understand it. If all of my graduating students in nuclear engineering understood radiation, I'd feel like my job was a success.

A. M. WEINBERG: Well, I'm not a professor, but I have thought about these things in a certain way. One of the very great difficulties in connection with the public's understanding of radiation and, indeed, in the assessments of health effects due to accidents is the fact that by its very nature, the effects of very low levels of radiation may be beyond the efficiency of science ever to really determine. Under the circumstances, one can take different positions. I think that Dr. Upton, in his remarks, is going to state what the upper limit will be. I think that the lower limit—and I now quote this from a report that will soon come out from the National Academy of Sciences—for the effects of very low levels of radiation could be zero. It was not possible to rule out the zero as the effect of very low levels of radiation.

This matter is really very central to these things. When Dr. Rasmussen says that he computes 45,000 cancers from the largest of his accidents and that that might be off by a factor of 3, and then some people say a factor of 10, you must realize that they're talking about very large numbers of people exposed to extremely low levels of radiation per person. At that level, there is no demonstration that anything happens; one must always keep that in mind.

J. R. LAMARSH: But to explain that to the general public is a very difficult thing to do.

UNIDENTIFIED SPEAKER: Yes, it sounds like you all also have your problems in understanding it yourselves.

H. LEWIS: Let me express some optimism here. Everything Dr. Weinberg said is absolutely right. The new academy report will, in fact, set an upper limit, which is what everyone has been calculating with, and a lower limit, which may well be zero. In fact, one of the reasons I included those curves in my *Scientific American* article was to make that point.

But when we speak of the public understanding radiation, understanding is a buzz word. Acceptance with familiarity sometimes masquerades as understanding. Some of us are old enough to remember when a lot of people didn't believe that flying was possible, and everyone now accepts flying. I'm sure others can go back to other technologies. When a thing has been around a long time, it becomes familiar and we say we understand it. In that sense, I'm rather optimistic about the ultimate fate of radiation.

T. H. MOSS: The other generality you might draw at this point is to observe that in chemical threats to health, we probably know even less than we know about radiation. Yet eventually we're going to have to have a public understanding of those threats too, enough to act rationally with respect to them.

M. MAGE (*Solar Transition Committee, National Institutes of Health, Bethesda, Md.*): My question is directed to Dr. Weinberg, and also parenthetically to the organizers of the conference. It has to do with Dr. Weinberg's comments on the level of spending likely to be accepted to improve the safety of nuclear reactors and how this will affect the perceived need for nuclear energy.

My specific question to Dr. Weinberg is, In view of the learning curve that is evolving for the changing technology and economics of solar energy, of energy efficiency, and of conservation, were you really being realistic when you spoke of a choice, in the mid-21st century, between having a nuclear electric system on the one hand or a fivefold generation of production of coal on the other hand? Is this a realistic statement of what our options are?

A. M. WEINBERG: I assume that in a rather indirect way, you were asking me my opinion of solar energy. Six years ago, I was the director of the Office of Energy Research and Development in the Federal Energy Office, which at that time was in the White House. One of the few things that our office did at that time was establish the Solar Energy Research Institute. My position on the matter is that I think solar energy is just great if we can really develop it, and develop it at a cost that society will consider to be acceptable. My view at the moment is rather agnostic: this may turn out to be the case, but it may not. Since it may not turn out to be the case, I don't think we should deny any of the alternative options at present.

M. MAGE: You didn't answer the question which was, Do you think it's realistic to view our options in the mid-21st century as having a breeder economy on the one hand or a fivefold increase in coal on the other?

A. M. WEINBERG: Perhaps I misstated my belief. Of course, I included solar energy as an option also by the mid-21st century, and perhaps it will prevail. I don't know.

THREE MILE ISLAND: ASSESSMENT OF RADIATION EXPOSURES AND ENVIRONMENTAL CONTAMINATION

Thomas M. Gerusky

Bureau of Radiation Protection
Pennsylvania Department of Environmental Resources
Harrisburg, Pennsylvania 17120

At 7:04 A.M. on March 28, 1979, the Bureau of Radiation Protection's duty officer, a nuclear engineer, was notified by the Pennsylvania Emergency Management Agency of an accident at Three Mile Island Unit 2 (TMI-2). He immediately contacted the control room to verify the notification and to receive detailed information. A site emergency had been declared because of high radiation levels detected on remote area radiation monitors in the auxiliary building.

Following receipt of technical information concerning plant status, he was informed that there had been no release to the environment. He, as is prescribed in our emergency plan, contacted other members of the staff and proceeded to our bureau offices.

The bureau office is located in the capitol complex in Harrisburg, only 12 miles north of the plant on the Susquehanna River. The main portion of the staff and the environmental radiation laboratory are also located there.

I was the first one to arrive at the office and immediately opened a telephone line to the Unit 2 control room. That line stayed open for the next month.

At 7:30 A.M., the plant declared a general emergency based upon in-plant radiation readings.

The plant health physicist, based upon a dome monitor reading of 800 R/hour, calculated a dose rate of 10 R/hour in Goldsboro, a community to the west of the plant. However, the calculation was based upon a containment leak rate of 0.02% per day. Since the containment pressure was only slightly positive and there were no on-site readings that would indicate such a massive release, we requested verification. A helicopter was used to provide quick information, and it was indeed verified that the readings in Goldsboro were normal. In the meantime, we had requested the Pennsylvania Emergency Management Agency to notify the county to the west of the site that evacuation might be necessary. That alert was called off when information was obtained from the helicopter.

Previously, at 7:35 A.M., the stack monitor had reached its alarm setpoint (2.8×10^{-7} μCi/cc or 0.3 μCi/cc release).

By 8:30 A.M., the Metropolitan Edison emergency response teams were finding some levels of detectable airborne radioactivity off site (1–3

0077-8923/81/0365-0054 $01.75/2 © 1981, NYAS

mR/hour). By 9:00 A.M., Metropolitan Edison teams were monitoring both east and west shore areas.

We were not notified until 10:00 A.M. that some off-site levels had been detected. At that time we dispatched two survey teams for verification purposes. We also requested assistance from the Brookhaven National Laboratory (BNL) emergency response team, which arrived by coast guard helicopter early in the afternoon.

By 11:30 A.M., on-site readings were from 5–10 mR/hour, with a high of 365 mR/hour on the western boundary of the site. Off-site readings began to show levels of 1–5 mR/hour and a high of 13 mR/hour at Kunkel School, west-northwest of the site and approximately six miles away.

Earlier, an air sample in Goldsboro on the west shore had shown an I-131 level of $1 \times 10^{-8}\,\mu$Ci/cc. [The maximum permissible concentration (MPC) is $1 \times 10^{-10}\,\mu$Ci/cc.] This was questioned because of high Xe-133 levels on the charcoal causing difficulties in using the single-channel analyzer. The sample was flown by helicopter to our lab in Harrisburg, where I-131 levels were found to be well less than MPC.

The Department of Energy (DOE) also sent a helicopter equipped with air sampling and radiation detection equipment, which was of tremendous importance in evaluating plume locations, types of radiation released, and levels over the entire episode. The DOE teams provided us with expertise, equipment, and personnel over the first month of the accident. Without their assistance, Pennsylvania would have been at a serious disadvantage in evaluating off-site exposures. They have been given little or no recognition in the official reports of the numerous investigating bodies. I wish to remedy that immediately.

During the next few days, Metropolitan Edison teams, the Nuclear Regulatory Commission (NRC), and DOE all provided us with routine information and data. We requested that DOE collate the survey data in daily briefings for all agencies involved.

The releases from the plant continued, with off-site readings varying from less than 1 mR/hour to 13 mR/hour on March 30 depending upon plant conditions. The 13 mR/hour reading was taken by Metropolitan Edison personnel on Friday, March 30, during the height of the venting of the makeup tank.

I wish I had time to describe in detail what really occurred on that Friday morning, but the Rogovin I report is fairly complete and will provide you with a summary of that day.[1]

On Saturday, March 31, the last of the significant off-site readings was made, with a high reading of 38 mR/hour just off-site to the northeast. Also, other federal agencies, including the Bureau of Radiological Health and the Environmental Protection Agency (EPA), began participating in the environmental monitoring program by placing thermoluminescence dosimeters (TLDs) throughout the area.

Early after the accident, a federal interagency Ad Hoc Population Dose Assessment Group from NRC, EPA, and the Department of Health,

Education, and Welfare (DHEW) evaluated the population doses from TMI. Their final report was published on May 10, 1979, and entitled *Population Dose and Health Impact of the Accident at the Three Mile Island Nuclear Station.*[2] It can be obtained from NRC or DHEW.

The president's commission also evaluated the exposures and their consequences.[3] The Task Force on Public Health and Safety published their report to the commission in October, 1979.[4] I would urge that both documents be mandatory reading for anyone interested in the details of exposure evaluations and the problems in performing evaluations because of the measurement systems involved.

The field survey data, with the exception of the data measured by DOE in the helicopter, are not very helpful in evaluating exposures. The noble gases that were released—Xe-133, Xe-133m, and Xe-135—have a variety of energies of both gamma and beta emissions. Geiger-Müller survey meters were used for field evaluations. The instruments probably overresponded to the Xe-133 gamma ray energy of 81 keV. In addition, many readings were taken with an open window, allowing for beta levels to be included in the overall measurements.

Field survey data were just that—field survey data. They did provide an estimate of the levels of radiation in the environment, but the numbers were just estimates, and all health physics personnel involved realized this.

The major tools for determining environmental levels were the thermoluminescence dosimeters that were in place before the accident and the additional dosimeters placed in the environment by NRC, EPA, and the Food and Drug Administration.

Dosimeters provided by Teledyne, Inc. were placed at 20 locations at distances ranging from 0.2 miles to 15 miles from TMI. Additional TLDs were supplied for 10 of these locations by Radiation Management Corporation (RMC) and at 4 of these locations by DOE and RMC. The Commonwealth of Pennsylvania deployed and recovered these dosimeters.

Beginning on the third day of the accident, federal agencies involved placed additional dosimeters around the site: NRC at 47 locations, DHEW-Bureau of Radiological Health at 173 sites, and EPA at 59 sites and on 54 persons.

The President's Commission Task Force on Public Health and Safety's analysis of the data reads as follows:

> The procedures for calibration, processing and reading these dosimeters were reviewed. Adjustments were made for estimated background values and energy dependence. Data from TLDs placed by NRC on the third day of the accident were rejected because their handling procedures were inappropriate for this evaluation. Because of their late deployment and distance from the source, the dosimeters placed by the other two federal agencies did not provide useful data.
>
> The population distribution used to calculate the collective dose is based on projections of 1970 census data to the year 1980, as given in the final

Safety Analysis Report for TMI-2. Adjustments were made to account for the fact that only one person is known to have been at the many summer cottage sites on the islands near the plant at the time of the accident.

The doses measured by the TLDs would be applicable to people who were outdoors all during the first few days of the accident. Because most people spent most of that time indoors, some protection can be assumed due to absorption of gamma radiation in the structural materials for houses and offices. It is estimated that the average dose received indoors is about three-quarters that of outdoors.

Persons within a 2-mile radius of the plant probably received the highest doses. The dose to the one person known to have been on the nearby islands, for about 9½ hours during the first few days of the accident, is estimated to be about 50 millirems (mrem). In addition about 260 people living mostly on the east bank of the river may each have received between 20 and 70 mrem. All other people probably received less than 20 millirem.

In estimating health effects of low doses to a population, it is important to know collective dose—the sum of the doses received by every person in the affected area. This is usually given in units of person-rems. The collective dose was calculated by multiplying the average dose at each of 160 areas surrounding the TMI plant by the population in that area and summing the products. The average dose in each area was estimated by interpolating between the locations at which TLD measurements were available. The collective outdoors dose to people within a 50-mile radius of TMI was calculated to be about 2,800 person-rems. Assuming that doses indoors were three-quarters of those received outdoors, the actual collective dose to the population is estimated to be 2,000 person-rems.[4]

A DOE estimate using aerial survey data was also 2,000 person-rems.[2]

The Ad Hoc Population Dose Assessment Group used four different methods of obtaining population exposure, resulting in doses of 1,600, 2,800, 3,300, and 5,300 person-rems.[2] The average of the four, or 3,300 person-rems, was considered acceptable.

The total radioactivity released during the accident was calculated using a stationary gamma monitor located at the base of and external to the stack. The stack monitor had gone off scale early in the accident. The total calculated release was 2.4 million curies. Using computer models of the release rate with time, taking into account meteorological and population distribution data, the president's commission task force estimated population exposure at 500 person-rems.[4]

It is also interesting to note that the beta component of the noble gases would have given a skin dose of as much as four times the gamma dose if any person were submersed in the plume. Because of clothing considerations and other variables, no real beta component of the exposure can be calculated.

Analysis of doses due to inhaled or ingested radioisotopes was also evaluated. Environmental sampling data from state and federal agencies were reviewed. Thousands of samples of milk, air, water, produce, soil, vegetation, fish, river sediment, and silt in the TMI vicinity were analyzed. The doses calculated by the president's commission task force

TABLE 1

RADIATION DOSES CALCULATED FOR THE THREE MILE ISLAND VICINITY[4]

Intake Mode	Organ	Dose (mrem)
Cow's milk ingestion	newborn thyroid	6.9
	1-year-old thyroid	4.7
	adult thyroid	0.6
	ovaries	0.00002
	testes	0.00002
	red bone marrow	0.00009
	total body	0.0003
Inhalation (off site)	newborn thyroid	2.0
	1-year-old thyroid	6.5
	adult thyroid	5.4
	ovaries	0.0002
	testes	0.0001
	red bone marrow	0.0007
	total body	0.003

are given in TABLE 1. This dose estimate is valid only for individuals living within three miles of TMI, because most of the sampling took place within that area.

An additional study was performed on 760 residents within five miles of the site. They were whole-body counted, with no reactor-produced isotope found.[5]

The Pennsylvania Bureau of Radiation Protection also evaluated film badge data in the general area—those badges worn by workers occupationally exposed to radiation. The data either verified the estimated doses in the area or suggested that the doses were too high.

The Federal Bureau of Radiological Health evaluated film that was in the area. Kodak analyzed the film and got no fogging in excess of 10 millirems.[6]

The Three Mile Island accident is far from over in the central Pennsylvania area. The problem of decontaminating the facility looms as an impossible task because of public concern over their radiation exposure. Many people just don't believe that their exposures were as I previously described. Individuals believe that they or their children will be dying of cancer within a few years. There has been a loss of credibility that is very difficult to reestablish by Metropolitan Edison, by the NRC, and by the state.

The road ahead looks as bumpy as the road behind.

REFERENCES

1. ROGOVIN, M., Director. 1980. Three Mile Island. A Report to the Commissioners and to the Public. Special Inquiry Report. Nuclear Regulatory Commission. Washington, D.C.
2. Ad Hoc Population Dose Assessment Group. 1979. Population Dose and Health Impact

of the Accident at the Three Mile Island Nuclear Station. Nuclear Regulatory Commission. Washington, D.C.

3. KEMENY, J. G., et al. 1979. Report of the President's Commission on the Accident at Three Mile Island. U.S. Government Printing Office. Washington, D.C.

4. President's Commission on the Accident at Three Mile Island. 1979. Report of the Task Force on Public Health and Safety. Washington, D.C.

5. Nuclear Regulatory Commission. 1979. Washington, D.C. (Unpublished data.)

6. Federal Bureau of Radiological Health. 1979. Department of Health, Education and Welfare. Washington, D.C. (Unpublished data.)

<hr>

DISCUSSION

UNIDENTIFIED SPEAKER: Were there any comparisons made between this accident and previous large accidents, such as the one in England? My recollection is that in the incident in England, more than 10,000 curies of iodine were released, which is obviously more than three orders of magnitude more than in this incident.

T. M. GERUSKY: No, not that I know of. The problem with the Windscale incident was that it occurred in 1957, and at least in the United States, it wasn't even well recognized as an accident for a long time. Nobody cared about those data. The public didn't care about them. There were many millions of gallons of milk destroyed, but there was a problem, I think, in terms of trying to relate TMI to iodine levels there. It would be easier to relate TMI iodine levels to the levels that were found in milk during the Chinese fallout than it would be to Windscale.

The other thing I failed to mention was that the iodine that was coming out of TMI was found not to be elemental iodine, and the cows were not on pasture—they were inside or they were out on nonpasture land; and so this was mainly inhaled iodine that was found in the milk. The numbers, if the cows were on pasture, could have been higher. However, all of the grass samples and soil samples and everything else did not reveal any significant levels of iodine on the ground, and the very few air samples revealed significant levels in the air.

D. P. SIDEBOTHAM (New England Coalition on Nuclear Pollution, Putney, Vt.): You mentioned samples taken at three miles and samples taken at five miles, and you said there was no body count detectable. Is that correct?

T. M. GERUSKY: Maybe I did say that, but what I meant to say was that whole counts of 760 people within five miles of the plant showed no detectable radioactivity as a result of fission. There was detectable radioactivity in some of the individuals. It was naturally occurring radon, and we had to go and do some well sampling as a result of that and, indeed, found radioactivity in the wells.

D. P. SIDEBOTHAM: I see. Well my next question is, Was there a substantial monitoring done at 10 miles, or were your figures calculations?

T. M. GERUSKY: Most of the monitoring was done within five miles of the plant, and most of the airborne iodine numbers were collected within three miles of the plant, although we had an air sampler in Harrisburg going all the time and did not find any significant iodine on the air sampler.

D. P. SIDEBOTHAM: That was just at Harrisburg?

T. M. GERUSKY: That's right.

D. P. SIDEBOTHAM: So there aren't specific exposures from monitoring at 10 miles, except at Harrisburg perhaps?

T. M. GERUSKY: They were calculated. No, excuse me. The Department of Energy helicopter did do studies out to 10 miles, and they did estimates out to 10 miles based on the DOE helicopter data. Also, there were some thermoluminescence detectors out beyond 10 miles, which were used and which did follow the ratio of diffusion that should have been found.

D. P. SIDEBOTHAM: Would it be fair to say that the primary body of your information is at three to five miles?

T. M. GERUSKY: Oh, yes.

D. P. SIDEBOTHAM: So as I understand it, a study is being done on the effects of people within a 5-mile area, with a 10-mile area as a control?

T. M. GERUSKY: No, that is not true. The studies that are being done on infant mortality, for example, are within a 10-mile radius of the plant and take into consideration every community whose boundary meets 10 miles from the plant. So it's even a bigger area than 10 miles from the plant. A report last week from the Department of Health on their first six months' data indicates that there is no difference in infant mortality rate in the 10-mile range than in any other place in the state.

D. P. SIDEBOTHAM: That's at variance to what we've seen in the—

T. M. GERUSKY: That's right. It's at variance with earlier reports on individual data, where data points were released and not rates. The rates are the same, and that's the important thing in evaluating epidemiological data.

D. P. SIDEBOTHAM: It's somewhat at variance to what I've seen recently in the papers in New England.

T. M. GERUSKY: Well, the papers in New England just didn't pick up the Health Department press release, which occurred after someone else had released the raw numbers to the press.

D. P. SIDEBOTHAM: I see. Thank you.

J. S. MILLER (*Northwood, N.H.*): Has the state done anything to follow up on the reports about farm animal deaths and birth problems? Also, has the state done anything to look at wild birds and wild animals for possible contamination?

T. M. GERUSKY: I doubt that we've done any wild birds and we haven't analyzed many animals, but we have done individual organs from animals. We have found no animals with any detectable contamination. The Pennsylvania Department of Agriculture has visited each of the farms that claimed to have problems, and a report on that was supposed to be out this week. Unfortunately, the person who was preparing the report left for private business, to set up his own veterinary office, so it will take a little longer to get the data out. But the information I have is that there's nothing abnormal at all in the Three Mile Island area, according to the Pennsylvania Department of Agriculture.

A. P. HULL (*Brookhaven National Laboratory, Upton, N.Y.*): I'd like to ask you a question. It's a little bit of a loaded question, but it's in the open literature (if the people have access to the literature), namely, in another one of the subdocuments of the president's commission task force, and you expressed some concern about what was happening in Washington. I'm not here to blow the DOE's bugle, but I've read the section describing a political discussion in the White House about the fact that EPA had ended up as the lead agency for making environmental measurements for the State of Pennsylvania and that this looked good to the public. So there was a political decision made at the White House that the DOE's profile would be as low as possible and that somebody else would look politically more acceptable—would be the lead agency as far as the public was concerned. I think it's in the volume on public information. You can read the discussion there for yourself, in case you doubt me. I wondered what your reaction is to the numerous agencies in this political finagling in Washington about who was going to be the king of the mountain. Would you care to comment on that?

T. M. GERUSKY: No. I've got to go back and work with these agencies, you have to remember. We have a brand new EPA spokesman coming in and setting up an office in Harrisburg to report all the data in case there is venting of 57,000 curies of krypton-85 from the stack. There is a problem of NRC credibility, and I think someplace there is a question—in the White House, in particular—about DOE credibility.The people who came in to help did a tremendous job, and they're not trying to hide any information. I'm afraid that the problem is at the higher levels. The accident was something I wouldn't want to live through again, but I sure hope that the federal government gets its act together before they come in like gang busters the next time.

I can see people down in Washington, in NRC, with this computer bank of information saying, Oh my God, what do we do now? Because none of them down there know anything about how to run a reactor, and they shouldn't try, especially the commissioners.

A. P. HULL: I think it's come up a couple of times: the decentralized or spontaneous epidemiology that is going on around Three Mile Island. Do you see the need for setting up very long-term, sustained epidemiological study of the region? Do you think that would help?

T. M. GERUSKY: Yes, and it's being done by the Health Department. I brought along with me a couple of copies of the *Report of the Governor's*

Commission on Three Mile Island, which includes some data and describes all of the studies that are going on around Three Mile Island. If anybody wants a copy of that report, they can write to me.

The one thing we would like to see in Pennsylvania is a tumor registry program, which we don't have yet and which would help in the areas around all nuclear power plants and in other locations in Pennsylvania where there are some radiation problems.

HEALTH IMPACT OF THE THREE MILE ISLAND ACCIDENT

Arthur C. Upton

Institute of Environmental Medicine
New York University Medical Center
New York, New York 10016

Health Effect of Ionizing Radiation

Nuclear accidents are distinguished from other kinds of accidents in their ability to release ionizing radiation, which is potentially harmful to health. The types, as well as the frequency and severity, of radiation injuries depend on the amount of radiation absorbed. Virtually any living thing may be killed outright if exposed at a high enough radiation intensity, but all forms of life have evolved in the presence of natural background radiation, which is present everywhere in small amounts (TABLE 1).

Whether health is affected by exposure at the low levels characteristic of natural background radiation is a matter of conjecture. Observations at higher radiation intensities, however, have implied that the risk of certain effects may be increased even at the lowest dose levels. These effects include (1) damage to genes and chromosomes, or *mutagenic* effects; (2) damage to the growth and development of the embryo and fetus, or *teratogenic* effects; and (3) damage to cells that increases the risk of their forming cancer, or *carcinogenic* effects.[1,2] While these effects are not known with certainty to be produced in human beings by natural background radiation levels, current theory attributes a small fraction of the natural occurrence of all mutagenic, teratogenic, and carcinogenic effects to background radiation. As yet, however, since these effects of radiation cannot be distinguished individually from similar effects produced by other causes, the effects of low-level irradiation are estimated only by extrapolation from observations at higher radiation doses and dose rates, based on tentative assumptions about the relevant dose-effect relationships. In the present state of our knowledge, such estimates must be regarded as highly uncertain at best.[3,4]

In contrast to the aforementioned effects, most other types of radiation injury occur only after doses many times higher than natural background levels. For example, sufficient depletion of blood-forming cells to cause fatal radiation sickness will not occur unless virtually the entire marrow is exposed to hundreds of rem,[1,2] but this type of injury can be one of the most serious consequences of a major nuclear accident (TABLE 2). Radiation-induced depletion of cells in the gastrointestinal (GI) tract, skin, gonads, and lymphoid tissues also can cause symptoms of injury within days after exposure;[1,2] however, serious such effects are

0077-8923/81/0365-0063 $01.75/2 © 1981, NYAS

TABLE 1

ESTIMATED ANNUAL WHOLE-BODY RADIATION DOSE RATES
TO THE POPULATION OF THE U.S.*

Source of Radiation		Average Dose Rates (mrem/year)
Natural		
Environmental		
Cosmic Radiation		40 (30–130)†
Terrestrial Radiation		40 (30–115)‡
Internal Radioactive Isotopes		20
	Subtotal	100
Man-Made		
Environmental		
Global Fallout		4
Nuclear Power		0.023§
Medical		
Diagnostic		72
Radiopharmaceuticals		1
Occupational		0.8
Miscellaneous		2
	Subtotal	80
	TOTAL	180

*From References 4 and 11.
†Values in parentheses indicate range over which average levels for different states vary with elevation.
‡Range of variation (shown in parentheses) attributable largely to geographic differences in the content of potassium-40, radium, thorium, and uranium in the earth's crust.
§Value for 1975.

likewise unlikely to occur unless hundreds, or even thousands, of rem are absorbed.

Radiation Doses from the Accident

The accidental radiation received by people residing in the vicinity of Three Mile Island (TMI) came almost entirely from xenon-133 (half-life,

TABLE 2

ESTIMATED CONSEQUENCES OF AN EXTREMELY SERIOUS
NUCLEAR POWER PLANT ACCIDENT*

Prompt Fatalities		3,300
Early Radiation Sickness		45,000
Thyroid Nodules (within 30 years)	8,000/year	240,000
Cancer Fatalities (within 30 years)	1,500/year	45,000
Genetic Defects (within 150 years)	200/year	30,000
Economic Loss Due to Contamination	$14 billion	
Decontamination Area	3,200 square miles	

*From Reference 12.

5.3 days), xenon-135 (half-life, 9.2 hours), and traces of radioactive iodine (principally iodine-131; half-life 9.0 days), which escaped intermittently from the plant as gases.[5,6] As these radioactive gases followed prevailing winds, they increased the level of ionizing radiation along their path; however, the increase was short-lived because xenon—which is relatively inert chemically and biologically—dispersed rapidly and because radioactive iodine was present only in barely detectable amounts to begin with. No release of long-lived fission products, such as strontium-90, cesium-137, and plutonium-239, was detected.

TABLE 3

VALUES ASSUMED IN ESTIMATING CANCER RISKS*

| | | | | Risk Estimate | |
Age at Irradiation	Type of Cancer	Duration of Latent Period (years)	Duration of Plateau Region (years)†	Absolute Risk‡ (deaths/10^6/ yr/rem)	Absolute Risk (% increase in deaths/rem)
In Utero	Leukemia	0	10	25	50
	All other cancer	0	10	25	50
0–9 Years	Leukemia	2	25	2.0	5.0
	All other cancer	15	30–Life	1.0	2.0
10+ Years	Leukemia	2	25	1.0	2.0
	All other cancer	15	30–Life	5.0	0.2

*From Reference 3.

†Plateau region = interval following latent period during which risk remains elevated.

‡The absolute risk for those aged 10 or more the time of irradiation for all cancer excluding leukemia can be broken down into the respective sites as follows:

Type of Cancer	Deaths/10^6/year/rem
Breast	1.5§
Lung	1.3
GI including stomach	1.0
Bone	0.2
All other cancer	1.0
Total	5.0

§This is derived from the value of 6.0 for women, corrected for a 50% cure rate and the inclusion of men as well as women in the population.

Based on all of the available measurements, it is estimated that the maximum cumulative whole-body radiation dose to anyone off site was less than 100 mrem, that the average cumulative dose to those within 10 miles of the plant was about 8 mrem, and that the average cumulative dose to those within 50 miles of the plant was about 1.5 mrem. Because these estimates make no allowance for shielding by shelter or other attenuation factors, they are generally considered to represent overestimates.[5,6]

Additional exposure of the population came from the beta radiation dose to the skin and from the inhalation dose to the lung. It is estimated that the total dose to the skin could have been larger than the whole-body dose by a factor as large as 3–4 if the protective effects of shelter and clothing are neglected.[5] In the case of the lung, the inhalation dose is estimated to have constituted no more than 3–7% of the dose to the whole body.[5]

Potential Impact of the Accident on the Risk of Cancer

The risk of cancer is generally assumed to be increased by low-level irradiation, but it is clear from observations at intermediate-to-high dose levels that the risk may vary depending on the type of cancer in question, age at the time of irradiation, the quality of radiation, and other factors. Such factors are therefore taken into account in efforts to estimate the risks of a given dose, as is illustrated in TABLES 3 and 4. Because the figures tabulated are based on a linear, nonthreshold extrapolation model, with no allowance for biological repair at low doses and low dose rates, they are regarded by many experts as being likely to overestimate the risks of low-level, low-LET radiation. For this reason, some experts prefer a linear-quadratic model, which yields estimates that tend to be 25–50% smaller than those in TABLE 3, depending on the type of cancer in question.[7-9]

If the risk coefficients presented in TABLES 3 and 4 are applied to the population (of about 2.2 million people) residing within 50 miles of Three Mile Island, they predict a lifetime risk (TABLE 5) of less than one

TABLE 4
ESTIMATED LIFETIME CANCER RISKS OF LOW-LEVEL RADIATION*

| | Risk per Million Person-Rem | |
Site	Fatal Cancers	Incident Cancers
Bone Marrow (leukemia)	20–50	15–40
Thyroid	10	100
Breast (women only)	50	100
Lung	25–50	25–50
Stomach ⎫ Liver ⎬ Colon ⎭	10–15 (each)	15–25
Bone ⎫ Esophagus ⎪ Small Intestine ⎬ Urinary Bladder ⎪ Pancreas ⎪ Lymphatic Tissue ⎭	2–5 (each)	2–10
Skin	1	15–20
Total (both sexes)	100–250	300–500

*From References 3, 4, 8, 9, and 13.

TABLE 5

PROJECTED IMPACT OF THREE MILE ISLAND ACCIDENT ON LIFETIME INCIDENCE OF
CANCER IN OFF-SITE POPULATION WITHIN 50 MILES*

	Estimated Numbers of Cancers Attributable to Radiation from the Accident†	Estimated Numbers of Cancers Occurring Naturally‡
Fatal Cancers	0.7 (0.15–2.4)	325,000
Nonfatal Cancers	0.7 (0.15–2.4)	216,000

*From Reference 5.
†Values cover extreme ranges of estimates presented in References 3, 4, and 9, with their geometric means (see Reference 5).
‡Based on the natural incidence for the U.S. as a whole.

additional fatal cancer and less than one additional nonfatal cancer (a central estimate of 0.7 case of each). These estimates compare with an expectation of some 541,000 naturally occurring fatal and nonfatal cancers in the population within its lifetime, based on cancer rates for the U.S. as a whole. The potential risk of cancer attributable to the accident thus represents an increase of 0.0002% of the natural incidence, or less than one case per million persons at risk.

Potential Genetic Impact of the Accident

It is generally assumed that irradiation can cause genetic damage in human germ cells that is transmissible to future generations in the form of various inherited diseases. However, since direct evidence of such radiation damage in human beings is lacking, the relevant risk estimates must be based on experimental data. From such data, it has been estimated that the incidence of genetic abnormalities in humans would be doubled by a dose of 20–200 rem.[3,4]

On the basis of this risk estimate, the major types of inherited diseases, their natural frequencies, and the extent to which their incidence can be expected to be increased by a given dose of radiation are shown in TABLE 6. Also tabulated are estimates of the numbers of descendants of the population within 50 miles of TMI who are likely to be affected by genetic disorders resulting from the accident. From the table, it will be seen that less than one offspring of the first and subsequent generations is predicted to be affected, in comparison with approximately 78,000 individuals naturally in the first generation alone.

Potential Impact on Growth and Development of the Embryo and Fetus

Although, as indicated above, teratogenic effects are customarily included among the effects of low-level radiation along with mutagenic

TABLE 6

PROJECTED IMPACT OF THREE MILE ISLAND ACCIDENT ON INCIDENCE OF HEREDITARY
DISEASE IN OFFSPRING OF POPULATION WITHIN 50 MILES

Disease Category	Natural Incidence (per million live births)	Incidence per Million Live Births per Rem to Both Parents*		Incidence per Generation Attributable to TMI Accident	
		First Generation	At Equilibrium	First Generation	150 Years
Dominant Diseases	10,000	10–100	50–500	0.01–0.1	0.05–0.5
Chromosomal and Re-cessive Diseases	10,000	slight	very slow increase	slight	very slow increase
Congenital Anomalies	15,000	1–100	10–1000	0.001–0.1	0.005–0.5
Anomalies Expressed Later	10,000				
Constitutional and De-generative Disease	15,000				
Total	60,000	12–200	60–1500	0.01–0.2	0.06–1.0

*From References 3, 4, and 9.

and carcinogenic effects, the risks of teratogenic effects on the human embryo and fetus are more difficult to estimate, owing to the paucity of relevant data.[3,4] The evidence at hand implies, however, that the risks of such effects are smaller per unit dose than are the risks of carcinogenic and mutagenic effects.[3,4] On this basis, it may be inferred that such effects are unlikely to result from the accident in view of the small magnitude of the radiation dose and the relatively small numbers of embryos and fetuses that were exposed (presumably fewer than 30,000 fetuses and embryos were exposed within 50 miles of Three Mile Island, based on the 1976 annual birth rate of 14.2 births per 1,000 people in the U.S.).

Mental Stress

Apart from the potential effects of radiation discussed above, the only known health impact of the accident is the anxiety and mental stress it has caused. These effects have been most pronounced in pregnant women and in families with teenage and preschool children living in the vicinity of the plant. Although transitory in many persons, the ultimate impact of these effects remains to be fully assessed, as does the degree to which the effects may have differed from those caused by other accidents or disasters.

Comment

The maximum radiation dose that any member of the public is estimated to have received from the accident—less than 100 mrem—is smaller than the dose normally received each year from natural background radiation (TABLE 1). By the same token, the average dose that a person residing within 50 miles of Three Mile Island is estimated to have received from the accident—about 1.5 mrem—is less than the dose normally received each week from natural background radiation. This is such a small amount of radiation that its potential carcinogenic, mutagenic, and teratogenic effects combined add up to only about a one-in-a-million risk of death. Other risks of this magnitude (TABLE 7) encountered in daily life customarily provoke little concern.

Because there is so little likelihood of radiation injury from the accident, the mental stress experienced by members of the population residing in the vicinity of the plant has been viewed as a more serious health impact.[10] Underlying the mental stress, of course, has been the fear that a larger release of radiation might take place, with consequences (TABLE 2) that could be disastrous. While we are thankful that such an event did not occur, we should profit from our experience at TMI by taking steps to minimize the risks of such accidents in the future.

Summary

The only health impact of the Three Mile Island accident that can be identified with certainty is mental stress to those living in the vicinity of the plant, particularly pregnant women and families with teenagers and preschool children. Although increased risks of cancer, birth defects, and genetic abnormalities are potential long-term consequences of low-level irradiation, few if any such effects of the accident are likely, because the collective dose of radiation received by the population within a 50-mile radius of the plant was so small. Estimates of the number of people in the population who may ultimately experience any such effects range from 0.4 to 10, in comparison with hundreds of thousands in the same population who can be expected to develop cancer, birth defects, or genetic abnormalities through natural causes.

TABLE 7

SITUATIONS INVOLVING A ONE-IN-A-MILLION RISK OF DEATH

Traveling 700 miles by air	(accident)
Traveling 60 miles by car	(accident)
Visiting 2 months in Denver	(cancer from cosmic rays)
Living 2 months in a stone building	(cancer from radioactivity)
Working 1½ weeks in a typical factory	(accident)
Working 3 hours in a coal mine	(accident)
Smoking 1–3 cigarettes	(cancer, heart disease)
Rock climbing 1½ minutes	(accident)
20 minutes being a man aged 60	(mortality from all causes)

*From References 14 and 15.

REFERENCES

1. UPTON, A. C. 1968. Effects of radiation on man. Ann. Rev. Nucl. Sci. **18**: 495–528.
2. UPTON, A. C. 1969. Radiation Injury: Effects, Principles, and Perspectives. University of Chicago Press. Chicago, Ill.
3. National Academy of Sciences Advisory Committee on the Biological Effects of Ionizing Radiation. 1972. The Effects on Populations of Exposure to Low Levels of Ionizing Radiation. National Academy of Sciences, National Research Council. Washington, D.C.
4. United Nations Scientific Committee on the Effects of Atomic Radiation. 1977. Sources and Effects of Ionizing Radiation. Report to the General Assembly, with annexes. United Nations. New York, N.Y.
5. BATTIST, L., J. BUCHANAN, F. CONGEL, H. PETERSON, C. NELSON, M. NELSON & M. ROSENSTEIN. 1979. Population Dose and Health Impact of the Accident at the Three Mile Island Nuclear Station. Preliminary Estimates for the Period March 28, 1979 through April 7, 1979. U.S. Nuclear Regulatory Commission. Washington, D.C.
6. GERUSKY, T. M. 1981. Three Mile Island: assessment of radiation exposures and environmental contamination. Ann. N.Y. Acad. Sci. (This volume.)
7. UPTON, A. C. 1977. Radiobiological effects of low doses: implications for radiological protection. Radiat. Res. **71**: 51–74.
8. Interagency Task Force on the Health Effects of Ionizing Radiation. 1979. Report of the Work Group on Science, Office of the Secretary. U.S. Department of Health, Education and Welfare. Washington, D.C.
9. National Academy of Sciences Advisory Committee on the Biological Effects of Ionizing Radiation. 1980. National Academy of Sciences. Washington, D.C. (As cited by FABRIKANT, J. 1980. Paper presented at the International Radiation Protection Association Congress, Jerusalem, March 9–14, 1980.)
10. KEMENY, J. G., B. BABBITT, P. E. HAGGERTY, C. LEWIS, P. A. MARKS, C. B. MARRETT, L. MCBRIDE, H. C. MCPHERSON, R. W. PETERSON, T. H. PIGFORD, T. B. TAYLOR & A. D. TRUNK. 1979. Report of the President's Commission on the Accident at Three Mile Island. U.S. Government Printing Office. Washington, D.C.
11. KEENY, S., et al. 1977. Nuclear Power Issues and Choices. Report of the Nuclear Energy Policy Study Group, sponsored by the Ford Foundation and the Mitre Corp. Ballinger Publishing Co. Cambridge, Mass.
12. NRC. 1974. An Assessment of Risks in U.S. Commercial Nuclear Power Plants. Report No. AEC WASH-1400. Nuclear Regulatory Commission. Washington, D.C.
13. JABLON, S. & J. C. BAILAR, III. 1980. The contribution of ionizing radiation to cancer mortality in the United States. Prev. Med. **9**: 219–226.
14. POCHIN, E. E. 1978. Why Be Quantitative about Radiation Risk Estimates? National Council on Radiation Protection and Measurements. Washington, D.C.
15. WILSON, R. 1978. Risks caused by low levels of pollution. Yale J. Biol. Med. **51**: 37–51.

DISCUSSION

M. FIREBAUGH (*Institute for Energy Analysis, Oak Ridge, Tenn.*): I understand that in some places in India and in Brazil, the natural background radiation is as high as 1,000 to 3,000 mR per year. Have there been any studies to try to correlate this variation of natural background with incidences of various cancers that we're seeing here?

A. C. UPTON: Yes, there are efforts to assess the effects of high levels of natural background radiation. To my knowledge, there have not been, thus far, any clear-cut epidemiological data. This is not surprising. The natural background levels that exist in Pennsylvania or most parts of North America—amounting to roughly a tenth of a rem per year or 100 mrem per year—given the risk models we've talked about, may be expected to contribute perhaps 1 to 2% of the natural cancer incidence. If one increases those levels by a factor of 10, then one might ascribe 10% of the cancer incidence to background. It would take a mammoth epidemiological study, very carefully controlled, to demonstrate an effect that small. So I don't expect that we're likely to find effects in the human population attributable to high background radiation levels, such as exist in Kerala or parts of Brazil.

There have been studies in plant materials that point to an increase in the frequency of chromosome aberrations, but some of my colleagues at New York University have been closer to this work than I have.

D. ARONSON (*Consultant, Upper Montclair, N.J.*): While you were talking about radiation from these nuclear incidents, we're exposed to quite a bit of it in normal activities or corrective activities. Is there any campaign going on to reduce such radiation? For example, I go to the airport, and my equipment is inspected by x-ray equipment that has very little effect. Yet, I go to a dentist and get an old-fashioned short-cone x-ray. Would it make a difference if more modern equipment were brought in for these purposes?

A. C. UPTON: There has been, in recent times, a concerted effort to do just what you're advocating. Early in 1978, President Carter asked Secretary Califano to form an interagency task force to look at the levels of radiation to which the U.S. population was exposed and to suggest steps by which radiation levels could be reduced. The work group on this particular problem area delivered a report in which it recommended a number of measures, perhaps chief of which was education. The largest single contributor to population exposure, above and beyond natural background, is medical radiation. The report mentioned that the population is estimated to get on the average about 100 millirems a year from natural background. It estimated that you get about 75 millirems a year, whole body, from medical and dental radiation, and much of it unnecessary. Physicians frequently order films simply for good measure, not because they think they're really necessary, but because they think they could be helpful and won't do any harm. In fact, given the models we're talking about, which postulate no threshold, we would assume that no x-ray is totally without some risk. So I think much can be done.

Baggage-inspecting apparatus at the airport, I'm told, scatters relatively little radiation to the operators or to passengers who have to go past. So I don't think that does constitute a major problem.

S. SEELY (*University of Rhode Island, Kingston, R.I.*): A lot of your discussion has had to do with background radiation, or gamma radiation. Have you done similar kinds of studies for the types of radiation that are absorbed by particular organs within the body and then serve essentially as point sources?

Let me be specific.

Some years ago, I tried to get some estimate of the whole-body radiation that was equivalent to about 10 millirems of plutonium ingestion. Ten millirems of plutonium ingestion is assumed to be a lethal dose. Hence, presumably, one will survive probably a week or two under that condition. That calculated to a whole-body dose of about 70 millirems. So that just talking about millirems whole-body dose, in my opinion, is very misleading because it does not take into consideration the selective action of the thyroid, of bones, muscles, and so on. Have you done this kind of work?

A. C. UPTON: Yes, those kinds of observations and computations go into the assessment that I presented. The best information we have about the risks of carcinogenic effects on the skeleton come from patients with high skeletal burdens of radium—the dial painters and patients who were injected with radium. So we do have some evidence of that sort for the skeleton. There are data for thyroid that come from the Marshall Islanders. There are other populations in which the best information we have comes from internally deposited radionuclides. I might mention uranium miners, of course, where we're dealing with radon daughters deposited in the trachiobronchial epithelium and in the regional lymph nodes and pulmonary tissue.

I don't want to imply that the analysis was valid only for uniform whole-body radiation.

W. MAIBEN (Columbia University, New York, N.Y.): Between March and July of 1979, the infant mortality rate increased 90% in Pennsylvania, and furthermore, the increase seems to be inversely related to the distance of the city from Harrisburg. I really don't think it's plausible to attribute that effect to cancer-inducing radiation because the radiation was too small. But I think it is plausible to assume an effect on the fetal thyroid such that birth weights could have been lowered and caused an increase in mortality. Do you want to comment on that?

A. C. UPTON: Yes. First, on the likelihood that doses to the fetal thyroid may have led to abnormalities in thyroid function, birth weight, viability, or vigor: there is no question at all in my mind but that profound effects on the thyroid can be elicited if one puts enough radioiodine in the gland—no question at all! In the Marshall Islanders, where the estimated doses to the gland were on the order of 1,000 rads or more, there were instances of hypothyroidism in children who were exposed under the age of 10.

As Mr. Gerusky has pointed out, the amounts of radiation that have been detected here are very, very small—a matter of a few millirems. To my knowledge, we should expect no abnormalities in thyroid function in the millirad range. I think we'd have to get up into the tens of rads or even hundreds of rads before we'd see abnormalities in thyroid function.

For cancer, I'm assuming no threshold, and the risks that I've projected on the board are based on the supposition that there's a strict proportionality between dose and risk. So we would get a hypothetical fraction of a case of thyroid cancer in the population as a result of one millirem delivered to the gland.

As to the frequency of infant deaths and fetal deaths, the question came up earlier. Mr. Gerusky gave me a release from the Pennsylvania State Health Department, and I would simply like to indicate—for the benefit of those in the audience who haven't seen it—some of the figures. Within 10 miles of Three Mile Island, the infant mortality rate for the period October 1978 through March 1979, six months before the accident, was 17.2 deaths per 1,000. Following the accident, in the subsequent six-month period of April to September 1979, the rate was 15.7. In the case of fetal deaths, the figure for the last half of 1978, that is, the six months preceding the accident, October 1978 to March 1979, was 23.7 within 10 miles. For the six months following the accident, the rate was 14.0.

Now, it is true that in the case of the infant mortality figures I gave you, the January to September time period showed 16.9 deaths within 10 miles of Three Mile Island as opposed to 13.5 for the whole state. But if we exclude Harrisburg, the number comes down to 13.4. Similarly, for the six months following the accident, April to September 1979, there were 15.7 deaths within 10 miles of Three Mile Island versus 13.3 for the whole state; but Harrisburg alone was 20.2. So I think that in these comparisons, the influence of variations from place to place, particularly urban versus rural, tend to confound the analysis. Looking at all of the information that's available at the present time, one really can't conclude reasonably that there has been an increase.

E. PENNA-FRANCA (University of Rio de Janeiro, Brazil): Since somebody asked for comments on the studies carried out in Brazil in areas of high natural background, I would like to comment a little bit. In one of our cities, the average exposure rate is about 600 millirems per year. It's about six times more than normal background. In the studies we conducted there, we found a slight increase in chromosome aberration rates. However, the sort of immunological studies that you mentioned were ruled out because—based on the risk estimates that Dr. Upton mentioned here—they would need a much larger population than we have, would need about 200,000 people, and about 20 years of careful research to be able to see if there were any significance. This type of immunological study is very, very difficult to conduct. It's a long-term type of study, and it's very expensive; and we decided that it was not worth doing.

Another thing I would like to comment on is the fact that that population is exposed from birth to death to 600 millirems per year. This is at least six times more than the most highly exposed people in the Three Mile Island accident. Based on present parameters for exposure of populations around power plants, that city should be phased out and the population evacuated. Instead, we have a population of about 30,000 living there, and ironically enough—as in any other place in the world where you have high levels of natural radioactivity—the place is considered a health center. People go there, and the medical doctor recommends that patients go there, to be exposed to radiation and to feel better.

UNIDENTIFIED SPEAKER: The question of very-low-level radiation, of background level or less, is very complex. Obviously, cancer is one of the factors that must be considered. There were animal experiments carried out several years ago that showed that the life expectancy first rises with very small radiation levels and afterwards starts to fall down. Could you comment on this.

A. C. UPTON: Yes, that's a very interesting question, whether a little radiation might be good for you. Dr. Penna-Franca mentioned that in Brazil, physicians are prone to recommend radiation as a tonic. It used to be a popular tonic in the Western world, in both Europe and the U.S., some years ago. There have been animal experiments that suggest that small amounts of radiation on the order of 1 rad per day for the duration of life may extend the average life span of the experimental population. I don't think this is a fluke; I think it's a reproducible observation that's been seen often enough. The question is, What does it mean? And if a little radiation extends the life of an animal population, would it extend human life? and why do we worry about these tiny doses?

It turns out that the animals in which a little radiation has been observed to extend life are animals that are prone to intercurrent mortality from infection. They're not in tip-top condition to begin with, and the daily radiation does not extend the life span of the longest lived animal, it extends the mean life span. Animals that live under favorable environmental conditions tend to have a life table that is almost rectangular. You start with young animals, 100% survive, they continue to survive, and then when they get old, suddenly they begin to die, and very shortly the whole population dies off.

Under such conditions, where the life table is almost rectangular, daily radiation at low dose rates does not extend life. But in populations in which the life table tends to look more the way it does in the wild, where there's some intercurrent mortality going on all the time, radiation tends to make the life table somewhat more rectangular. Why that should happen remains a mystery. Whether radiation, by killing cells, heightens the immune response, we don't know. That's one of the theoretical explanations. I don't think it has relevance for the human population, at least not in the U.S. We live in a very favorable environment.

UNIDENTIFIED SPEAKER: Could I have a point of clarification? Maybe I misunderstood something. I thought Dr. Upton said that there had been an increase in the fetal mortality at Three Mile Island, but that there would have been a decrease if you eliminated the city of Harrisburg. My question is, Why would you eliminate the city of Harrisburg? Presumably, it had been used in other calculations of state mortality. Why would you make a selective exclusion? This reminds me of that recent Atomic Industrial Forum study of reactor efficiency, where they left out Brown's Ferry, Tennessee Valley Authority coal generators, and Three Mile Island, and came up with a very good figure. I can understand why you would want to exclude Harrisburg to get a lower figure, but I don't understand the methodology here and the science behind it.

A. C. UPTON: Please allow me to apologize. I'm sure I must have jumbled the figures in presenting them, so let me do this once more. During the six months before the accident, the figures for the radius within 10 miles were 17.2; and during the six months following, 15.7. During the latter period, the figure for the whole state is 13.3. If one were to look during the last six months at mortality rates within 10 miles and compare them to mortality rates within the state at large, one would say, Aha, the mortality rate is higher within the first 10 miles, and it must mean something has happened there to increase the rate. In fact, the rate has gone down since the accident. That it is higher than in the state at large is attributable to the inclusion of Harrisburg within that rate. Harrisburg is traditionally higher than is the state at large. So, I think it points up the problems of epidemiological methodology. You have to use a valid comparison in trying to arrive at any inferences as to what may have happened within 10 miles of Three Mile Island.

UNIDENTIFIED SPEAKER: I'll repeat the question. Was Harrisburg not included in previous calculations of state mortality? So the question is, Why drop it out now?

A. C. UPTON: I don't have the data for the whole state in the six months before the accident. In the six months before the accident, those data for the area within 10 miles show 17.2. The data for the six months after the accident show 15.7. There has been no increase within 10 miles during that period.

In the latter period for the whole state, the figure is 13.3; and for Harrisburg, it's 20.2. I'm talking about the latter period, the period April 1979 to September 1979—13.3 for the whole state, 15.7 for the first 10 miles, 20.2 for Harrisburg.

THREE MILE ISLAND UNIT 2:
DECONTAMINATION PROGRAM

Robert C. Arnold

Metropolitan Edison Company
General Public Utilities System
Middletown, Pennsylvania 17057

I'm very delighted to be with you today to share with you some of the experiences that we've had at Three Mile Island (TMI) since the accident on March 28, 1979. I hope to describe to you what we see as the major activities that must be carried out to accomplish the cleanup of Unit 2, including some discussion as to what we see as the major technical and institutional issues that have to be resolved before that cleanup is, in fact, accomplished.

For the first few weeks after the accident, those gathered at Three Mile Island to deal with the situation had three major objectives toward which all the efforts were oriented. First was to stabilize plant conditions and bring the plant into what was generally described as cold shutdown; second, to contain the radioactive fission products that were released from the core by the accident; and the third was to carefully monitor for releases of radioactive materials from the site.

These objectives were fulfilled very successfully during those initial few weeks and on into the first two or three months of the postaccident period, and in somewhat modified form they remain our objectives today. As progress was made towards the first objective, the focus gradually shifted to designing and installing modifications that would give us redundant capabilities to maintain the plant in a stable cold shutdown condition. Similarly, as the assurance of containment of the radioactivity released from the core became more certain, focus shifted to re-collecting and immobilizing those fission products. The need to continue careful monitoring of all emissions from the plant has continued and clearly will be a requirement throughout the cleanup period.

This situation is summarized in TABLE 1. We have achieved a stabilization of the reactor and are almost in position to shift to the system we envision using for long-term cooling of the core, including during fuel removal. We call this system the mini–decay heat system. This system is basically the same from a conceptual standpoint as the original plant equipment for decay heat removal, but it is much smaller (approximately 5% of the capacity of the original system). We have made substantial progress on decontamination of the auxiliary building, the building other than the containment building most significantly affected by the accident. We also have almost completed decontamination of the surfaces in the fuel handling building and the emergency diesel generator building,

0077-8923/81/0365-0076 $01.75/2 © 1981, NYAS

TABLE 1

PLANT STATUS

- Reactor stabilized, and cooling to be changed to long-term mode
- No unintentional activity releases
- Auxiliary building partially decontaminated
- Some waste processing capability in place
- Containment radiological status established

the other two buildings at the facility that were affected by the accident. We have in place some capability to process the radioactively contaminated water, and I'll talk about that in some detail, and we have made substantial progress in identifying the radiological status inside the reactor containment building.

As indicated in TABLE 2, the specific plant parametric conditions are as follows: there is a current decay heat generation rate of about 150 kilowatts, the plant has been brought down in pressure to approximately 100 pounds, and the average reactor coolant temperature above the reactor core is 140°F. Cooling is being accomplished by natural circulation of the reactor coolant water through the core and through the A loop. The secondary side of the A loop steam generator is lined up to the condenser, creating a vacuum on the secondary side of the steam generator so that boiling occurs at about 135°F. The low-pressure, low-temperature steam is transported through normal system piping to the condenser, where the steam is condensed and eventually pumped back into the steam generator, as would be the flow path for normal system operations.

The auxiliary building and the fuel handling building are approximately 90% decontaminated with regard to surface contamination, and what we mean by that is that the activity of loose surface contamination is

TABLE 2

TMI-2 PLANT STATUS

Reactor	150 kW
	100 psi
	140°F
	cycle natural circulation, "A" OTSG*
Auxiliary and Fuel Buildings	90% decontaminated
	~1,000 DPM/100 cm^2
	~1 mrem/hour
	systems decontamination in progress
Contaminated Water	auxiliary and fuel buildings, ~60% processed
	containment building, 0% processed

*OTSG = once through steam generator.

below 1,000 disintegrations per minute (DPM) per 100 square centimeters, and the radiation level from any fixed contamination is less than one millirem per hour. The decontamination of the remaining surfaces and the decontamination of the interior of the systems in those two buildings are in progress and proceeding quite smoothly.

In addition, contaminated water is being processed. We started out in October 1979 with about 425,000 gallons of water that we considered to be contaminated to an intermediate level, and which was in storage in tanks in the auxiliary building and the fuel handling building. The auxiliary building tankage was part of the original plant design. The fuel handling building storage totals 110,000 gallons and was installed after the accident in one of the spent fuel pools to increase the capacity of contaminated water storage in the unit. About 60% of the 425,000 gallons has been processed through a system we call EPICOR 2, and I will talk in more detail about that system later.

We have not started processing any of the highly contaminated water in the containment building or the reactor coolant system. It probably will be the latter part of this year before we have a system available for that purpose.

TABLE 3

TMI-2 CURRENT ACTIVITIES

• Prepare for long-term effort
• Maintain reactor and radiological safety
• Process and manage contaminated water
• Decontaminate auxiliary and fuel handling buildings
• Plan and obtain data for containment decontamination
• Engineer longer term facilities

TABLE 3 summarizes activities currently under way at TMI-2. We are doing the technical planning, staffing, and organizational development necessary to prepare for the long-term cleanup effort. We are maintaining the plant in a stable condition from a criticality and thermal viewpoint as well as from a radiological viewpoint. We are proceeding with the processing and the management of the contaminated water, including provision for storage of the processed water that results from the cleanup of the contaminated water. We're continuing with the decontamination of the auxiliary and fuel handling buildings. We're undertaking the planning and data-gathering efforts necessary for decontamination of the containment building, and we have initiated the engineering of the long-term facilities that must be in place to support major cleanup activities inside the containment building.

I prepared TABLE 4 to give you some idea of the scope of the water cleanup effort. I've identified the water that's in the plant in terms of

TABLE 4
TMI-2 WATER STATUS

Source/Type	Location	Quantity (gallons)	Activity (μCi/ml)					
			Gross	Tritium	Cs-137	Cs-134	Sr-39	Sr-90
1. Nonaccident	BWST	450,000	0.002	0.01	9×10^{-4}	5×10^{-4}	LLD*	LLD
2. Accident	RCBT	190,700	48	0.5	35	8	12	3
	AB & FHB	109,300	30	0.03	15	3.1	2.1	1.6
	RB sump†	400,000	270	1.0	180	40	41	3
3. Primary System	RC system‡	90,000	171	0.2	59	10	75	25
4. Processed Water	EPICOR 2	19,000	1×10^{-5}	0.02	LLD	LLD	LLD§	LLD§
	"B" SFP‖	156,830	5×10^{-5}	0.03	4×10^{-5}	8×10^{-6}	5×10^{-6}§	4×10^{-6}§

*LLD = lower limit of detection.
†RB = reactor building.
‡RC = reactor coolant.
§Estimated values.
‖SFP = spent fuel pool.

nonaccident water, the water that was affected by the accident, the primary coolant system water, and the water contaminated by the accident that has been processed. Starting with the nonaccident water, we have the borated water storage tank (BWST), which is part of the original plant design and which holds approximately 450,000 gallons. That tank capacity may well be used in the management of processed water to reduce the pressure to dispose of the water in the near term. The accident water is currently contained in a variety of tanks, as I indicated to you. The three largest tanks are the reactor coolant bleed tanks (RCBT). They are part of the original plant design and normally provide inventory of and storage capacity for reactor coolant quality water. Those three tanks have a total capacity of about 225,000 gallons and currently hold about 190,000 gallons of water to be processed. In the auxiliary building (AB) and fuel handling building (FHB), we have a variety of smaller tanks, but the only ones that currently contain accident water are the ones that were installed in the fuel pool. The table also shows the concentrations of the radioactive isotopes of major interest.

TABLE 5 shows how extensively the auxiliary and fuel handling buildings were contaminated by the accident and the levels of surface contamination subsequent to decontamination efforts. For reference, the 305-foot elevation is the grade level. The 282-foot level is below grade and is where the majority of the equipment that directly supports the reactor coolant system is located and, consequently, was the location of the highest levels of surface contamination.

I listed at the bottom of TABLE 5 the exposure history for the personnel involved with the cleanup to date. We've been very, very pleased with the ability to control the exposure. We've been able to utilize a large

TABLE 5

TMI-2 FUEL AND AUXILIARY BUILDINGS DECONTAMINATION STATUS

Elevation	Source	April 27, 1979	February 26, 1980
282 feet	Surface, DPM/100 cm^2	15×10^5	<1,000
	Area, mrem/hour	1,000	<5
305 feet	Surface, DPM/100 cm^2	7×10^5	<500
	Area, mrem/hour	80	<1
328 feet	Surface, DPM/100 cm^2	4×10^5	<500
	Area, mrem/hour	10	<1

Exposure History:	April 27, 1979–February 26, 1980	
	Total Personnel (sum of quarters)	611
	Average Quarterly Exposure,	
	millirem	84–517
	Total Exposure, man rem	84.7
	Maximum Individual Exposure, rem	1.585
Remaining Work:	Hot Spots	
	Tankage	
	Systems	
	HVAC	

TABLE 6

CONTAINMENT RECOVERY PROGRAM

1. Preentry planning
2. Containment purging
3. Containment sump water processing
4. Containment recovery service building
5. Hands-on decontamination
6. Reactor pressure vessel head removal
7. Fuel removal
8. Requalification of nuclear steam supply system
9. Refurbishment of containment systems
10. Requalification of containment integrity
Goal. Commercial operation

number of system employees for cleanup work by having them come to the plant on temporary assignment for a period of about two weeks and then return to their normal work location. In some cases—in fact, more accurately, in many cases—we've had people come back on a voluntary basis for a second, a third, and even a fourth assignment for decontamination work. We've gotten very good support from our employees and a high level of interest in assisting in the cleanup work.

Below the exposure data is a list of work activities that have yet to be done: cleanup of localized hot spots, flushing and cleanup of the inside of the tankage and piping of the systems themselves, and to some extent there is still work to be done on the heating, ventilation, and air-conditioning equipment (HVAC) to remove surface contamination inside those systems.

So the situation we have at this point is that the major cleanup effort other than the containment building is well under way and essentially should be completed by the end of the year. Cleanup of the containment building is yet to be started.

TABLE 6 shows the 10 major activities that we see comprising the containment recovery program. The first 7 of those 10 get us to the point where the plant has effectively been restored to the preaccident conditions. The last 3 are the steps necessary to return the plant to service. Of these first 7 items, number 1 is under way. We're proceeding with planning on the basis of being able to make an entry before containment purge, if that appears to be what is necessary. The containment purge proposal has received a lot of discussion today, and it continues to generate a lot of discussion in the communities around Three Mile Island. I think that substantial progress is being made in providing reassurance to the public that the venting of the containment building atmosphere is both safe and appropriate.

Just clarifying a couple of the items, number 3 includes the processing of the reactor coolant system water as well as the water that floods the basement of the containment building. Number 4 is a substantial structure that we have to design and build to support major activities inside

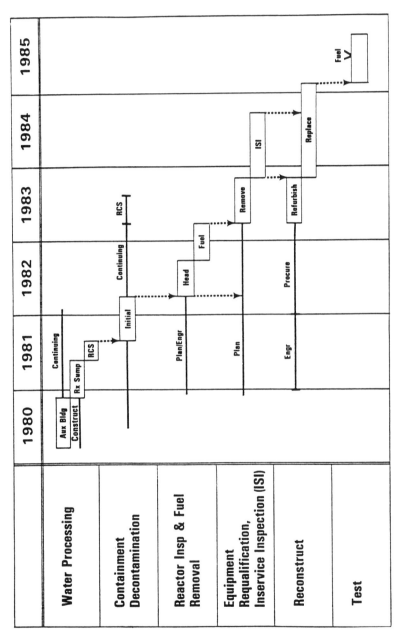

FIGURE 1. TMI-2 recovery program critical path schedule.

the containment building. It will be constructed immediately adjacent to the containment building and will provide access to the building through the large equipment hatch that was part of the original containment building design.

FIGURE 1 provides a summary schedule of the extent to which items in TABLE 6 lie on the critical path for completing the cleanup and recovery of the plant. It projects fuel removal by mid-1983 and an additional six months to complete the cleanup of the reactor coolant system and to inspect the plant to determine its suitability for service. We project another two years of reconstruction and restart activities before the plant could be returned to service. With that overall scenario in mind, let's look at what we think are the major technical issues that will be encountered in the course of accomplishing the cleanup (TABLE 7).

We anticipate that the containment building decontamination will pose some substantial technical difficulties. We don't know at this time

TABLE 7

TMI-2 MAJOR TECHNICAL ISSUES

- Containment building decontamination
- Containment and reactor coolant system water processing and cleanup
- Reactor core removal procedure
- Radioactive waste classification, accessibility to disposal grounds, resin treatment
- Reusability of major reactor system components, i.e., pressure vessel, OTSG, etc.

just what they'll be; and because of the uncertainties, we identify this as a major technical issue. We think we have to assume for the present, though, that decontamination techniques used on previous occasions may not be adequate for all the decontamination work we have to do and that new techniques will have to be developed to complete the decontamination of the surfaces inside the containment building. We have a tremendous amount of surface area, a wide variety of surface conditions, and an uncertain time frame between the time of deposit and the attempted removal of the materials. We believe that the longer the delay in starting that effort, the more difficult the overall effort will be. One consequence of that delay is a higher occupational exposure to the people in the course of accomplishing that work.

Our experience with the cleanup of the auxiliary building water would indicate that it's appropriate for us to be optimistic about the cleanup of the containment building water and the reactor coolant system water, but clearly, processing that water and the disposal of the waste associated with it are substantial technical items.

The reactor core removal will undoubtedly involve development of some new inspection techniques, some new handling techniques, and the adaptation of previous container methods to the situation we have at

Three Mile Island. There is a lot of uncertainty as to just what will be involved with core removal, since we don't know the geometry of the damaged core. I think it would be fair to say that we do not expect to have to develop new technology for the core removal. However, extension of current technology may be necessary in the sense of development of specific tools and specific procedures for our particular situation.

A very large issue—it may well become the dominant one of the cleanup program—is the disposal of the radioactive waste. Questions exist as to how it should be classified, particularly the material that results from the processing of the water; what will be our continuing access to the waste disposal sites for routine material as well as material that's more unique to the cleanup; and what we will have to do to solidify or otherwise immobilize the resin material to make it suitable for long-term storage, or suitable for additional processing if that turns out to be desirable. These items are all under extensive discussion between the Nuclear Regulatory Commission and ourselves, and undoubtedly will receive a great deal of public discussion in the coming months.

Although our analyses all indicate that we have no basis for believing the major reactor system components will not be reusable, that has to be

TABLE 8

TMI-2 Recovery: Institutional Factors

| ● PUBLIC COMMUNICATIONS |
| ● REGULATORY INTERACTION |
| ● FINANCIAL |

an important open issue pending the ability to inspect the equipment directly. That ability won't exist until after the core has been removed and the reactor coolant system has been decontaminated.

Table 8 identifies what we feel to be the major institutional issues that will impact on the completion of the cleanup, both in terms of schedule and in terms of the specific methods chosen to conduct the various cleanup activities. I think that good communication with the public—by the company, by the regulatory agencies, by elected officials, and by the various public interest groups that can contribute to the public's understanding of the situation at TMI-2—is absolutely essential. We believe that TMI-2 must be cleaned up. To not decontaminate the facility and remove the core from the reactor vessel simply would be an unacceptable situation. We think that it can be done safely and that it can be done in a timely way. But there are major impediments to accomplishing that, one of which is the public's perception of what has to be done.

The public's perception of what happened during the accident at TMI-2 and what will be involved in the cleanup of the unit has greatly affected the response of the various regulatory agencies to the problems presented by the accident. At both the state and federal level, regulators are faced with a variety of unprecedented situations. Regulations implementing the Atomic Energy Act and the National Environmental Policy Act never contemplated the circumstances existent at Three Mile Island. The situation calls for large measures of imaginative and pragmatic application of regulatory procedures. We recognize the challenge facing the regulators, and we are dedicated to working cooperatively and constructively with the regulatory agencies to increase the public's confidence that the cleanup will be accomplished safely and expeditiously.

Last summer we developed, in conjunction with Bechtel Power Corporation, a preliminary cost of $400 million for the cleanup. The company has $300 million of property damage insurance coverage. We are currently working on a new cost estimate based upon the schedule that I outlined earlier, and it appears that with the additional information available to us since last summer's cost estimate, there will be a substantial increase in our estimated cost for cleanup. The ability of the General Public Utilities System to meet additional financial requirements is an open issue, which will be determined primarily by the action of our state utility commissions.

Our present perception is that these institutional factors will dominate the cleanup effort at least during its initial phases. We believe the unit can be successfully decontaminated, and we know of no technical reason at this time why the plant cannot be recovered. However, our first objective—and our only current priority—is completion of the cleanup safely and expeditiously.

BACKGROUND OF THE THREE MILE ISLAND NUCLEAR ACCIDENT, II: GENERAL DISCUSSION

Moderator: Thomas H. Moss

Subcommittee on Science, Research, and Technology*
U.S. House of Representatives
Washington, D.C. 20515

H. ETZKOWITZ (*In These Times, Chicago, Ill.*): With respect to the proposed venting of krypton gas, a recent *New York Times* report stated that community leaders from the Three Mile Island [TMI] area suggested that there might be riots from the public if the gas were vented. Do you have any comment on the public reaction in the area to the proposed venting? And is there any level of negative reaction from the public that would lead you to reconsider the proposal?

R. C. ARNOLD (*Metropolitan Edison Company, Middletown, Penn.*): I don't think there is any question that there is anxiety on the part of some members of the public in the vicinity of Three Mile Island with regard to the venting approval. We have a problem there that clearly is not going to go away. It's one that we're going to have to solve. What we have taking place now within the vicinity of Three Mile Island is a process that is very desirable. This is a time in which the public has the opportunity to hear about the various trade-offs that are available and about the risks that are understood to be associated with the available options. The report that you indicated I think was based upon a meeting that six local residents had with NRC [Nuclear Regulatory Commission] in which they asserted that they represented a substantial portion of the population in that area. Since then, a number of members of the public and local public officials have come forward who would challenge whether those six really represented the majority of the people in the vicinity of Three Mile Island in regard to their anxiety over the venting. I think that while it is a matter of concern on the part of some of the people in the area, and consequently has to be addressed very constructively and very openly, it's not the predominant attitude.

T. H. MOSS: Let me ask, Mr. Arnold, how you would describe your options if you are not allowed to vent the krypton. Decontamination?

R. C. ARNOLD: I think the best way to generalize the options is that they all involve a collection of the krypton by some mechanical method and storing it on site. The three basic processes are liquification of it by a cryogenic system; passing it through very, very large charcoal beds in which krypton does have a slight affinity for absorbing to the charcoal; or storing it in piping or pressurized containers, having pumped it out in the containment building with compressors. None of these processes would

---

*Dr. Moss is science advisor to this committee, not a member of Congress.

0077-8923/81/0365-0086 $01.75/2 © 1981, NYAS

appear to present an opportunity for complete collection of the krypton. All of them would take a considerable period of time for design, construction, and operation of the facilities—on the order of two to three years. During that time period, there would continue to be the vulnerability to inadvertent and uncontrolled releases of the gas, either from the containment building, from the systems while they're being operated, or while the gas is in storage. As we look at those potential, inadvertent releases, they could lead to substantially higher doses to individual members of the public than would a controlled venting.

A. C. UPTON (New York University Medical Center, New York, N.Y.): Dr. Moss, I haven't been involved personally in the study of the release problem; I don't know the amounts of radioactivity that remain in the containment chamber or the number of curies and projected doses to the population that might result if the krypton were vented. Just in the session this morning, reference was made to a fraction of a millirem per person on the average. I think there are several difficulties here.

First of all, there is an enormous credibility problem. I think people are concerned about the reliability of the estimates that the experts might offer them; they'd rather not take any chance if they can avoid it.

But beyond that, I think there is a tremendous lack of understanding as to what a millirem is. If the estimates are anything at all correct, we're talking about almost infinitesimally small doses of radiation that people are likely to receive. It seems to me that some comparisons might be helpful to put this problem into perspective.

If you burn natural gas or coal, you're going to release some radiation inevitably; and there are probably people in the audience who can make a back-of-the-envelope calculation and say how much coal you'd have to burn to release the same amount of radiation—granted that it isn't in the form of krypton—as you'd have to vent if you bled off the contents of the containment vessel. People are used to seeing smoke come out of smokestacks. It doesn't worry them nearly so much as does this mysterious quantity of radiation. Would you happen to have that kind of comparison at your fingertips?

R. C. ARNOLD: Let me say that the amount of krypton in the building is 57,000 curies. The method that we propose for venting of it, we calculated, would result in less than a tenth of a millirem whole-body exposure to a member of the public, and less than five millirems of skin dose, beta dose. The NRC's environmental assessment used average annual meteorological data in place of our selected meteorological conditions per discharging, and their result was two-tenths of a millirem under those assumed conditions, and perhaps as much as 16 millirems of beta skin dose. But, they acknowledged that those could be reduced by a factor of two or three with the appropriate selection of meteorological conditions.

M. EISENBUD (New York University Medical Center, New York, N.Y.): I can say that the amount of radium and uranium and thorium that is emitted from a coal-fired power plant—when corrected for the relative risks of the radium series and the actinides, as opposed to the noble gases—is greater than that from the nuclear reactors. Of course, this is under normal operating conditions, and this was what my comment was going to relate to.

I think that there's a basic fallacy here, if I might say so—a basic failure particularly on the part of the federal government. It's my understanding with respect to both the liquid wastes that you were prevented from discharging into the Susquehanna at the very peak of the episode (as I recall, you were releasing it, and then you were stopped) and this particular required krypton release, that these releases were to have been permissible under the normal operating ground rules for a licensed reactor under normal conditions.

It appears that in the hysteria that has developed, you're prevented during this period of recuperation from doing what you would be allowed to do if the reactor were operating normally. I believe you could stagger the 57,000 curies in such a way as to comply with your license. I would think that there should be a ruling, at the level of the NRC, that anything that you propose to do that would have been permissible under normal operating conditions is permissible under these emergency conditions.

R. C. ARNOLD: You're absolutely right. The discharge of industrial wastewater that was interrupted temporarily at the time of the accident would have met all of our technical specifications for contamination levels for discharge. The krypton could be done within normal technical specifications quite readily. What we have in the area up there, it seems to me, is an absolute vacuum in terms of the Nuclear Regulatory Commission coming forward to the area and explaining the basis for the regulations and the reasons why they have confidence that the regulations are adequate for protecting public health and safety. Not having that for a reference point, I think the company, the local citizens, and other groups as well are at sea; there's no reference except people's intuition and how they make their own independent evaluations on risk. Its a real problem.

T. H. MOSS: I might ask your comment on what I think is a fairly common theory of the phenomenon there now, that is, the straw that broke the camel's back theory. Many members of the public felt that they went through a period of events that were out of their control and that caused them great upset, concern, and worry. Now, at last, there is an event that is possibly within their control, that hasn't happened yet; and you're not permitted to do it. They are determined to see that it doesn't happen. Have you, in your plans for decontamination, thought about the merit of not pushing this issue, that is, at least trying to show the public that there is something that they can control about the events there and that will be respected even if, in purely technical terms, it's an irrational wish?

R. C. ARNOLD: We very definitely have adopted that posture where it has not interfered with the ability to get on with the cleanup. We have not felt it was appropriate for us to take that type of an approach where the cleanup itself was hindered. For example, we had no difficulty at all in settling the suit that was initiated by the City of Lancaster with regard to discharge of the water that was cleaned up. We agreed that the water would not be discharged from the plant pending a completion of an environmental impact statement or until the end of 1981, a two-year period. So we were quite willing to adopt that type of a posture, as if to say: It's not necessary for us to discharge that water to proceed with the cleanup, and we recognize the anxiety associated with it. We'll voluntarily set that aside.

With regard to the krypton, though, we're not in that posture. It's certainly our judgment that the alternatives would involve more likely and higher off-site doses to individuals than would be incurred by a controlled venting, and very clearly they involve higher exposure to the workers and would incur a delay in the completion of the cleanup. All of those things, we feel, are impediments to solving the problem and should not be voluntarily undertaken.

UNIDENTIFIED SPEAKER: I was wondering where the money is going to come from to pay for the cleanup operations—whether it's going to be passed along to customers or whether there is some other method of funding such a large operation.

R. C. ARNOLD: The cost of the cleanup itself is estimated at this time to be $275 million. We have $300 million of property insurance. To the extent that costs, in fact, go beyond the property insurance coverage, then we would anticipate that it would be recovered in base rates at some point, probably over some period of amortization. But obviously that's a subject of future rate making.

UNIDENTIFIED SPEAKER: I guess what I'm asking you is whether you think that the customers should take on whatever additional financial costs there are in this, or whether some blame should be assessed and financial responsibility set up that way.

R. C. ARNOLD: Let me talk about that from several aspects, because I think it's a very important issue and not a very simple one.

One has to start with the understanding that the customers were the ones that benefitted from the lower costs of energy from the nuclear facilities that General Public Utilities operates. The stockholders did not gain any benefit from that: whether their investment was in nuclear plants, coal plants, utility poles, or wires, their return on that investment was the same and their opportunity for return was basically the same.

The stockholders have suffered dramatically in terms of costs of the accident. Our stock was trading at about 18½ the day before the accident; I guess it's down now around the range of $4.00 a share—at 60 million shares, that's probably around $850 million. There has been a reduction in dividends through two payment periods of $0.45 a share to $0.25 a share, and the last quarterly dividend was passed. So, there's very clearly a tremendous impact on the stockholders from this.

One of the things that we learned from the accident is the need to be able to spread the cost of such an accident over a broader base in an individual company. The liability insurance was available to the company and covered that. The property damage insurance was available; replacement energy costs at that time were not. I think it would be desirable if there were some way in which we, as a nation, could recognize that. Our government, our regulators, and our industry did not see the potential need for replacement energy costs and provide that kind of insurance.

So I think that a spreading of the costs will be necessary, and I think that our customers, clearly, are going to suffer from it. Our stockholders have suffered, and our company employees. However, at this point, even very minimal efforts to spread the costs to some extent—such as by having the state tax surcharge applying to all utility sales not be applied to the interchange sales—have not been able to gain support in the legislature. So these are not easy issues to resolve.

J. W. JACOX (*Nuclear Consulting Services, Inc., Columbus, Ohio*): My company and I have been considerably involved in the cleanup effort from about the first week of the accident on, so I'm quite aware of some of the incredible roadblocks—political and bureaucratic—as far as getting any technical, engineering, and practical work done. I'm unaware of any publicity or any public statements by Med Ed, General Public Utilities, or any other body relating the specific costs to the consumer, the stockholder, or whomever of these, at times, technically irrational delays—such as the krypton venting and having to store water that's already been cleaned up. I question the reason—directed specifically to the utility representative here—why these very substantial costs are not being publicized so that, hopefully, a more rational decision could be made.

R. C. ARNOLD: With regard to the water, it's because it's a rather minimal amount in terms of total cost. We may be talking about $1 million or $1¼ million. This may not be insignificant, but in terms of a percentage of total cost, it did not seem to us to be worth pursuing—and for basically the reasons that Dr. Moss indicated. There are advantages that come from taking a more compassionate, a more understanding, and a more receptive attitude toward the community.

With regard to the krypton, the issue there, in our judgment, is that it's the safest thing to do. There are many who have charged that the reason we want to proceed with the venting is because it's the cheapest. Well, it is the cheapest, but it is also the safest thing to do. So rather than muddy the argument, we have chosen not to try to make an issue out of the additional costs associated with prolonging the cleanup.

UNIDENTIFIED SPEAKER: I've heard of a method of cleaning up that is called cryogenic, where you actually refrigerate the whole thing; and I'm wondering whether that is ridiculous. Is it practical? Is it expensive? Is it less or more safe than any other? Secondly, does your insurance pay for any of the cleanup, or is that only for damage suits and things of that sort?

R. C. ARNOLD: Let me take the latter one first. We carry two kinds of insurance. Under the Price-Anderson Act, we carry liability insurance. It has a maximum of $560 million, and that is not available to the company at all. It's strictly for off-site liability. We carry $300 million in property damage insurance from private insurers. We have this, and it will go to offset the cost of cleanup.

A cryogenic type of system was one of those that we have investigated and analyzed as alternatives to the venting. Those of you from the Pennsylvania area may be familiar with an editorial that was run in a Philadelphia newspaper, chastising us for turning away an offer from Philadelphia Electric to use a cryogenic system that they have at their Limerick station, which is under constuction. I think that was picked up in a couple of other stories. The alternative of using a cryogenic system was, in fact, based upon utilization of that system. However, as Mr. Denton identified very clearly in his press conference with Governor Thornburgh on Thursday, March 27th, it is not realistic to expect a cryogenic system to work reliably—to collect, as he gauged it, any more than 60% to 70% of the krypton gas. Frankly, 30% or 40% can have as great or greater off-site doses, depending on when it is released, than the controlling venting would have. And the system that they used at a plant at the Brunswick station in the south has experienced a number of detonations, and that system has just been shut down. It is not operating at this time.

We are at a point where we're looking at the state of the art, and the state of the art is not consistent with the reliability and performance that would be necessary to achieve something that approached zero releases.

T. H. Moss: Would you like to comment on the radiation levels you'd have from a liquified stored krypton, and the occupational hazard that would cause?

R. C. Arnold: Well, I think that we could construct whatever degree of shielding would be necessary, as far as the specific radiation levels from the stored material. Our estimate of the occupational exposure that we would incur under the more optimistic projections of how the equipment would work, how much maintenance we'd need, and those kinds of things was about 25–30 person-rems of exposure, compared with the off-site exposure from controlled venting of less than 1 person-rem to everyone within the 50-mile radius.

Unidentified Speaker: I'm having trouble with little bits of sensational, intriguing information that keep coming in, and I wonder if I could ask for a clarification on just a few points.

On one of the television networks last night, there was reference to radiation unexpectedly found in some tunnels being dug near the site, with descriptions of—I think the phrase was—potential disaster. So, first, I would like to ask Mr. Arnold if he knows anything about that.

Also, Dr. Upton, I know you made a statement that the amount of radiation seems much too small to cause hypothyroid problems. There are these data floating around of 13 cases. I'm wondering if investigations have verified this or have shown some alternative explanation that might be plausible.

And Mr. Gerusky, I also would like to refer to the question of whether radiation levels were higher 10 miles away than closer in. I wonder if some reconciliation might be available?

R. C. ARNOLD: Let me talk about last night's news article, since it was based on a news release that I signed out. We have a monitoring program to insure that we can identify whether there is leakage of the water in the basement of the containment building out through the containment building walls. As part of this, we've dug, over the last few weeks, seven wells around the Unit 2 plant. Those wells don't go right immediately against the containment building; they're 75 to perhaps 250 feet away and pretty well cover the compass sector, depending on the degree of interference of other structures.

We put in an eighth well at the north end of the island that would not be expected to be influenced by anything happening at either Unit 1 or Unit 2.

When we took the first water samples from those wells, the results of the analyses indicated higher than expected tritium levels in three of the eight wells. The one closest to the borated water storage tank, which is over on the east side of the plant, showed about five times the normal level of tritium. That's a concentration that's still about 10% of the Environmental Protection Agency guidelines for tritium levels for drinking water; but clearly, it's higher than we expected. In the two other wells, we saw about two times the level that we would expect.

We think that the borated water storage tank is the source of that tritium. We are continuing to investigate the problem to try to isolate what the source is for sure.

Obviously, the wells were put in for the purpose of monitoring leakage from the containment building, so one cannot state categorically at this point that the water did not come from there. But we think it's extremely unlikely that that's the source of it. We believe we would see other material there as well as the tritium if it were coming from the containment building. However, until we're able to determine the source—or if it turns out not to be from any leakage at all, if it goes away—we can't say categorically that it's not from the containment building.

A. C. UPTON: I'll try to tackle the question on the thyroid, Dr. Moss. I think we have to distinguish between effects that may occur through damage to one gene, which can be transmitted in the germ cells to future generations, or damage to one cell, which may go on to divide and grow into a cancer cell years later, and the kind of damage that gives rise to abnormalities in thyroid function, where we really have to injure many cells.

I'm not aware of any evidence, experimental or clinical, that thyroid dysfunction can occur without the delivery of hundreds of rems to the thyroid gland, damaging many cells. At the same time, I emphasize that our theoretical models all lead us to predict some finite risk of cancer, even with doses in the millirem region, although the risks become very, very small as we do go down to the small doses.

That there have been instances observed in which children or others have hypothyroid conditions in Pennsylvania doesn't lead me to relate those to the radiation. There's too large a difference between the actual possible dose and the doses that would be needed to cause those effects. I'm more inclined to think those effects, if they've been observed, relate to other causes. Before they began to iodize table salt, hypothyroidism was a common disease in our country, and there are areas of the country where thyroid abnormalities still tend to be prevalent. But I have to confess that I haven't seen the report, I've just heard about it. I don't know the data, so I'm really not in a position to comment on them knowingly.

T. H. MOSS: Before we leave the monitoring, I wanted to ask Mr. Gerusky, Are you monitoring on site, or is the Environmental Protection Agency monitoring on site, in addition to the test wells Mr. Arnold mentioned?

T. M. GERUSKY (Bureau of Radiation Protection, Harrisburg, Penn.): Right now there is a combined environmental monitoring program that consists of on-site and off-site monitoring with a variety of agencies. The Environmental Protection Agency is the lead agency and NRC, the Department of Energy, Health, Education, and Welfare, and the Commonwealth of Pennsylvania are all being involved in a postaccident environmental monitoring program. That includes on-site monitoring of water discharge in our discharge pipe.

There are a lot of samples that are being collected. We're spending most of our environmental monitoring time on Three Mile Island and not on the other reactors, unfortunately. In relation to the highest dose being found 10 miles away, I read that this morning also, and I went through the references. I had not heard that number before. I have heard some other numbers before. So, I went through the documents I had, including the President's Health Physics Task Force Report, and I haven't been able to find any verification of any numbers 10 miles away that have been reported. I don't know where that came from. I'd like to know, because I've been looking for data that would disprove the numbers. We have been searching everywhere and asking for all the information that we could to find out if those numbers were correct, the off-site numbers. So far, we haven't been able to find any data that would indicate that the numbers, the off-site doses, were higher than indicated in any of the studies.

As for the hypothyroidism cases, the Health Department is studying each case. First of all, last year was the first year of any hypothyroidism reporting in Pennsylvania. Therefore, baseline data in Pennsylvania are not available. In Lancaster County, the numbers for the rates are small but are higher than the state average. The numbers also are higher in counties farther away from the plant.

Individual cases already have been looked at, and many of the cases have been discounted as being hereditary, and not related to radiation in any way. The final results of that analysis will be completed, and a report issued by the Department of Health.

About a week ago, a special task force put together by the Department of Health reviewed the environmental radiation data to determine whether or not they should investigate the hypothyroidism and the environmental radiation data any further. They concluded that there was no reason to go into the radiation data any more. They believe that it was acceptable. This was not our office, it was an independent group looking at the data for the Health Department.

T. H. MOSS: That there hasn't been, until now, any long-term collection of hypothyroid data is an important issue. If we don't have that kind of baseline health data, we can hardly detect any sort of public health event.

T. M. GERUSKY: That program was not started because of Three Mile Island. It was started to determine whether early detection of hypothyroidism could save children a lot of problems. The reason that it was done was that the Health Department felt that such an early warning program was vitally important, and it was started prior to Three Mile Island.

WHAT THE PRESIDENT'S COMMISSION LEARNED ABOUT THE MEDIA

David M. Rubin

Task Force on the Public's Right to Know
President's Commission on the Accident at Three Mile Island

and

Department of Journalism and Mass Communications
New York University
New York, New York 10003

Most discussions of the performance of the news media during the accident at Three Mile Island (TMI) begin, and often end, with the question, Did the media sensationalize the accident? Behind the question is the thesis that the image of nuclear power is suffering not because of the actual events at TMI, but because of the way in which the news media portrayed those events.

The sensationalism thesis was being advanced even a year after the accident. The Subcommittee on Energy Research and Production of the House Science and Technology Committee stated in a report released April 3, 1980, that mental stress among the population near the plant was "greatly exacerbated" by news media coverage.[1] Subcommittee Chairman Mike McCormack, known for his pronuclear position, criticized CBS and a New York City newspaper and recommended that the government deluge the public with information on how safe nuclear power really is.[1]

Congressman McCormack's view of media performance is at odds with the findings of both the President's Commission on the Accident at Three Mile Island and the Nuclear Regulatory Commission's (NRC) own internal investigation, yet his criticism cannot be easily dismissed. The suspicion persists that Three Mile Island was a media creation—that the reality was much less serious than the image. Thus, any discussion of media performance must first deal with the question of sensationalism.

The subject is a knotty one because there is no clear and accepted definition of the term, and it is a difficult concept to operationalize for purposes of quantitative research. To some, sensationalism is the mismatching of a headline or broadcast "teaser" with the story it accompanies. To others, it is the selection of pictures. Sensationalism can also be seen in the choice of evaluative words to describe situations; in the choice of persons to quote; in the inaccurate presentation of facts; in the "play" given a story, either by making it the first item in a newscast or by placing it on the front page with a large headline; or in the tone of voice of a newscaster as he or she delivers the story.

Nor does the same news item necessarily seem sensational to different members of the audience. Nuclear power is an issue that has

0077-8923/81/0365-0095 $01.75/2 © 1981, NYAS

polarized the public. Information on the status of the reactor that is alarming to a nuclear opponent might be unremarkable to a nuclear engineer. Information on radiation releases that would concern an antinuclear activist might reassure a nuclear proponent. All this makes reasoned discussion of sensationalism very difficult.

Three separate news stories appearing during the accident have frequently been criticized as sensational. These three are (1) the NRC press conference on the third day that produced a United Press International (UPI) story on the chance of a meltdown;[7] (2) the "CBS Evening News" broadcast of the third day, which Walter Cronkite opened by saying, "The world has never known a day quite like today";[11] and (3) the Associated Press story on the evening of the fourth day that said the hydrogen bubble in the containment might explode.[12] As a first step in studying more closely the charge of sensationalism, each of these stories will be examined in turn.

The NRC Discusses the Possibility of Meltdown

On Thursday night March 29, the second day of the accident, it became clear to both Metropolitan Edison and the NRC that a large volume of hydrogen gas had collected at the top of the reactor vessel.[2] The danger posed by this gas bubble, however, was a matter of disagreement between Met Ed and NRC personnel.

NRC personnel were concerned that a decrease in pressure in the reactor could permit the bubble to expand and cover the nuclear core, impeding the flow of vital cooling water and causing further damage to the fuel. Roger Mattson, director of the NRC's Division of Systems Safety, was alarmed enough to recommend to NRC Chairman Joseph Hendrie that a precautionary evacuation be undertaken.[3] He worried that if one of Met Ed's suggested maneuvers for removing the gas bubble turned bad, a core melt could start in a matter of hours.[3] But in the TMI control room, Met Ed engineer Gary Miller thought the bubble could be taken out of the system without danger.[4]

On Friday afternoon the public and the news media were not yet aware that a meltdown was anything more than a theoretical possibility, although NRC concern about a core melt had been expressed to Pennsylvania Governor Richard Thornburgh, White House Press Secretary Jody Powell, Energy Secretary James Schlesinger, and others.[3] The public position of the NRC since the beginning of the accident, however, had been that a meltdown was not a legitimate scenario.[5]

The NRC's concern was first publicly acknowledged at a press briefing Friday afternoon in which staffers Dudley Thompson and Brian Grimes discussed the state of the reactor with reporters. In response to the question "Is a meltdown possible?" Grimes said, "Yes, there is a possibility. . . . I think everyone technical would have to concede that if the reactor was depressurized, and the hydrogen bubble expanded, that there was a possibility, not a likelihood, but a possibility that things could go wrong, or all the cooling systems quit."[6] Thompson said much the same thing later in the briefing.

In that this was the first public discussion of the possibility of a meltdown, it is not surprising that the UPI story resulting from the briefing began with this information. Three key paragraphs of that nationally-distributed story read as follows:

> Dudley Thompson, a senior official in the NRC Office of Inspection, said the threat is posed by a steam bubble inside the reactor that could increase in size as pressures in the reactor are lowered, leaving the core without vital cooling water.
>
> "We are faced with a decision within a few days, rather than hours, on how to cool down the core," Thompson told reporters at an NRC news center.
>
> "We face the ultimate risk of a meltdown, depending on the manner we cope with the problem. If there is even a small chance of a meltdown we will recommend precautionary evacuation."[7]

Given his experience with the press, Frank Ingram, who works in the NRC's Office of Public Affairs, knew that the meltdown predictions would be the most important information from the briefing, and Ingram did not disagree with this news judgment.[8] No matter how it is couched, a discussion of meltdown by official sources during an accident is bound to have a "sensational" effect on the public.

Met Ed responded by maintaining its optimistic posture, stating that "reports of a meltdown are unfounded."[9] The five NRC commissioners chose to undercut Thompson and Grimes by issuing a press release, drafted in committee, that stated in the first paragraph that "there is no imminent danger of a core melt."[10] In the fourth paragraph, however, the commissioners admitted in somewhat more technical language that should the gas bubble expand, "some of the fuel would fail to cool and further damage to that fuel could occur."[10]

The press release was designed to diminish the impact of the UPI story without specifically denying the possibility of a partial meltdown. This was accomplished by use of the word "imminent" and by creating the impression that the media were being alarmist. In contrast, however, Chairman Hendrie later told Jody Powell that the chance of a meltdown was "pretty small . . . but you can't rule it out."[3]

This episode illustrates the industry's reluctance to discuss the subject of meltdown. It is not an example of media sensationalism, but rather of the difficulty in communicating scientific uncertainty to the public in a manner that maximizes understanding and minimizes alarm. None of the parties involved—the utility, the NRC, the news media, and the public—were experienced at this in relation to a nuclear power plant accident.

Years of disaster communications in the area of hurricane forecasting, for example, have created a common language for experts and the public, so that communication can occur with a minimum of misunderstanding and panic. The TMI experience is an important step in building that history for nuclear power plant accidents. To blame only the news media for the impact of the UPI story is to disregard the obligation (and failure) of the utilities and the NRC to develop language that would clearly communicate the likelihood of a meltdown.

The UPI story is in fact a relatively accurate account of what the NRC in Bethesda thought about the chances of a meltdown on the afternoon of the third day of the accident. Since, one year after the accident, it is still not clear how much of the fuel did melt, it seems unreasonable to expect journalists to have handled this information in any other way. Indeed, if a complete core melt *had* started that day, the performance of UPI, Thompson, Grimes, and the NRC would be viewed rather differently. The solution is not for officials to refuse to answer questions about a meltdown—questions that are inevitable in such circumstances—but for officials to develop a more sophisticated and understandable (and truthful) language for discussing the state of a reactor during an accident.

The "CBS Evening News" Broadcast of March 30, 1979

The news report that is singled out perhaps most frequently as an example of sensational journalism was the "CBS Evening News" of Friday, March 30.[11] By any measure, Friday was a disturbing and confusing day for those living near the plant. The utility had released radioactive gas early that morning in a manner indicating that utility personnel might not be in complete control. The release had prompted Governor Richard Thornburgh to recommend that individuals living within 10 miles of the plant stay indoors and that pregnant women and young children within 5 miles of the plant leave the area. This was followed by the Thompson/Grimes briefing in Bethesda (discussed above) concerning the possibility of a meltdown.[6]

When CBS prepared its evening newscast, it did *not* know that the NRC had recommended to the governor that an area out to 10 miles around the plant be evacuated because of the release of radioactivity that morning (and the probability of further releases). Nor was the network aware of the growing concern in Bethesda that the hydrogen bubble posed problems beyond those detailed by Thompson and Grimes. Nor did the network have access to NRC's Roger Mattson, who was strongly urging a general evacuation. Had all this been available to CBS that evening, the broadcast might have been even more pessimistic.

Cronkite's introduction to a series of reports from CBS correspondents has become the most celebrated portion of the broadcast and is worth quoting in its entirety:

> The world has never known a day quite like today. It faced the considerable uncertainties and dangers of the worst nuclear power plant accident of the atomic age. And the horror tonight is that it could get much worse. It is not an atomic explosion that is feared; the experts say that is impossible. But the

spector was raised [of] perhaps the next most serious kind of nuclear catastrophe, a massive release of radioactivity. The Nuclear Regulator— Regulatory—Commission cited that possibility with an announcement that, while it is not likely, the potential is there for the ultimate risk of a meltdown at the Three Mile Island Atomic Power Plant outside Harrisburg, Pennsylvania.[11]

This introduction is an accurate summary of what was known then about the course of the accident. The "horror" of the situation getting much worse did not materialize, but Cronkite can hardly be blamed for that. He was only reporting concerns voiced by members of the nuclear establishment. The events of the day *were* unique in reactor history. The industry has since conceded that TMI was the worst civilian nuclear power plant accident. While the entire first sentence and such words as "horror," "specter," and "catastrophe" are clearly evaluative, they must be judged according to how events appeared when they were written. Against that standard, they do not seem particularly alarmist.

The introduction was followed by a tape of Grimes and Thompson repeating for the cameras what they had said to reporters at the Bethesda briefing, by a report on the radiation release and evacuation recommendations, by interviews with evacuees at the Hershey Park Arena, by an interview with Dr. Paul Milvy in New York about the health effects of the radiation release, by a review of the many conflicting statements coming from Met Ed and NRC personnel, and by quotes from congressional figures on the likely impact of the accident on the nuclear industry. One might take legitimate exception to two aspects of this reporting, both related to the radiation release.

First, a CBS reporter said that the release Friday morning was "ten times the amount considered safe for the general public to be exposed to in a full year."[11] He failed to state that the release as measured at the stack of the power plant was not the amount residents were exposed to. Nor did he discuss how long a person would have to be exposed to a 1,200 mR release to absorb a yearly limit. He did not say what happened to the radioactive plume after the release or suggest that it might have dissipated. In short, the reporter overstated the case on the seriousness of the release.

The second criticism concerns the wisdom of permitting Dr. Milvy to predict on camera an "increased incidence of cancer" 20 or 30 years following the accident.[11] In the absence of the detailed studies necessary to determine the amount and type of fission products released, such a prediction seemed premature.

These two examples illustrate the news media's problems in reporting radiation information and the media's often unnecessary preoccupation with speed in attempting to provide perspective. But the charge that CBS sensationalized the accident in this newscast is not supported either by treatment of the events as they appeared to journalists at the time or as they now appear to those familiar with the accident. The anxiety level among some NRC officials was clearly as high as CBS indicated, and some very disturbing events—such as the NRC's evacuation recommendation to Governor Thornburgh—were not reported at all. Reaction to Cronkite likely divides along pro- and antinuclear lines and is more a recognition of his power and credibility than a statement about content of the broadcast.

The Associated Press and the Hydrogen Bubble

On Saturday March 31 at 9:02 P.M., the Associated Press (AP) sent out a story quoting NRC officials as saying that the hydrogen bubble "showed signs of gradually turning into a potentially explosive mixture that could wreck the already damaged reactor. One NRC source, who asked not to be identified, said that critical point could be reached within two days."[12] The story described the view of some NRC officials in Bethesda that oxygen was slowly increasing in the system and that at some point it would form an explosive mixture with the hydrogen to threaten a breach of containment.

This AP story caused great upset among residents of the Harrisburg area and among reporters and NRC officials at the site. It led to a White House order to close the NRC's Bethesda press center, from which the story originated. All information from the NRC was, from that point on, centralized in Harold Denton at the site. From the perspective of those at the site—who did not consider the bubble potentially explosive—the story was a media creation that unnecessarily upset the local community at a sensitive time.

The AP story was, in fact, incorrect. Oxygen was not collecting in the system, and the bubble could not have exploded. But the story was an accurate reflection of what the AP reporter was told by the NRC's Frank Ingram (in public affairs), Edson Case (Harold Denton's deputy), and a third, unnamed source. It also accurately reflected the concern of Roger Mattson and of Chairman Hendrie, in comments the latter made at a Saturday afternoon press conference. Tapes of telephone conversations between Case and Ingram and NRC officials on site, and with the AP reporter, prove that the reporter checked the story line by line and that the story accurately reflected the mistaken views of the NRC in Bethesda.[13] The tapes even reveal Case and Ingram defending the story to their colleagues on site, who did not share this alarming view of the bubble's chemistry.[14]

What the AP story really illustrates is the confusion among sources that existed throughout the accident. It is highly unlikely that reporters would ever have the expertise in this area to second-guess the sources talking to AP on Saturday night. Indeed, the public has a right to know what responsible officials are saying, even if (especially if) these sources turn out to be incorrect. AP had an obligation to publish their views. If they had turned out to be correct about the bubble's explosiveness, there is little doubt that AP would not have been called sensational for sending out the story to its client media.

This example does, however, raise a number of other important questions about news coverage of nuclear accidents. Should there be a single source of information about the reactor? Or multiple, competing sources? Who should speak for the NRC and the utility? Who are credible sources? And how are reporters to distinguish them from the false crisis generators in any organization? How much fact checking is required of a reporter before a story such as this one is moved?

It is possible, of course, that while the news media may not have behaved in a sensational manner in these three instances, their reporting overall was sensational in tone. To test this hypothesis, the task force examined one important dimension of sensational reporting: the kinds of evaluative statements about various aspects of the accident (such as the threat of radiation exposure or the threat of a hydrogen explosion) that journalists chose to include in their stories. Categorizing these evaluative statements into two groups, either alarming or reassuring, helped to determine whether reporters were presenting an alarming or reassuring view of the accident. For the charge of sensationalism to hold up, it was hypothesized that the number of alarming statements would be larger than the number of reassuring statements for most aspects of the accident.

The task force content analyzed all news coverage about the accident during the first week in the following 10 news media: AP and UPI; ABC, CBS, and NBC; New York Times, Washington Post, Los Angeles Times, Philadelphia Inquirer, and Harrisburg News.

The hypothesis that alarming statements would outnumber reassuring statements was not supported. (A full report of this study appears in the task force report, pages 198–207.)[16] TABLE 1 shows that 1,168 evaluative statements about 14 aspects of the accident were coded. Of these, 56% were reassuring, only 39% were alarming, and 5% were neutral. For 9 of the 14 accident-related issues coded, there were more reassuring than alarming statements. These 9 included the ones most directly related to the health and safety of the public: the status of the accident, the threat of danger, the threat of radiation exposure, the threat of hydrogen explosion, the threat of the bubble growing larger, and the need to evacuate. On only one issue related to the public health and safety were there more alarming than reassuring statements coded: the risk of meltdown. Here the split was approximately 60–40 in favor of alarming statements.

To the extent that alarming and reassuring statements selected by reporters for readers and viewers are a measure of sensational treatment of a subject, the major news media in this analysis did not portray the accident at Three Mile Island in an alarming manner.

Given this analysis, why is the view that the press was sensational held by some members of congress, by residents of the TMI area, and by at least one member of the president's commission? Their condemnation of media performance can better be understood through examination of a half-dozen other aspects of the press-public and press-source relationship. These are discussed below.

TABLE 1

FREQUENCY OF ALARMING VERSUS REASSURING STATEMENTS
ON ACCIDENT-RELATED ISSUES

Issues	Reassuring		Neutral		Alarming		Total	
	F*	%	F	%	F	%	F	%
1. Accident Status	55	47.4	56	48.2	5	4.3	116	10
2. Threat of Danger	99	88.4	—		13	11.6	112	10
3. Radiation Exposure	79	82.3	—		17	17.7	96	8
4. Hydrogen Explosion	29	55.8	—		23	44.2	52	4
5. Hydrogen Bubble	136	91.3	5	3.3	8	5.4	149	13
6. Meltdown	55	39.6	—		84	60.4	139	12
7. Management of Accident	18	56.2	—		14	43.8	32	3
8. Information	1	1.0	—		113	99.0	114	10
9. Evacuation	70	85.4	—		12	14.6	82	7
10. Local Reactors	15	88.2	—		2	11.8	17	1
11. Nuclear Power	47	75.8	—		15	24.2	62	5
12. Future of Nuclear Energy	12	33.3	—		24	66.7	36	3
13. Evacuation Preparedness	7	31.8	—		15	68.2	22	2
14. Citizen Reaction	34	24.5	—		105	75.5	139	12
TOTAL	657	56.3	61	5.2	450	38.5	1,168	100

Meaning of Terms		
Reassuring	Neutral	Alarming
1. Improvement/Crisis Over	No Change	Deterioration
2. No Danger		Danger
3. No Health Threat		Health Threat
4. Not Serious Possibility		Possibility
5. Shrinking	Same	Growing
6. No Serious Possibility		Possibility
7. Prepared,		Not Prepared,
Knowledgeable,		Not Knowledgeable,
Competent		Incompetent
8. Reliable,		Unreliable,
Clear		Conflicting
9. Not Necessary		Necessary
10. Not as Dangerous		As Dangerous
11. Necessary,		Unnecessary,
Desirable		Undesirable
12. Optimistic		Pessimistic
13. Adequate		Inadequate
14. Positive		Negative

*F is the frequency of number of statements coded in that category.

Unrealistic Public Expectations of Media Performance

The social responsibility theory of the press asks that journalists present stories in a context that gives them meaning.[15] The public has a right to expect that the news media make the confusing understandable and help resolve conflict. In covering TMI, however, this proved to be an impossible task. The story changed from hour to hour. News sources were themselves confused about, or ignorant of, what was happening.

Reporters often received directly contradictory information from Met Ed and NRC personnel on the status of the reactor or off-site radiation releases.

When this happens and when reporters have no sure way of determining which information is correct, they present the contradictory statements to their audience and let them make up their own minds.[16] Indeed, it is a function of a free press to present contradictory views from credible sources on matters of public importance. It is easy to understand, however, why the audience, particularly in the area around TMI, might not find this helpful in trying to make an evacuation decision. Anne Trunk (a member of the president's commission), speaking for many of her neighbors in Middletown, says this stream of conflicting messages on the status of the accident forced her to conclude that the media were part of the problem at TMI.[17]

Nevertheless, given the extraordinary degree to which reporters were hostage to their sources in this story (not being able to really see any of it first hand), it is difficult to see how they could have avoided creating this emotional roller coaster without intentionally suppressing pessimistic information from the NRC and other sources. Such suppression would not be in the tradition of American journalism and, in the long run, would not serve the interests of the public. Confusion in reporting this type of story is probably inevitable.

Inaccurate Reporting and Ethical Lapses

Most of the important errors in media coverage of the accident are traceable to inaccurate sources of information—but not all. Where reporters erred, such errors should be admitted and corrected. Media credibility is seriously damaged when a columnist such as Jimmy Breslin of the New York Daily News writes that "evil" steam "laced with radiation" is coming from the TMI cooling towers, or that he is suffering an attack of radiation poisoning, or that the radioactive gas in the reactor "is the steam we see rising from the cooling towers."[18] The barrage of photographs and television clips showing the massive cooling towers no doubt misled the public about the relationship of the towers to the accident. (There was none.)[16]

The comparatively alarming (although not generally inaccurate) manner in which two New York newspapers—the Daily News and the Post—played the story may have led many in this media capital to conclude that papers all over the country were adopting headlines such as "Nuke Cloud Spreading," or "Race With Nuclear Disaster," or "H-Blast Danger Fades at Plant."[16] Given the high visibility and circulation of these papers, the impression lingers that this coverage was typical, when in fact the vast majority of newspapers treated the accident in a sober, careful fashion.[16]

The Pressure to Get a Story

Journalists do not have the luxury of *not* filing a story because events are confusing and unclear, or because news sources are not available. One of the cardinal rules of journalism is to "get the story." In the first days of the accident, neither the NRC nor Met Ed had a credible, efficient information plan in place. Since journalists could not get what they needed from these sources, they were forced to consult a wide variety of alternative sources, including Met Ed workers (who may not have been in a good position to know what was happening in the control room), antinuclear spokesmen, academicians, executives from other utilities, state government officials from outside Pennsylvania, and residents of the area.[16] Some of these sources proved more reliable than others.

Journalists must, of course, always cross-check what they are hearing with individuals outside official channels, but these sources were not an adequate substitute for the flow of information that should have been coming from the NRC and the utility. Had NRC and Met Ed recognized that the "media goat" has to be fed regularly during a story of this magnitude, much of the pressure on reporters to rely on sources with little direct knowledge of events would not have been present.

Media Inexperience and the "What If?" Question

While many experienced science writers covered the accident, the majority of reporters were not familiar with nuclear technology. Nor was the press corps stable in its composition, with reporters leaving and arriving daily. This meant that there was no shared base of information and questioning of sources often was repetitive and simplistic. For the inexperienced reporter, the most common question (and the easiest to frame) concerned projections of worst-case scenarios: "What if this happens? What will that mean for the reactor?"[16] While such questions were certainly legitimate given conditions in the reactor, they forced sources to address the question of meltdown perhaps more often than conditions warranted. The overemphasis on worst-case scenarios could well have increased anxiety in the media audience.

In addition, the confusion during mass press conferences and the lack of sophistication in the questioning meant that some stories of great importance were overlooked. One good example was the error made by some NRC staffers in Bethesda concerning the explosiveness of the hydrogen bubble (referred to above in relation to the AP story of Saturday evening). The fact that the panic over the bubble was caused entirely by NRC miscalculation should have been an important follow-up story in the media. It was *not* because Harold Denton skated over it quickly in one of his press conferences and reporters did not question him about it.[19]

Given that inexperienced reporters will likely continue to form a substantial portion of the press corps at nuclear accidents, officials must be prepared to provide background information and answer "what if?" questions in a manner not guaranteed to alarm the public unnecessarily.

Weaknesses in Radiation Reporting

The information the public most needed to know during the accident concerned off-site radiation releases. Such information bears directly on evacuation decisions. This information was not being made available to the public or the news media in any systematic manner, and reporters compounded the problem by making basic errors in radiation reporting. A content analysis of radiation reporting by the president's commission uncovered a number of common errors. An amount or unit of radiation was often not accompanied by a rate, that is, a radiation dose delivered over a period of time and measured in a unit of time. Often it was not clear when a release took place and how long it lasted. The location of a release or an indication of where the measurement was made was often not indicated. (This was particularly troublesome, for example, with the 1,200-mR venting on March 30, a release that was measured directly above the plant's stack.) The news media did not always state which types of fission products were being released, and comparisons of releases with x-rays generally proved more confusing than helpful.[16]

In sum, the public's need to know in this area was not well served. The blame lies in part with officials who did not make this information available in any systematic manner and with reporters who did not understand the subject. The confusion over releases contributed significantly to the public's frustration with the flow of information and bears on the perception of sensationalism.

Pack Journalism

The psychological impact on a small community of the arrival of 300 or 400 reporters can be significant. A common concern voiced by some members of the president's commission was that the larger number of reporters impeded management of the accident. The task force examined this hypothesis but did not find any evidence that reporters interfered with management of the accident or with the delivery of health care. Nevertheless, the constant interviewing of evacuees and local officials by reporters and the confusion at mass press conferences added to the impression of a press corps bent on sensational treatment of the accident.

While the charge of sensationalism is a convenient shorthand for criticism of the press at Three Mile Island, it is not an accurate reflection of coverage in most of the nation's news media, and it does not explain the real reasons for the type of news coverage that resulted. Like any complex institution, the news media cannot be understood by retreating to clichés. The dependence of reporters on ill-informed sources, the difficulties of reporting developing events, the problems of communicating scientific uncertainty, the overall complexity of the story, and the lack of background of many reporters all contributed to the quality of coverage.

A public more sophisticated about how the news media work and about the limitations of media performance will be better able to interpret and evaluate the information being made available during crisis situations. To that end, unfounded charges of media sensationalism are not productive.

REFERENCES

1. New York Times. 1980. April 4: A8.
2. MATTSON, R. 1979. Chronology of TMI-2 Hydrogen Bubble Concerns: 1. Supplied to the President's Commission on the Accident at Three Mile Island. Nuclear Regulatory Commission. Bethesda, Md.
3. NRC. 1979. Transcript of closed NRC meeting, March 30: 68, 85, 130, & 189. Nuclear Regulatory Commission. Bethesda, Md.
4. MILLER, G. 1979. Task Force on the Public's Right to Know interview with Gary Miller (Tape 2): 5.
5. MILNE, D. 1979. Task Force on the Public's Right to Know interview with David Milne: 47.
6. GRIMES, B. 1979. Task Force on the Public's Right to Know interview with Brian Grimes (Tape 3): 1.
7. UPI. 1979. Wire service story of March 30, 4:02 P.M., from Washington, D.C. United Press International. New York, N.Y.
8. INGRAM, F. 1979. Task Force on the Public's Right to Know interview with Frank Ingram (Tape 2): 2.
9. Met Ed. 1979. Press statement of March 30, 5:30 P.M., Hershey, Penn. Metropolitan Edison Company. Reading, Penn.
10. NRC. 1979. Press release, March 30, 6:30 P.M. Nuclear Regulatory Commission. Bethesda, Md.
11. CRONKITE, W., et al. 1979. CBS Evening News with Walter Cronkite. TV program aired March 30. Columbia Broadcasting System. New York, N.Y.
12. AP. 1979. Wire service story of March 31, 9:02 P.M., from Harrisburg, Penn. Associated Press. New York, N.Y.
13. IRC/NRC. 1979. Telephone log transcripts, March 31, 04-283-CH7/25 HF 2-4. Nuclear Regulatory Commission. Bethesda, Md.
14. IRC/NRC. 1979. Telephone log transcripts, March 31, 04-237-CH6124 LFR 9-12 and 04-283-CH7/25 HF 11-13. Nuclear Regulatory Commission. Bethesda, Md.
15. SIEBERT, F. S., et al. 1963. Four Theories of the Press: 87–88. University of Illinois Press. Urbana, Ill.
16. Public's Right to Information Task Force. 1980. Staff Report to the President's Commission on the Accident at Three Mile Island: 174-176, 207-217, 225-230. U.S. Government Printing Office. Washington, D.C.
17. TRUNK, A. D. & E. V. TRUNK. 1981. Three Mile Island—a resident's perspective. Ann. N.Y. Acad. Sci. (This volume.)
18. New York Daily News. 1979. March 31.
19. DENTON, H. 1979. Transcripts of press conference, Middletown, Penn., April 2: 4. Nuclear Regulatory Commission. Bethesda, Md.
20. DENTON, H. 1979. Transcript of press conference, Middletown, Penn., April 3: 2. Nuclear Regulatory Commission. Bethesda, Md.

THE PRESS AND NUCLEAR ENERGY

David Burnham*

The New York Times
Washington, D.C. 20036

I'd like to begin with a little bit of history, because to understand what happened at Three Mile Island and the reaction of the media—and, by extension, the reaction of the public—you really have to look at what has happened in the last say 20 years.

It seems to me that the nuclear establishment, perhaps more than any other powerful technology, has lied to or misled the public about what was happening. There are repeated instances of untruthfulness on the part of the industry and the government. These repeated instances have made reporters who have worked in the area very suspicious of any public statement and have led the public to be very suspicious also.

The skepticism is based on a number of real instances and is going to be very hard to overcome. It seems to me that it's perhaps the critical problem of our age: How can we have a powerful technology? Can the engineers, the scientists, and the managers who want to install this technology deal with the public? Can they inform the public truthfully enough so that the public will trust them?

I'll give some examples. In 1964, Brookhaven National Laboratories was asked by the Atomic Energy Commission (AEC) to do a study on the potential consequences of a serious accident at a reactor the size of Three Mile Island (TMI). The study—which was completed, written—found that the accident might result in damage equivalent to a "good sized weapon," with various levels of radiation affecting an area—and I think you'll remember this phrase—"equal to that of the state of Pennsylvania."[1] That was the phrase that Jane Fonda picked up on in the China Syndrome.

Internal AEC memos, which were made public a number of years later under a Freedom of Information request, showed that the AEC was fully aware that the publication of such a study might affect reactor siting and construction. Further, the memos show that the Atomic Industrial Forum met with the AEC and suggested that the study "not be published at this time."[1] The forum thoughtfully suggested the exact wording of a brief, totally uninformative letter that the chairman of the AEC should write to the joint committee of Congress.

On June 13, 1965, following the instructions of the forum to the last period, Glen Seaborg, then AEC chairman, sent the brief letter to Congress. The study was not made public until eight years later under a

*Present affiliation: Aspen Institute for Humanistic Studies, Washington, D.C. 20036.

Freedom of Information request. The industry said that the study never had been finished. That was just not true.

In 1965, the AEC staff discovered that a company called Nuclear Materials and Equipment Corp. (NUMEC) in Apollo, Pennsylvania, was unable to account for a large amount of highly enriched uranium. After an incomplete and bumbling investigation (those questioned were not required to do so under oath), the FBI was scared away. It was decided that about 200 pounds of highly enriched uranium could not be located. Though some in the AEC felt there was a good chance that the uranium had somehow been obtained by Israel, the AEC secretly informed Congress that there was "no evidence that would lead us to believe or suspect that the material had been diverted."[2] Several years later, the CIA and then the Defense Intelligence Agency concluded that Israel had the bomb—probably built from materials from NUMEC. Documents showed that Richard Helms told Lyndon Johnson about the matter and that Lyndon Johnson responded, and I quote, "Don't tell anyone. Don't tell McNamara, Don't tell Rusk."[3]

This case, among other things, is a marvelous example of the terrifically difficult problem of dealing with nuclear energy. This matter was literally too hot to handle. Nobody could look at it. It has continued to plague the Ford administration, the Nixon administration, and the Carter administration. Just two years ago, Lee Gossick, the top executive officer of the Nuclear Regulatory Commission (NRC), told Congress that there was no evidence of a diversion. About six months later, the NRC was forced to admit that this was not the whole truth. Shortly thereafter, Robert Fri, then acting head of the Energy Research and Development Administration, told Congress that there was no evidence of a diversion. It was investigated by the Department of Energy and the inspector general. They said it appeared to be a deliberate lie, and it was sent to the Justice Department for treatment as perjury. The Justice Department did not act.

Dr. David Rubin has told you that on Friday, March 30th, the White House press secretary put out a statement to the press. The first paragraph went as follows: "The Chairman of the Nuclear Regulatory Commission, Joseph M. Hendrie, said this afternoon that there is no imminent danger of a core meltdown at the Three Mile Island Nuclear Plant."[4]

As Dr. Rubin suggested, the problem was the word "imminent." At the very same moment that the White House was putting out Hendrie's statement to the public, Victor Stello, Roger Mattson, and other senior staff members were briefing members of Congress. They were privately telling Congress a different thing than the president was saying to the public. They were saying, Hey, this is very serious. One congressional staff man had a note on his desk that read in big letters, "BAD," with three high exclamation points, from his conversation with Stello.[5] Another one remembers Mattson saying that there was a good chance of a meltdown.[6] I'm not saying that Mattson and Stello were right, but I am saying that there was one statement being put out to the public and to the press and another, different one being put out privately to Congress. As a reporter, I really resented that conflict.

One final example. About a year ago, the Atomic Industrial Forum published a survey on the cost of nuclear energy.[7] It showed that allegedly the cost of generating a kilowatt-hour of electricity from oil was three cents; from coal, two cents; and from nuclear energy, one cent. It was very impressive. General Electric, Westinghouse, and others have used that figure over and over and over again. About three months ago, Komanoff (I guess he's an antinuclear economist) started looking at that survey. It turned out that somehow, just by accident, the six most expensive reactors had not been counted in the survey. And just by accident, the coal plants run by the Tennessee Valley Authority and by another large utility—the most efficient coal plants in the United States—also had not been included in this survey. The Atomic Industrial Forum denied—no, they asserted affirmatively that they had never deliberately left out any figures.[8] They did not admit that these plants had been left out.

The point is that over and over again, the government and the industry have either lied to or misled the press and the public about nuclear energy. And the press and the public don't trust the industry. They don't trust the government. Whether that trust can be rebuilt is a very very tough question for our society.

REFERENCES

1. BURNHAM, D. 1974. The New York Times (November 10): 1.
2. SEABORG, G. 1966. Letter to Congressman Chet Holifield, chairman of the Joint Committee on Atomic Energy. Atomic Energy Commission. Washington, D.C.
3. NRC. 1978. Inquiry into the Testimony of the Executive Director for Operations, Office of the General Counsel, and Office of the Inspector and Auditor 3: 178. Nuclear Regulatory Commission. Washington, D.C. (Interview with Carl Duckett, former associate director of the Central Intelligence Agency for science and technology.)
4. Office of the Press Secretary. 1979. Notice to the Press, March 30. White House. Washington, D.C.
5. Personal interview with congressional staff member.
6. Personal interview with Paul Leventhal, staff director, Senate Subcommittee on Nuclear Regulation.
7. AIF. 1979. 1978 Economic Survey Results. Atomic Industrial Forum. Washington, D.C. (Press release, May 14.)
8. KOMANOFF, C. L. 1980. Power Propaganda. Report prepared for the Environmental Action Foundation. Washington, D.C.

REPRESENTING THE NUCLEAR
REGULATORY COMMISSION

Richard Vollmer

Nuclear Regulatory Commission
Washington, D.C. 20555

A little bit of background, I think, would be necessary here to give you an idea of my involvement with the accident. On the 29th of March, I took a small group of specialists from Washington and arrived at the Three Mile Island site around 11 in the morning. Basically, we were to be the team to try to find out what was going on and what the prospects looked like for the immediate and longer term future of the accident.

What we found when we got up there was sort of a nightmare in terms of communications. At that point on Thursday, you might recall, things were pretty quiet. The reactor coolant pump had been turned on the day before. It appeared that things were pretty well in control. We had not yet discovered the hydrogen bubble, and we hadn't had any of the Friday morning releases at that point.

It was very difficult for me to get any information that I could use, for a number of reasons. I don't think the utility was covering up any information, but it just was difficult to piece together because there was so much information—so much information that it was very difficult to put it together into a comprehensive and logical story, not only to describe what had happened and the extent of the accident, but also to tell what might happen in the future.

That, I think, was one of the big problems in reporting to or communicating with the public about the accident. It was a very technologically complex accident. It was not one where the consequences, causes, or anything in between was immediately apparent. It was a difficult scenario to piece together, a difficult one to project in terms of future consequences.

Though it was not so unusual in that such technological incidents do occur, the problem was its duration, its extended scenario, and the many things that had to be pieced together to produce a full picture.

I believe that as questions were asked by the press, answers were given as truthfully as could be given at the time. In some cases, though, answers were given without enough qualification. For example, the hydrogen bubble situation was an example of answers being given without a good technical basis at the time, without enough qualification; this caused a significant concern that spread throughout the public— more concern, in retrospect, than any single event that happened during the accident.

What can be done to correct that type of situation? I've been asked by Harold Denton to give you some of his thoughts on the causes and

0077-8923/81/0365-0110 $01.75/2 © 1981, NYAS

possible remedies. The first obvious one is that when an event like this occurs, we're talking as specialists to the press and the press is supposed to be responsible to the public. But there are large gaps in between.

The matter of better communication from the specialists to the press is an important one. We should be educated on how to deal with the press.

Secondly, it would be helpful to educate the press on how to deal with complicated technical issues.

Thirdly, the press should understand how to project not only facts of the story but also risks that the public might be exposed to, so that the people themselves can make intelligent decisions. It's clear that in the first few days of the accident, the public was in no position to make an intelligent decision on whether to evacuate or not.

Another part of the gap question is the use of a single spokesman. This has been criticized on the grounds that if you use a single spokesman, he could more or less dominate the news and, if you will, lead the press and the public whichever way he wants.

Well, it was clear in the first couple of days of the accident that there were too many conflicting stories. And a great deal of press was given to those as well as to the facts of the situation. That was one of the reasons why Harold Denton was sent by the president to be a single spokesman at the site.

I might mention that when he came up on Friday afternoon, he found that we at the site—who were closely following the actual radiation measurements—were completely unaware of the talk of a possible evacuation. We were aware of the 1,200-millirem release measurement, but we also knew that it had been measured by a helicopter well above the plant and that the actual dose rates on the ground were quite low and would not have prompted an evacuation. So there was a big problem in that staff officials at the site—both from the Nuclear Regulatory Commission and from the utility—and staff officials in Washington held rather divergent views.

There was a similar problem with the hydrogen bubble situation. Vic Stello and others were with us at the site looking at the hydrogen bubble problem. We did our own independent homework and did not feel that there was any possibility for a buildup of oxygen that could lead to burning or explosion of the bubble. We were caught quite unawares when the story broke that the bubble was possibly explosive.

So it's important to have a spokesman, one who has some sort of an integrated network from which he can draw the best and most responsible facts and make the best conclusions.

One of the problems that I saw with the media coverage was—as Dr. Rubin mentioned previously—the "what if?" problem. In other words, being asked, What if this or that happens? We attempted to be scientifically accurate in response to those questions, to interpret the facts of the situation as well as we could, and to convey them to the press. However, it resulted in a discussion of, for example, meltdown, hydrogen explosion, and radiation dangers without a real characterization of the chances of their occurrence or of the real risk involved. It may be a difficult thing to do, but I think the public is owed some characterization of the risks being faced.

Secondly, there was always the demand for very prompt news. It was typical during the first few days for a reporter to come up and ask if there had just been a release. Of course, since the press was listening in on all of the telecommunications or radiocommunications between our staff in Trailer Village and the site, they knew as well as we did that there had been a release. So we would say, Yes, there was one. And they would ask, How bad? And we'd have to say, We don't know. This "don't knowing" is a difficult thing, it appears, for press to understand, even though we would try mightily to come up with some factual information to characterize how bad it was.

A good example of that was an incident on February 11th. There was a minor leak—some tubing ruptured, spilling about 1,000 gallons of primary coolant onto the floor in the auxiliary building. There wasn't really any particular problem with that, except that 300 mCi of krypton also got out. This is a trivial or inconsequential release by any standards. And the initial news that we gave to reporters—who knew about it a few milliseconds after it happened, somehow—was that we saw no increase in activity of releases from the site, which was a factual statement. After pouring over the charts later, Metropolitan Edison discovered that there was a slight increase, and they could integrate it out to be something like 300 mCi over a two-hour period; this is three-tenths of a curie. The plant itself at that time was releasing on the order of 75 or 80 curies a month anyway just from background sources. So it wasn't really clear or easy to see this very minor release on the charts. Yet the local papers and the Washington papers made a great deal about its being a significant release.

Last, there seemed to be a need for getting "punch" to the stories. That did not occur so much during the accident. I would generally agree with some of the other people here who characterized the reporting during the accident, except for a few instances, as generally factual and nonsensational. But after the accident, when the news media were searching for other ways to keep the story alive, if you will, it appeared that they were really reaching for some of the issues. For example, the krypton issue that we're faced with today: there are 57,000 Ci of krypton in the reactor building. Scientific analysis would show that if it were released over a period of time, the maximum whole-body dose to any individual would be one-tenth of one millirem—a very, very minor dose. Yet I've seen stories and editorials describing the possible release of that krypton as a potentially very hazardous, unhealthy situation that would create cancers, stillborn babies, and so on.

That is irresponsible. Even the part of the scientific community that would be characterized as antinuclear has felt that such a release would not be a public health and safety concern if indeed the numbers are correct. And I believe those numbers are being confirmed by some of the antinuclear groups.

I'd just like to finish up by saying that we learned a lot from the accident, and not only from a technical point of view. Clearly, we've learned that there are other things that we need to attend to, and one of those things is communication with the public. The Nuclear Regulatory Commission is taking a number of steps to try to educate the media on basic reactor technology and radiation technology. We're also taking steps to assure that if there should be another incident, there will be a mechanism by which prompt and accurate reporting of information can be made to the media in order to eliminate the things that really went wrong during the Three Mile Island event.

REPRESENTING THE CONGRESS

Henry Myers

*House Committee on Interior and Insular Affairs**
U.S. House of Representatives
Washington, D.C. 20515

I'd like to emphasize that inaccuracies in the press resulting from inaccurate reporting were, it seems to me, very small in comparison to inaccuracies resulting from inadequate and incorrect information provided by the Nuclear Regulatory Commission (NRC) and utility officials. There's been some focus on the third day of the accident, when it seemed to dawn on people that maybe the accident had been worse than had been initially reported. But the Rogovin inquiry now has concluded that information available in the control room at nine o'clock in the morning on March 28th indicated that the situation was sufficiently uncertain that it would have been reasonable to advise state officials as to the following:

That the core has been badly damaged and has released a substantial amount of radioactivity.

That the plant is now in a condition not previously analyzed for cooling system performance. Presuming that the full high pressure injection flow is turned on, advise the state that if the cooling systems do not function adequately, portions of the core could begin to melt, which could lead to significant off-site releases in a few hours.

If the cooling systems are successful, evidence of that success should be available in a few hours.

Begin a precautionary evacuation of the first few miles around the plant, with an alert for a larger radius, 10-mile evacuation that may follow. Evacuees from the inner zone of a few miles radius should be removed to locations at least 20 miles distant.

This advice—based on information that was available in the control room—should have been provided to state officials.

Instead, whatever was told to the state officials led Lieutenant Governor Scranton to announce at 10:55 A.M. that he had been advised that everything was under control and that there is and was no danger to the public health and safety.

Later in the afternoon, somewhere around one or two o'clock, Metropolitan Edison vice president Jack Herbein was briefing a crowd of reporters at the Three Mile Island (TMI) visitors' center. There seems to be no transcript for this press conference, but Richard Lyons reported on it in the *New York Times* on March 29th.[1] He said that Herbein had spent almost an hour describing the accident and had emphasized that

---

*Dr. Myers is science advisor to this committee, not a member of Congress.

114

the main safety system in the almost-brand-new plant had worked well enough to prevent a very serious accident or indeed a catastrophe. Herbein had said that there had been some minor fuel failure. And, in a later piece by John Fialka, who was at this press conference, Herbein was reported to have closed the conference by saying, "This is not what I would call a serious accident."[2] This all led to the stories that appeared on March 29th, the day after the accident, which had the general tone that this was the worst accident that had ever happened in a commercial nuclear generating facility in the United States, but that it seemed to be something that was more or less under control.

It was not until late Friday that there started to be discussions about evacuation. The lead story of the *New York Times* on Saturday, March 31, stated that the Nuclear Regulatory Commission told Congress that day that the risk of a reactor core meltdown had arisen at the crippled Three Mile Island atomic plant at Middletown—an event that could necessitate the evacuation of the immediate area.

Now, this pessimistic assessment did not appear until late Friday, March 30, even though the situation was more serious on the 28th, when neither state nor federal officials appreciated the magnitude of the problem. In sum, lacking important information available to control room personnel and indicative of a significant potential for a major radiological release, the major newspapers' initial accounts of the TMI accident generally understated its severity. The picture was distorted because those who had the information did not provide it to the press.

REFERENCES

1. LYONS, R. 1979. New York Times (March 29): A-1.
2. FIALKA, J. 1979. Washington Star (June 11): 1.

SHARED RESPONSIBILITIES OF THE UTILITY AND THE MEDIA IN CRISIS SITUATIONS

William B. Murray

Communications
General Public Utilities Service Corporation
Parsippany, New Jersey 07054

Three Mile Island (TMI) more than any accident in modern history pointed up pertinent lessons for both the media and industry during extended crisis situations. Utilities and the media share a large responsibility in applying these learnings so as to improve public understanding and insure public safety.

Reporting the early detail on the accident to a demanding media at a time when the utility staff itself was still struggling to determine the exact status of plant operations produced serious and long-term problems of credibility. The protracted time period of the accident's development and the growing understanding of the utility staff over that period were often interpreted by the media not as increased knowledge and understanding of what was happening in the plant, which indeed it was, but rather as some cover-up or doctoring of the information being provided.

This extended period of real-time involvement of the public probably is an inherent characteristic of serious nuclear plant accidents. Appreciation of this characteristic by both the utility and the media is an important communications lesson of TMI. On the utility side, it means that the best possible information available must emanate from the plant. Where the accident situation has unknowns, where the information available is not complete, where full understanding is lacking, the utility should candidly communicate this to the public along with the known definitive facts. This in turn places a heavy responsibility on the media. Concentrating coverage on divergences of opinion and sensationalism must be avoided if the public is to be truly served. The necessary responses and decisions of civil authorities with emergency planning responsibilities must not be preempted, nor the public confused by media handling of the available accident information as it progresses. The aim of our very powerful modern communications techniques in such situations should be one of assisting and reinforcing the messages of the civil authorities who are coping with the emergency.

A key issue is the selection and role of spokesmen to the media and the public. Problems still being felt today at TMI arose from the presence of too many spokesmen during the accident, particularly unofficial commentators. Some believe the solution lies with the designation of a single government official. Other investigations have pointed out the danger of that course. I am convinced that the utility must retain a spokesmanship role in situations such as the TMI accident. But this

116

0077-8923/81/0365-0116 $01.75/2 © 1981, NYAS

should be a shared role. On a practical basis, I have urged that press conferences, for example, should present three spokesmen to the media: a utility spokesman to provide plant status data, a senior emergency representative in the area for information on public impact and necessary instructions for public welfare, and a regulatory agency spokesman responsible for the involved plant. The concurrent availability of these three spokesmen could do much to reduce the serious consequences of media speculation on differing comments on the same subject delivered at different places at different times by different spokesmen. Media questions at joint press sessions—held as frequently as the situation demands, perhaps two or three a day—could then be directed at the spokesman most expert on that particular query and commented on immediately, if desired, by the other two speakers. I believe this would do much to control the problem of conflicting stories with the resulting loss of credibility of all concerned that was so damaging at TMI. The media by nature seek out points of conflict, discrepancies, rumors, and suspected cover-ups. In major crises, the public is best served if such concerns can be aired and settled quickly.

Much has been written about the need for media representatives with detailed knowledge of nuclear systems who can better interpret accident situations to the public. This is an educational matter of paramount importance. Here again, the utility and the media share a responsibility.

The utility must adopt a more aggressive role in providing understandable information to the print and electronic media, particularly those in the vicinity of its nuclear installations. The utility's educational process must take place before the accident occurs. Special short courses, seminars, plant tours for journalists, and extensive question-and-answer sessions should be arranged and offered by the utility.

Such programs should go well beyond covering the simple operation of a nuclear plant. We would probably all agree that one of the least understood areas needing special emphasis is that of radiation. Radiation is without doubt the single greatest concern in the public's mind. Some of its aspects are not agreed on even by the experts. The misinformation and lack of perspective on radiation sources and radiation effects remain high today.

We are beginning to see the emergence of such expanded educational efforts. Nonetheless, radiation remains the most easily sensationalized and the most emotional topic associated with nuclear power.

The utility also must provide education on less technical subjects that nevertheless impact heavily on public understanding. One such topic is that of utility economics and, in particular, rate making and risk capital. Initially, these subjects may sound somewhat far afield of accident crisis communication. However, our company has learned from the TMI accident, with all its resultant financial fallout, that the public's understanding of the dollar-and-cents issues heavily colors their thinking on nuclear safety and future accident risks. Utility economics is not a simple subject. Its treatment by the media often has been uninformed, and the utilities must move to provide a clearer understanding of how these pocketbook issues operate and affect the public.

Effectiveness in communicating from the nuclear plant control room to the press room has been hindered by technical jargon. We all know this, but in many cases we all continue to be guilty. The engineer will tell you that there is a level of nomenclature and expression below which you cannot go without losing preciseness of meaning. The reporter will respond that preciseness is useless if there is little or no understanding. Examples from other developing technologies over the years have shown that there is a middle course, and the utilities and the media must work to achieve that level. For some years, the University of Missouri at Columbia has offered a course in nuclear energy, specifically directed toward journalism majors. Utilities should encourage the expansion of such efforts in the academic communities they serve. It might also be worthwhile to expand dialogue between the utilities, the engineering faculties, and those involved in teaching the university courses outside of nuclear engineering. Misinformation is frequently presented in academic fields—such as environmental, biological, and social studies—that must be corrected.

I am encouraged that we have come quite a distance since March of 1979. It is only occasionally now that I see pictures of cooling towers labeled as nuclear reactors. However, the turnover of personnel in the journalism field necessitates a continuing emphasis on improved education and the reduction of jargon.

Two situations arise in the nuclear accident scenario that create great difficulty for both the media and the utility. I am referring here to the handling of the "what if?" question and the "maybe" answer.

There is no doubt that during the days of the TMI accident, the posing of "what if?" questions, the responses received, and the handling of those responses by the media were probably the cause of a great part of the public's concern. The utility must have an appreciation of the press' need to ask the "what if?" type of question. It would be foolish and wrong to argue that because of their impact on the public's emotions and even their safety, such questions should not be asked. The media, on the other hand, should also recognize their responsibility in reporting the responses to such questions. They must recognize that the scientifically trained spokesman, in his efforts to be precise and cautious, will almost always qualify himself. He will refuse to absolutely rule out almost any reasonable hypothetical eventuality posed by a reporter's question. This often leaves an open field for sensationalism and scare stories.

The "what if?" questions are further complicated by the editorial distaste for "maybe" answers. Highly qualified reporters have told of the pressure from their editors not to take a "maybe" answer but to press on, sometimes with the result that an extremely low probability occurrence can be reported as a near fact. We will probably never completely solve the problems inherent in these "what if?" and "maybe" situations. Our hope lies in better understanding on both sides, not only of the technical details of nuclear power but also of the often differing mental processes and objectives of the journalist and the engineer.

The credibility problem in nuclear accident situations has been much discussed. The mechanics of the utilities' reporting to the regulators, the politicians, the media, and the general public are a key to the maintenance of credibility. In the past, the operating rule has often been to report to the public only those events at a nuclear plant that by their nature must be reported to the regulatory agencies. Such events, often under regulatory procedures, require reporting only on a 30-day basis, not as they occur. Following such a procedure may be legally correct and may satisfy the regulations, but it is obviously not workable from a public communications standpoint.

New regulatory procedures since the TMI accident have broadened the definitions of emergencies that must be reported. As a utility that was severely hurt under the earlier policy, we have, on our own, expanded even further the definition of which events to report from the technical innards of the plant to public officials, to the media, and to the regulators. We now report any event that could have potential public interest. Obviously, such a system is only as good as its interpretation and use by those working with it, but I believe this is a significant step forward in responsible reporting, certainly under today's climate of public fear and concern.

Efforts to avoid sensationalism in the media during extended crises that impact on large numbers of people are everybody's responsibility. The degree of sensationalism in TMI accident reporting was surveyed by a task force of the Kemeny commission. The general conclusion was that, with very few exceptions, the media were more reassuring than alarming, and in areas where they could be accused of sensationalism, they were only reporting what they were told. My own opinion is that this is an overly optimistic evaluation.

Overall, in the year since the accident, it appears that through better understanding of what happens in a nuclear accident, the most extreme examples of sensationalism have declined. In my opinion, however, too little is being done to objectively inform the public. This contributes to misinformation and unnecessary emotional concern regarding nuclear accidents. I am referring here specifically to the fictional movies made for television.

I do not argue against the value of fiction as a method of examining controversial subjects. The disturbing aspect of several such news-related movies for television is that they lack objectivity. They seem to adopt what the arts editor of the *Wall Street Journal* recently called the "Snidely Whiplash" formula in handling complex social issues. This technique is simply to arrange all incidents and dialogue in such a way as to make one side look saintly and the other evil. Without, I hope, sounding paranoid, the evil side is almost always represented by business and industry, often by a utility. Advocacy journalism is an honorable and useful undertaking. Honest advocacy, however, requires much effort and soul-searching. The more complex the issue, the more difficult it is. Certainly nuclear energy, nuclear accidents, and their relative risks to the public are complex issues technically, ethically, and morally.

Post-TMI accident surveys by psychologists report finding a direct relationship between the degree of risk perceived by laymen and the frequency with which a potential risk is simply mentioned in news reports. This being the case, one can only surmise the impact of overdramatized and single-sided fictional portrayals of such risks.

We have here another area of shared responsibility requiring close reexamination by the media and the utilities with the goal of providing the public with evenhanded, informed reporting on the complexities of nuclear power, how it works, its benefits, its risks, and, finally, the accurate interpretation of nuclear accidents.

There are many lessons from TMI for the utility in the pure mechanics of handling the influx of an extremely large number of press and electronic media personnel, as well as the solid stream of telephone inquiries. Utilities with nuclear plants are making the physical changes necessary to alleviate such problems. However, I think it is necessary for the media also to recognize the problems inherent in maintaining all of the niceties desired for ease of good reporting when numbers of media personnel approach 1,000 on a continuing basis, and for accident probabilities of low frequency. Nevertheless, nuclear utilities must be prepared to expand rapidly their handling of the off-normal levels of media interest that TMI showed will be generated during the course of a major nuclear accident.

It has been pointed out that modern man has been better able to handle the complex technical problems of space exploration than the more difficult human relations problems of our decaying central cities. Perhaps a parallel exists here—the communication problems involved with nuclear accidents may be ultimately more difficult compared to the correction of the technical deficiencies indicated in the area of nuclear plant safety. There certainly is little question but that all concerned must work as diligently as possible in applying the many communication lessons learned from the TMI experience.

REPRESENTING PUBLIC INTEREST GROUPS

Arthur Tamplin

Natural Resources Defense Council
Washington, D.C. 20006

I've been involved in nuclear power controversy since about 1969. And over that period of time, I think, the media have done an excellent job of presenting both sides of the nuclear power controversy to the public. Moreover, I think the media's role in this respect was absolutely essential. If I could, I'd like to explain the reason for that by quoting Bertrand Russell. He said:

> Whether men will be able to survive the changes of environment that their own skill has brought about is an open question. If the answer is in the affirmative, it will be known someday. If not, not. If the answer is to be in the affirmative, men will have to apply scientific ways of thinking to themselves and to their institutions. They cannot continue to hope, as all politicians hitherto have, that in a world where everything has changed, the political and social habits of the 18th century can remain inviolate. Not only will men of science have to grapple with the sciences that deal with man, but, and this is a far more difficult matter, they will have to persuade the world to listen to what they have discovered. If they cannot succeed in this difficult enterprise, man will destroy himself by his halfway cleverness.

Now without the cooperation of the media, the men of science would find it next to impossible to communicate their concerns to the public. If the only approach to the public were through paid advertisements, the world's nuclear industries and bureaucracies would have had it all their own way.

At the same time, during the Three Mile Island (TMI) accident, I think both the media and the nuclear critics felt constrained to operate in a different mode. For example, I was asked by a number of reporters, "Do you think the situation is dangerous?" I was hesitant and felt somewhat intimidated, but I responded that if I lived in the area and had a pregnant wife or young children or both, I would get them out of the area—and that if I had no compelling reason for staying, I would leave myself.

The situation there was just too confusing, to say the least. Even those who were on the scene and should have known what was going on, didn't know.

The TMI incident is still not ended. There remain important problems relating to the venting of the containment, to the decontamination of the plant, to the start-up of the other reactor at the site, and then to who will pay for the accident and its aftermath. And what about the next nuclear power plant accident? Are the new emergency plans for the 10-mile evacuation zone adequate? I doubt it. It seems that the electronic media will be integrated into these emergency plans. Will these media be or feel constrained to announce only official bulletins?

I think that the overall media response was quite responsible in reporting the TMI accident, and additional constraints would certainly be unwarranted.

Here in New York, there are immediate questions concerning the fate of the Indian Point reactors. Should the operation of these reactors be suspended until the known safety deficiencies are corrected and until the lessons learned at TMI are implemented? I think they should be.

But the broadest possible segment of society should participate in this decision. To participate, they must be informed; and for their education, the media are absolutely essential.

THE ROLE OF THE MEDIA: PANEL DISCUSSION

Moderator: Merril Eisenbud
Panel Members: David Burnham, Henry Myers,
William B. Murray, David M. Rubin, Arthur Tamplin,
and Richard Vollmer

D. M. RUBIN (*New York University, New York, N.Y.*): I would like to comment on something that you said earlier, Dr. Eisenbud, about the misleading story in the Middletown, New York, paper on radioactive industrial wastewater being dumped into the Susquehanna. I think that is a misplaced example of press criticism. The main problem with that story was that if the Nuclear Regulatory Commission [NRC] and the State of Pennsylvania had gotten their act together and had, in a timely and clear fashion, announced on Thursday what was happening, why it was happening, and why there shouldn't be any concern—if they had notified downstream communities and done it in an open press conference rather than delaying eight hours, haggling over who was going to take responsibility for it, trying to duck the political implications, and ending up with a press release that was issued after midnight at the beginning of Friday morning, thereby raising the suspicions of an already sensitive press corps that yet another fast one was being pulled—the story would have gotten the little attention it deserved rather than the major attention. I think that this story is an example of how you cannot look at a headline or a picture or a single story or any piece of media content and attempt to state that it is an example of poor press performance unless you know how it got there, why it got there, and who said what to whom. This isn't to say that there aren't examples—after you've gone through that process—that the press can be criticized for. However, the question of industrial wastewater, I think, is not one.

M. EISENBUD (*New York University Medical Center, New York, N.Y.*): Of course, thinking in retrospect, it's possible that whoever had to make the decision assumed that it would be alright to discharge wastewater in concentrations that are permissible under normal operating conditions, since their license allowed them to do it. And in retrospect, you might say they should have understood the sensitivity of the subject and dealt with it. But people have certain frailties in judgment.

D. M. RUBIN: Yes, but that wasn't the reporters' responsibility. That was the utility and the government and the NRC.

A. TAMPLIN (*Natural Resources Defense Council, Washington, D.C.*): I think I will carry on in that vein. You know, I've been in this nuclear controversy since 1969. And every time I've heard nuclear proponents speak about a thing like this, they say, Well, you know, we have to learn how to communicate with the people and with the press. But the one fact they seem never to have learned in the whole 11- or 12-year period is that if they never lied to the press, the press would understand them. Yet they have continually fabricated. Many times reporters have asked me, This

123

0077-8923/81/0365-0123 $01.75/2 © 1981, NYAS

guy from nuclear industry said this; and what do you think? And generally I know that they don't believe him. Usually I have to say that he's wrong or that he's lying. If the nuclear industry and nuclear bureaucracies want to communicate with the public, what they need to do is be honest. Going into some kind of a ritual about a picocurie being one-tenth or one-twelfth of a curie is ridiculous because a picocurie is a dangerous amount of radiation. It just happens to depend upon where it is.

W. B. MURRAY (*General Public Utilities Corp., Parsippany, N.J.*): Since we've been throwing words like lying around here rather loosely in the last hour, let me just point out two things that were concluded by both the Kemeny commission, the Rogovin commission, and the more recent supplement of the Rogovin commission. I don't have the books right in front of me, but they conclude that the utility in the case of the Three Mile Island [TMI] accident did not willfully withhold information from the authorities during the course of the accident. And I think that's worth looking up again, if we're going to continue to talk about this question of verity.

D. BURNHAM (*The New York Times, Washington, D.C.*): As a footnote to that, sir, I think that question is being investigated by Mr. Stello now. It's still open, and he has said that there may be more finds on that specific question of whether information was withheld from the government improperly.

H. MYERS* (*U.S. House of Representatives, Washington, D.C.*): Governor Babbitt [of Arizona, a member of the Kemeny commission] also had a footnote to the Kemeny report on this very point, where he said that there appeared to be "evidence to indicate that Met Ed technicians had understood within a few hours of the accident that the nuclear core had been uncovered, and that this specific information had been transmitted to supervisory personnel at the plant early Wednesday. There seems to be little question that the technicians who took the temperature readings that morning understood what they found. The real question is what happened to this information and whether it was transmitted to the appropriate management personnel. It certainly did not get transmitted to responsible public officials, including Lieutenant Governor Scranton, during a meeting with Met Ed that afternoon." It is also the case that Frampton and Rogovin have backed away from saying that there is no evidence that information was intentionally withheld. Frampton [deputy director, Nuclear Regulatory Special Inquiry Group] has said that there is evidence and that while he concluded that the weight of evidence did not support intentional withholding of information, some people may look at it and conclude that it was intentionally withheld. This is what Stello is looking into now.

R. VOLLMER (*Nuclear Regulatory Commission, Washington, D.C.*): I'd

*Dr. Myers is science advisor to the House Interior Committee; he is not a member of Congress.

like to get back to the radiation issue. We've all conceded and stated over and over again that any amounts of radiation pose a risk, and I think that's inherent in our regulatory structure and in the principles by which the government regulators operate. The problem is, How do you characterize how big the risk is or how little the risk is? I think it's a slanting of the way the things are stated and the perception that comes out of them that cause the problems. Again, a good example of that is the one that I mentioned dealing with these very small releases of krypton. We have taken every step we can as regulators, and the utility takes all reasonable steps it can, to prevent any releases from the facility. Since the accident, releases have been extremely low compared to any normally operating plant. Yet anytime anything is released, it creates a public concern. It's played up in the press, and in some cases, rather gross estimates are given—exaggerated estimates—of the potential harm of these releases. That's the type of education and communication that I think is important: not to say that radiation is not a risk, that's not the point; it's just trying to communicate what is the risk.

C. STARR (*Electric Power Research Institute, Palo Alto, Calif.*): I've listened with real interest to the panel this morning, and there are some general themes that go through it. I'd like to comment on these themes very briefly.

There are really three separate points that have been raised. One is the intent of communication—whether it's to inform or to mislead. There's some confusion in the presentations as to which the panel is addressing. Obviously if the matter is one of informing, then it depends upon the information that the informant has. And as evidenced from all the papers you've heard, the information at TMI was pretty vague at the time, and a certain amount of confusion resulted. If the intent was to mislead, that I think is something of a different nature and really ought to be segregated out.

The second point is one of competence. Who is competent to really provide information in a situation of this sort—a very complicated technical situation? I can tell you that from my point of view, the Nuclear Regulatory Commission was incompetent to do that at this particular occasion; and the evidence supports that point, as history developed. I'll come back to that later.

The third point is one of process. What is the process of communication? That's been briefly mentioned here at the panel; but in fact, that may be one of the most important points because the process is obviously inadequate.

I want to address specifically some questions in those general areas to Dr. Rubin, who gave I think a very proper and a sound analysis of the instances he picked. But he did not mention one issue that is of real concern to everybody in the technical community, not just in the nuclear area, but in every other area. That's the quality of treatment. The hydrogen bubble was a technical fiasco on the part of the NRC. Everyone in the technical profession who knew anything about the subject knew instantly that a hydrogen explosion was impossible. That information gradually seeped into the NRC. It gradually got admitted by the NRC and in a very roundabout way. Now, the hydrogen bubble was given front-page treatment by the *New York Times* and everybody else. But when the NRC found it had made an error—I suggest you go back to the news media to find out how much coverage it got and compare that coverage with what the hydrogen bubble got. The lack of equality of treatment when a major error is observed and a scare issue turns out not to be scary, that is one of the big issues.

I want now to address David Burnham because I find his presentation an example of journalism that could stand a little improvement, and in fact, that is what I would call bad journalism, though maybe the *New York Times* would not. For example, Mr. Burnham, you mentioned the NUMEC situation. Now, I have followed the NUMEC situation for years, as I have most of the other things in this business. The NUMEC situation is really very uncertain. In the case of NUMEC, the situation—at least as far as the NUMEC management was concerned—was one of accountability. They claim, and the government looked into it, that the difficulty of accounting in terms of sewage disposal, waste disposal, etc.—that this process accounted for the lack of record on what happened to this material. After NUMEC's management was changed, it turned out that the unaccountability was just about the same. The fact is that no evidence has ever been published that enriched material was stolen from that plant. Now it's true that nobody can say it wasn't, but there's no evidence that it was. And in your statement this morning, you left the innuendo that material went to Israel. I'd be very interested in the *New York Times* publishing a factual analysis to prove that that's the case.

You mentioned the Brookhaven report. I'm familiar with that report. Any technical report that makes a study on a prediction is based on the premise that goes in the report. Brookhaven was asked at that time to look at what would happen if about half of the fission and radioactive products from the plant were released to the public without any containment, without any control, without any evacuation procedures, with no protection. So there was an extreme analysis based on a set of premises that were very extreme. Now you did not mention that. So I can understand, incidentally, why the Atomic Energy Commission [AEC] at the time soft-pedaled it. I don't disagree with you on the fact that this was a bad policy. But when you talk about something like that, you don't leave one shoe hanging. Drop the other shoe and give the whole story.

You also talked about the Komanoff studies on cost. Our institute for years has followed Komanoff analysis, and we've published reports on it. Why didn't you look into it? The Komanoff analysis is wrong. And if it weren't wrong, the industry would not be ordering—as they still are ordering—additional nuclear plants. The fact is that you ended up with an incomplete statement, and that's bad journalism.

I think the issue that you're following here is one of prediction versus one of reporting what happened. Whenever you get into the issue of prediction, you're going to end up with a situation that's confusing.

I want to make one last comment and that's on Henry Myers' comment. I think that it's very easy to show on hindsight that the information that was given by any agency or anybody else at a time of crisis is inadequate information. But at the time when the information is given, most of the stuff that comes out in an area like this is very uncertain. I think—as Dr. Eisenbud mentioned—that one of the big issues in all of this is how to transmit uncertain information.

D. BURNHAM: I would like to respond to a couple of things. On the NUMEC matter, Carl Duckett—who was the deputy director of the CIA for science and technology—wrote a report, which has been released, that said that the CIA believed the material was stolen. I have seen a Defense Intelligence Agency statement that said that same thing. There is no question that Mr. Helms went to see Mr. Johnson and told him this, and Mr. Johnson said that. As far as the credibility of the AEC in this investigation, you can look at the minutes from a meeting on February 14th, 1966, where the then chairman commented on the desirability of stressing the theory that no diversion had taken place. Mr. Brown, who was then the general manager, said that maybe they ought to make a more thorough investigation of this, that they couldn't be sure it was true. The objection to this thorough investigation was that the impact on both NUMEC itself and the nuclear industry in general would be, to say the least, traumatic, that is, they were not going to look at the possibility of diversion because it would hurt the nuclear industry. And that's from the minutes of an AEC meeting.

Now as far as Komanoff goes, the only study I mentioned was one that looked at the cost analysis put out about a year ago by the Atomic Industrial Forum. Komanoff charged that they left out—I think it was 11 out of the 12 most expensive nuclear plants. They also left out of their calculations the coal plants of the Tennessee Valley Authority and a large utility in the Midwest whose name escapes me. I went to the Atomic Industrial Forum and asked them. They acknowledged that those plants had not been included. They said, "We never said our survey was accurate. We never made any claims for our survey." Well, you go look at their press release, and they do. So I find you less than persuasive, sir.

C. STARR: Well, obviously we disagree.

M. EISENBUD: I'm going to rule the NUMEC matter out of order. It hasn't been settled in 12 years. We've got 10 minutes.

E. J. FENYVES (University of Texas, Dallas, Tex.): My feeling is that the nuclear controversy is actually such a gap of understanding and lack of information that sometimes I feel that it is almost impossible to over-bridge this. But it reminds me today of the story about two Texan cowboys who were shouting at each other from a very large distance so that they couldn't understand each other. Then a third cowboy came and told them, why don't they use a telephone. Maybe they could communicate better by means of this technical system. I'm not saying that you can so easily solve problems. But in some ways, technical solutions are bringing people closer to each other. I would like to mention here very briefly a system we are working on in the University of Texas at Dallas. We are building a continuous environmental radiation monitoring system around our nuclear power station, which will start operation next year. The system will be like a meteorological system, and it will be able to give data every day, every hour, every week. The public could be informed about the radioactive conditions around the nuclear power plant, if required, on TV or on radio. Then anybody can see the data and check the radiation levels. We are now measuring the background radiation before the power station is in operation. So later, we can compare the measured data with the background—water, soil, air, etc.—data. I feel that under the circumstances, we can make a small step forward in respect to mutual understanding. This is a state university; we have no interest in misrepresenting the facts to anybody—to the press or to the nuclear industry—to anybody. The meters are there; if somebody wants to read them, he's welcome to read them. This is something that we feel is certainly a small step in overbridging this very big gap between people, scientists, concerned citizens, etc.

D. BURNHAM: Could I make just one brief comment on the measurement? Of course, accurate measurements are very helpful. But there was a little incident last week that shows the problem. Dr. McCloud—who was fired as the secretary of health of Pennsylvania—obtained some information showing some birth statistics around Three Mile Island. He just had the hard numbers of some deformities—I think it was deformities, I don't remember exactly what it was—and that there had been an increase. He did not have the rate. McCloud talked to the *New York Post*, and the *Post* had a scare story. And it was based on hard numbers. If you talked to Dr. McCloud, it sounded very reasonable—the number of neonatal deaths had increased. When you got into the rates, or at least the state says, there have been no changes at all. So the hard numbers, once again, can be very misleading to the press and to the public. Dr. McCloud at least had some standing as a scientist. He's a doctor; he's a professor at the University of Pittsburgh. And the state didn't want to talk. I had to badger them to come up with the rates and develop them. You know we were able to write a more or less balanced story, but it's very hard to deal with that problem. The hard numbers aren't enough; we need more than that.

L. BERKOWITZ (*Westinghouse Electric Corp., Pittsburgh, Penn.*): Dr. Rubin has argued eloquently that the media coverage was not sensational. Mr. Burnham, who has said that industry has misled and lied, has argued with equal eloquence that media coverage has not been sensational. The Kemeny commission recommended that efforts be made to improve communications from the industry, from the utilities, and from the commission. I don't think there is anybody who could argue against that recommendation. But somehow or other, I have a deep feeling that we have a situation here where there is a fox watching the sheep—with representatives of the press arguing that media coverage was not sensational—when we spent the last hour-and-a-half discussing the public's right to information, and we have ignored the prior history of the media coverage of the industry. I find it rather interesting that no mention was made by Dr. Rubin or Mr. Burnham of the Media Institute's study of prime time network TV coverage of the subject of nuclear power over the last 10 years. That comes to a conclusion that is quite the contrary. I also find it extremely interesting that no mention has been made either this morning or yesterday of the findings of the Kemeny commission's technical staff with regard to some of the very serious "what if's" of Three Mile Island. In fact, when you go into the technical staff assessment of alternative event scenarios, you find that the consequences could not have been nearly as grave as anyone was given the right to believe—either at the time of the accident or subsequent to it. I don't think there's any question in anyone's mind but that Three Mile Island was a serious accident. But the going-in supposition for every investigation that has been held was that it was a disaster. And people believe it. It was certainly serious, and there were certainly serious psychological effects of the accident. But was it a disaster? and in the sense that the DC-10 crash was a disaster? and in the sense that the collapse of the oil rig in the North Sea two weeks ago—which was out of the press in two days—where 123 men died was a disaster? I really question this. I wonder whether Dr. Rubin and Mr. Burnham would like to comment.

M. EISENBUD: I'd like to focus that a little bit more. I think it's a very important point because I think in Dr. Rubin's presentation, he discussed the possibility of a meltdown as a black or white situation—go or no go. He gave the impression that if there had been a meltdown, there would have been what the public understood as a China syndrome. I wonder if anyone on the panel would care to comment on that? How about you, Mr. Vollmer, since I think the Kemeny commission did address that question.

R. VOLLMER: Yes, and the Rogovin special inquiry group also addressed that. I think they're the ones that were quoted as saying that if nothing else had been done at the point when the additional cooling water was reinstituted, there was something like 30 minutes to an hour to the initiation of core melting. But they did go on to say that at that point in the accident, they felt the impact on the public would likely be not significant because they questioned whether or not such a meltdown would penetrate the containment. So it's certainly true that you could talk about a meltdown, but that doesn't necessarily mean disaster. The whole story has to somehow unfold when you talk about such things.

M. EISENBUD: Dr. Myers, do you want to comment on that?

H. MYERS: Yes, Rogovin was ambiguous on that point. He says any analysis of how far it could have gone—that is, if there were a meltdown or whether the reactor building would have held it—is speculative and depends on complex analysis with many necessary assumptions. He then goes on to say that one thing the analyses *do* seem to agree on, though, is that even with a core meltdown, there is only a small probability that the consequences of TMI would have been catastrophic to public health and safety. Now there's a contradiction in those two sentences.

M. EISENBUD: Yes, but the point is that there was not a black and white situation.

H. MYERS: No, not black and white, but also not a clear conclusion that had there been a meltdown, it all would have been contained.

A. TAMPLIN: The Kemeny commission stated also, as I recall, that their conclusions were based upon just what they had analyzed and that there were other scenarios possible. They didn't analyze the problem completely, and one of the things they didn't take into consideration was additional operator errors during the course of the accident. So clearly it was still possible. Fortunately, it didn't happen.

D. M. RUBIN: Dr. Eisenbud, since I was attacked it seems to me I ought to respond. Mr. Berkowitz, I think you don't mean that it's the fox watching the sheep—you mean that the sheep are watching the sheep, because you mean that we didn't do a tough job on the press. I can only respond to that by saying I'm not a member of the working press. I'm not a journalist, I'm an academician. If you want to look at my past writing about the press at a number of places, you will see that I'm hardly a press supporter. I'm a vigorous critic. I think David Burnham would support the fact that I'm basically *persona non grata* at the *New York Times* for what I've written. I hope that gets rid of your first point.

Second, on past coverage of nuclear accidents, I think the proper criticism to be made is that the press didn't give enough space to things like Brown's Ferry, for example. If past accidents had been covered with anything near the same size press core and degree of assiduousness, this accident might not have stood out in the bold relief that it did. It seems to me that the main test is how much coverage is given to successive accidents. I think Crystal River got more than it might have because of Three Mile Island. So the proper criticism is that there was not enough coverage of previous accidents.

Third, the Media Institute study—I've heard of it. They have not bothered to send it to me. If they would, I'd be happy to read it. Let me point out that these kinds of content analyses, unless they are done by people who understand content analysis procedure, are generally methodologically weak and support nothing more than the biases of the people who did them. I'm not saying that's the case with this one, but I'd like to see it and find out if this is the exception to the rule.

Fourth, on the question of whether this is a disaster or not, you're from Westinghouse. Obviously you don't think it's a disaster. And antinuclear people do think it's a disaster. You are never going to get an agreement on whether Three Mile Island was or wasn't a disaster. Everyone is going to have their own perspective on it. It's not a matter that's up for a final conclusion; it depends on your perspective. I think that it is misleading to try to indicate that there is some bottom line on whether this was or was not a disaster.

Finally, the North Sea question. I have no difficulty understanding why that was not a major media story for weeks and weeks and why Three Mile Island was. There's no difficulty understanding what happened. The risks and benefits of that sort of mining are somewhat clearer than they are on nuclear power; there's been more public debate about it. There isn't any great mystery as to how they died or why they died or how others may die in similar circumstances. The simple fact that more people died there than died at Three Mile Island completely beggars the question and misses the point about the confusion of the two technologies. There isn't any doubt in my mind that the new judgment that that sort of a disaster is not in the same league with this sort of a disaster—if you want to use the word disaster—is an accurate one.

A. TAMPLIN: I'd like to add one thing to that, too. Mr. Berkowitz mentioned the DC-10. I think if you follow the press coverage of the DC-10, it was quite extensive until the DC-10s were grounded. Well, one of the reasons why the press coverage on nuclear power is continued is that they haven't grounded the other nuclear power plants.

W. BASSOW (*Center for International Environment Information, New York, N.Y.*): One comment and a brief question. I assume that the desired end result of this morning's discussion is to address the question, How can the media do a better job in reporting on nuclear energy and what can be done to help the media do a better job? The panel did not dwell sufficiently, I believe, on some of the problems faced by the news media in this dramatic situation where 500 reporters descended on Middletown—most of them probably coming right off the city desk covering the police beat, maybe pulled off the sports beat, real estate, who knows. But they were the bodies that were available at the time to cover a story. Now these people had to work against incredible deadlines, with a lack of technical information and know-how about the nuclear process. They didn't know where to turn for authoritative and useful information. Now, it seems to me that what we should be looking at are—as this gentleman from Texas described it—the first small steps. What are the first small steps that can be taken to help the media do a better job?

We at the center have been concerned with this problem for over two years now, and last year we published a guide to energy specialists for the news media that lists about 2,000 of the country's leading experts in every known energy technology, including nuclear. These people are drawn from government, from industry, from environmental organizations, from the universities, and so on. Everyone listed in it has agreed as a public service to respond to frantic telephone calls from editors and reporters working on energy stories.

When Three Mile Island broke, this guide was in the hands of about 800 daily newspapers—at that time, we hadn't completed distribution— around the United States. We checked to find out whether this guide was actually used to identify authoritative sources of information relating to Three Mile Island. And we found that, despite what you want to say about the way some reporters and editors carry out their professional activity, it was indeed used. And it was indeed used especially for the more esoteric kind of questions about nuclear insurance, the economics of nuclear construction, and so on. So that's my comment. The question is, What do you gentlemen think should be done to help the media to do a better job in reporting on nuclear energy?

D. M. RUBIN: I'm familiar with those guides. I think those guides are useful. I think that the main question is whether or not media owners and managers are willing to invest the time and the money to train reporters—to give them a chance to attend seminars, to make them specialists, and so on. I don't think anybody was grabbed from sports or real estate. But you're quite right, a lot of reporters were sent without the necessary background. We recommend in the commission's report that media managers do in this area what the Kerner commission recommended in the 60s on race relations reporting—and there was an improvement in race relations reporting for a while, though that seems now to have gone by the boards. But the question in this, as so often with the media, is the bottom line and whether that sort of investment is going to be made. My own guess is that it is not going to be made and that unless organizations like yours forcibly distribute these kinds of publications, run seminars, and dragoon the reporters into them, I don't know how much better it will get, except by experience. The more accidents you cover, the more familiar you are with what's going on.

D. BURNHAM: One brief comment. I'm aware of your publication, and it's helpful. But the problem isn't so much that we need more information. If you look across the press of the United States, it seems that the problem is a lack of skepticism on the part of the press. The press tends to be very accepting of what's handed to them. If you think about the coverage of nuclear power in the last 20 years, the problem has been the lack of skepticism, the acceptance of anything that's given without thinking about it.

S. KANDER (*Harvard University, Cambridge, Mass.*): Mr. Murray from General Public Utilities stated—and it seems to be the general consensus—that credibility has been lost vis-à-vis anybody who's been informing the public to date. He also called for educating the public over the whole industry on exactly what everything means, which is a nice idea. I have two questions. Who do you propose should be responsible for educating the public, in light of the fact that credibility has been lost from anybody that we've been listening to so far? So who should be responsible for that education? And then, how should the public know the difference between what is education and what is public relations and advertising? I address the whole panel.

W. B. MURRAY: As I said in the beginning, we have a shared responsibility, and I was addressing myself to what the utilities should do. We certainly don't accept that we're going to have this credibility problem for ever and ever. We're out there today. We've had literally thousands touring the plant during the cleanup. We have invited reporters on tours, and we're going to continue to have these; we've had them there during the venting. We are doing everything we can to build this credibility back up by trying to show confidence and trying to be very candid. You have problems, you have lapses. You don't do this overnight by any manner or means. But then, there has to be a role by others. This question of What's propaganda? What's public relations? What's fact? and Who do you believe? is basic to any question. It's not one you can just give a pat answer to. A question came up before that I don't think was ever addressed: "You should never believe anybody that has a vested interest." Well, the person who doesn't have a vested interest many times is someone who has little knowledge of a subject. On the other hand, there are many that have a lot of knowledge and may not *appear* to have a vested interest. I would say that many times a public interest group has a vested interest. Just because you're not a manufacturer or utility doesn't mean you don't have a vested interest. So it's a very difficult thing. You have to satisfy your own mind and do the kind of digging that we talked about here to verify what's being put out—whether it's factual or whether it isn't.

BEHAVIORAL AND MENTAL HEALTH EFFECTS OF THE THREE MILE ISLAND ACCIDENT ON NUCLEAR WORKERS: A PRELIMINARY REPORT

Rupert F. Chisholm,* Stanislav V. Kasl,†
Bruce P. Dohrenwend,‡ Barbara Snell Dohrenwend,‡
George J. Warheit,§ Raymond L. Goldsteen,‡
Karen Goldsteen,¶ and John L. Martin‖

This study of nuclear workers was conducted by the Task Force on Behavioral Effects of the President's Commission on the Accident at Three Mile Island (TMI). The study was one of four that the task force had conducted to examine effects on "the mental health of the public and the workers directly involved in the accident at TMI-2."[1] Determining the behavioral responses of nuclear workers under stress during the accident was of particular interest.

Examining the potential impacts of the TMI accident on nuclear workers involved considering a wide variety of factors. To begin with, disaster research (e.g., References 2–4) clearly demonstrates the power of disaster situations to cause symptoms of psychological distress among high proportions of affected populations.[5] Hence, the disaster literature provided one theoretical base for the research.

It was also necessary, however, to consider the special status of nuclear workers in this situation. Since the nuclear workers were permanent members of the organization that experienced the accident, they had both a psychological and an economic stake in its outcome. And, to varying degrees, these workers were directly involved in the incident and had responsibility for controlling the plant and bringing it to a safe condition. Consequently, organizational behavior literature on key dimensions of employee reactions under stress (e.g., perceived workplace hazards, job-related stress, and job security) served as the second major conceptual base of the study. In short, nuclear workers were involved in the accident situation both as members of the general population and, more directly, as employees of TMI. Therefore, it was essential to examine key aspects of likely worker experiences from both perspectives.

*Graduate Program of Public Administration, Pennsylvania State University, Capitol Campus, Middletown, Penn. 17057.
†Yale University, New Haven, Conn. 06510.
‡Columbia University, New York, N.Y. 10032.
§University of Florida, Gainesville, Fla. 32601.
¶Capitol Area Health Research, Inc., Columbia, Md. 21045.
‖Graduate Center, City University, New York, N.Y. 10036.

0077-8923/81/0365-0134 $01.75/2 © 1981, NYAS

METHOD

Sample

The sample of workers included all employees on the payroll of Metropolitan Edison Company who were assigned permanently to the Three Mile Island plant as of March 1, 1979. These permanent employees included bargaining unit employees, supervisory employees, and nonexempt employees. The survey did not cover contractor personnel or company employees who were assigned temporarily to TMI during the accident.

Nuclear workers at the Peach Bottom plant of the Philadelphia Electric Company served as a comparison group. This plant is located approximately 40 miles from TMI. All employees permanently assigned to Peach Bottom comprised this comparison group.

Names of employees from each plant were arranged into combined alphabetical lists, and order of contact was determined by a random number system. This procedure was designed to assure randomness in the order of contact in case time prevented making calls to all workers. A total of 324 TMI workers and 298 workers from Peach Bottom participated in the study. These represent response rates of 60.8% at both plants (324 of 533 employees at TMI; 298 of 490 at Peach Bottom). The nonrespondents include both interview refusals and individuals who could not be reached by the cutoff date for collecting data despite several attempts to reach each individual. However, results from a separate study of nonrespondents indicate that the findings would be substantially the same if completion rates were as high as between 70 and 80%.

Procedure

Data were collected during a one-hour telephone interview with each participant. This procedure was based upon the growing body of evidence from health-related studies that indicates that telephone interviews can provide as valid and reliable information as can face-to-face interviews or mailed questionnaires (e.g., References 6–8). Another recent study of several Pennsylvania communities found that respondents will provide detailed and complex information on a variety of personal topics and that this information is comparable to that obtained in person.[9] Underrepresentation due to excluding nontelephone households was not a problem in the present study since virtually every nuclear worker on each plant list had a phone.

Interviews were conducted by trained interviewers who used a standard guide for introducing and asking questions and for recording data. Random follow-up phone calls were used to verify that respondents had been contacted and to monitor the quality of interviews.

The survey was conducted between August 20 and September 29,

1979. Due to the end-of-September cutoff date, which was necessary to meet the deadline of the task force report to the president's commission, response rates were lower than desired. However, an analysis of major demographic characteristics indicates general similarities between TMI and Peach Bottom respondents (see TABLE 1).

As the data in TABLE 1 indicate, however, differences exist between respondents and nonrespondents at both nuclear plants. For example, disproportionate numbers of supervisors were interviewed at TMI and Peach Bottom. Consequently, a separate study of nonrespondents was conducted in early October to assess the impact of nonresponse on study

TABLE 1

COMPARISON OF WORKERS INTERVIEWED AND NOT INTERVIEWED

	Site			
	TMI		Peach Bottom	
Worker Characteristic	Interviewed (%)	Not Interviewed (%)	Interviewed (%)	Not Interviewed (%)
Supervisory Status				
Supervisor	34.0	26.3	38.6	13.0
Nonsupervisor	66.0	73.7	61.4	87.0
College Graduate	22.3	not obtained	22.6	not obtained
Sex				
Male	89.8	90.0	97.3	98.0
Female	10.2	10.0	2.7	2.0
Age				
Less than 30	30.9	27.8	22.9	24.0
30–39	49.4	47.8	46.1	46.8
40–49	13.0	12.9	19.5	19.3
50 or more	6.8	11.5	11.4	9.9
Distance of Home from TMI				
5 miles or less	37.3	49.3	0.3	not obtained
Over 5 miles	62.7	50.7	99.7	
Total Respondents*	(324)	(209)	(298)	(192)

*Bases for percent vary somewhat from totals shown because of missing data for some respondents for some variables.

results. Small, representative samples of TMI and Peach Bottom workers who had refused earlier interviews or who could not be reached previously were phoned again and asked to participate. This follow-up effort resulted in 28 out of 50 interviews with former TMI nonrespondents and 30 of 75 interviews with Peach Bottom employees who had not been interviewed before. As mentioned earlier, analysis of data from these interviews indicates that the results reported in this paper would be substantially the same even with considerably higher completion rates than those obtained.

Variables

Because of the unique nature of the TMI incident, the study was highly exploratory. However, as noted above, the disaster literature and written material on stressful events in the workplace provided conceptual underpinnings for the study. Variables were selected based upon the findings of previous studies, preliminary results of other TMI studies, and general knowledge of the accident situation. Insofar as possible, the questionnaire was constructed from scales that had been tested in previous studies. Naturally, in many cases, the wording of items had to be adapted to the study. In several instances, items or scales were developed to attempt to tap some special aspects of nuclear workers' experiences. Due to the exploratory nature of the study and the extremely unusual nature of the TMI incident, the questionnaire was designed to cover as many potentially significant areas of employee attitudes and behavior as feasible during a one-hour phone interview. This paper reports findings of the data that have been analyzed to date. The APPENDIX contains a detailed description of questionnaire items used for this paper.

The major measures of objective threat stemming from the accident were:

1. Place of work—working at TMI instead of the Peach Bottom plant when the accident occurred;
2. Proximity of residence to TMI—living within versus beyond a five-mile radius of the TMI nuclear plant;
3. Family status—having (or not having) preschool-age children in one's family.

Measures of overall health, attitudes, and behavior included:

1. Recall of degree of upset before, during, and after the crisis period;
2. Demoralization since the accident;
3. Perceived threat to physical health;
4. Uncertainty about the future of their occupation;
5. Perception of hostility from the community;
6. Trust in authorities.

Several of these dimensions were identical or similar to those used in studies of the other populations conducted for the Task Force on Behavioral Effects of the president's commission. Therefore, for these variables, it is possible to compare responses of nuclear workers from the two sites with those of the general population.

Dummy regression analysis was used to conduct the statistical analyses.[10] This technique allowed for assessing the effect of one factor while holding other relevant factors constant. All the effects reported were statistically significant at the 0.05 level or better using one-tailed tests.

RESULTS

Degree of Upset

To determine how upsetting the accident was, respondents were asked whether they had experienced periods of extreme upset during three time spans: six months immediately before the accident, during the crisis period (3/28/79–4/11/79), and at the present time. TABLE 2 gives responses to this item.

Results indicate that similar proportions of TMI and Peach Bottom employees experienced periods of extreme upset during the six months preceding the accident. In contrast, a much higher proportion of TMI than Peach Bottom employees reported having been extremely upset during the immediate crisis period (28.1% vs. 15.2%). However, at the time of interviews, the difference between workers at the two sites had diminished considerably.

TABLE 2

PERCENT OF WORKERS AT TMI AND AT PEACH BOTTOM WHO REPORTED PERIODS OF EXTREME UPSET BEFORE AND DURING THE ACCIDENT AND AT PRESENT

Location of Plant	Time of Periods of Extreme Upset		
	During the 6 Months before 3/28/79	Anytime during Crisis (3/28–4/11/79)	At the Present Time
TMI	10.3	28.1	16.9
Peach Bottom	10.0	15.2	12.8

Demoralization

Demoralization is a common distress response when individuals experience a serious predicament from which they see no escape. Findings reveal that the average TMI worker was somewhat more demoralized than was his typical counterpart at the Peach Bottom plant (13.1 vs. 11.2). Since almost 90% of TMI respondents and over 97% of those interviewed from Peach Bottom were male, results from these two groups were compared with those of male respondents in the general population study. Based upon interviews conducted mostly in July, male respondents from the general population had an average demoralization score of 9.7, which is significantly lower than that for TMI employees (but not that for Peach Bottom workers).

It is also revealing to consider results in the context of when data were collected. Early studies of relatively small samples of male and female heads of households who lived within a 20-mile radius of TMI had been conducted in April and May 1979. These studies found an average demoralization score of slightly over 19 for April respondents, contrasted with slightly above 15 for those interviewed in May. In mid-July, the average male-female response had declined only a slight degree to below 15, and the average male response to 9.7. Data on the workers were not available before August 20, with the bulk of interviews conducted in September. Hence, it is impossible to trace changes in the profile of nuclear worker responses from soon after the accident to the time of interviews. Nonetheless, it is important to note that even in late August and September, TMI employees remained considerably more demoralized than did men in the general population.

Differences also emerged between levels of demoralization of supervisors and nonsupervisors at TMI. The average level of demoralization among TMI supervisors was 10.7, which is identical to that of supervisory personnel at Peach Bottom and only slightly higher than the average of males in the general population. Nonsupervisory personnel were substantially more demoralized, with an average demoralization score of 14.3. This score also is significantly higher than that of nonsupervisors at Peach Bottom (11.6). Thus, the demoralizing effect of the accident occurred only among nonsupervisory employees.

Several demographic characteristics (education, age, sex, marital status) also related significantly to levels of demoralization. In general, college graduates were less demoralized than were those with less formal education. In terms of age, workers under 40 were the most demoralized, those over 50 were the least demoralized, and those 40 to 50 were at an intermediate level. Consistent with the findings from other studies, women were more demoralized than were men, and unmarried individuals more demoralized than were the currently married.

Threat to Physical Health

This scale was designed to indicate the extent to which nuclear workers felt that their health had been threatened by the TMI accident. Scores ranged from 1 (low concern) to 3 (high concern). The objective presence of threat did reflect in greater perceived threat among TMI employees than among Peach Bottom workers ($\overline{X}$ = 1.58 and 1.28, respectively). However, the fact that both means were below the uncertainty point of 2 suggests that workers did not have a generally high concern about threats to their physical health. In this respect, workers' responses are similar to those of respondents from the general population, where scores on an index of concern (a somewhat different scale) about the impact of the accident on physical health also were relatively low ($\overline{X}$ = 1.68).

Uncertainty about Occupational Future

It seemed reasonable to expect that after the accident and the resulting negative public reaction to nuclear power plants, TMI employees would have a high level of concern over the security of their jobs. Consequently, a scale was used to assess the extent of workers' insecurity about the future of their occupation as nuclear workers. Scale scores range from 1 (most certain) to 5 (most uncertain). While the average score of TMI workers (2.12) is higher than that of employees at the Peach Bottom plant (1.61), the general level of nuclear workers' concern about the future of their occupation is surprisingly low.

Hostility from the Community

These items were designed to determine the extent to which workers felt that the public was critical and unappreciative of their work. Scores on the scale range from 1 (indicating greatest hostility) to 10 (greatest appreciation). The first and third questions tapped perceived hostility, the second question the extent to which the hostility was justified. TABLE 3 gives average responses (means) to these items. As these results show, responses to items one and three are almost identical for employees at the two plants: both groups of workers felt that the public had a somewhat negative view of their performance. On the other hand, responses by the two employee groups to the second item are quite different ($\overline{X}$ = 4.54 for TMI, 5.55 for Peach Bottom; $p < 0.001$). This difference indicates that TMI employees felt that the public view of their performance was less justified than did their counterparts at Peach Bottom. Thus, in general, workers at the accident site perceived that the somewhat negative public image of their performance was unjustified to a significantly greater extent than did nuclear workers at the Peach Bottom plant.

TABLE 3

PERCEIVED HOSTILITY FROM THE COMMUNITY

	Site	
Item Content	TMI ($\overline{X}$)	Peach Bottom ($\overline{X}$)
1. Community view of nuclear worker performance during accident	4.61	4.64
2. Degree to which community view of performance was justified	4.54*	5.55*
3. Public appreciation of nuclear workers' work during accident	4.06	4.07

*Difference in means significant at the 0.001 level.

TABLE 4

PERCENT OF WORKERS AT TMI AND AT PEACH BOTTOM WHO EXPRESSED TRUST IN
INFORMATION RECEIVED FROM STATE AND FEDERAL OFFICIALS DURING THE ACCIDENT

	TMI	Peach Bottom
No	46.6	42.6
Yes	23.7	33.3
Don't Know/Maybe	29.7	24.1

Trust in Authorities

Two questions focused upon workers' trust in authorities. The first asked whether individuals felt information from state and local officials was truthful. Responses to this question were similar to those of respondents from the general population. Results in TABLE 4 indicate that almost half said they did not trust the information from state and federal officials, a quarter were uncertain, and between slightly less than a quarter and a third expressed trust.

The second question (which turned out to have no relationship to the first) asked whether their employers kept them fully informed about dangers or unhealthful conditions during the accident period. A substantial majority of both TMI (73%) and Peach Bottom (82.1%) employees responded positively to this item. However, workers at TMI showed less trust in their company than did those employed at the Peach Bottom plant. The general expression of trust in the utility companies contrasts sharply with the distrust that general population respondents expressed towards these organizations.

DISCUSSION AND CONCLUSION

One basic finding clearly emerges from results of the study: TMI employees reacted differently to the accident than did nuclear workers at the Peach Bottom plant. Significant differences showed up on six of the seven variables examined. TMI employees experienced more periods of being upset during the accident, were more demoralized, perceived greater threat to their physical health, were less certain of the future of their occupation, expressed that the public view of their performance during the incident was less justified, and felt that their employer was less trustworthy in providing information about the accident than did their Peach Bottom counterparts. Consequently, we conclude that in general, working at the site of the accident had a major impact on the psychological experiences of nuclear workers.

On the other hand, TMI/Peach Bottom differences do not exist for the two items indicating experienced hostility from the community and trust in public officials. These dimensions deal with workers' general perceptions of their relationships to the total society, while the variables cited above concern more immediate personal impacts of the accident on workers. Thus, it appears that for nuclear workers, a differential effect occurs between more immediate personal psychological reactions and more general attitudes about relationships to the larger community. This is a tentative conclusion that will require much more thorough analysis of the data for documentation.

The difference between demoralization of TMI workers and that of the general population is noteworthy. Not only were these nuclear workers more demoralized than were men in the general population surrounding the plant, but their psychological distress persisted after that of male members of the general population appeared to have abated. These findings indicate that the accident caused longer lasting psychological reactions among employees who were involved at the accident site than among the general population. The lack of relationships between living within five miles of the accident site and having preschool children and demoralization for TMI employees (contrasted with relationships for the general population) gives further evidence of sharp differences in the predicament of nuclear workers contrasted with that of members of the communities that surround TMI.

ACKNOWLEDGMENTS

This paper is based upon findings from the "Report of the Task Group on Behavioral Effects" to the President's Commission on the Accident at Three Mile Island (GPO No. 052500300732-1, U.S. Government Printing Office, Washington, D.C.). The authors also gratefully acknowledge the work of Brenda Eskenazi, Ph.D., for assistance on further analysis of the original data.

REFERENCES

1. Task Group on Behavioral Effects. 1979. Report to the President's Commission on the Accident at Three Mile Island. Report GPO No. 052500300732-1. U.S. Government Printing Office. Washington, D.C.
2. LINDEMANN, E. 1944. Symptomatology and management of unit griefs. Am. J. Psychiatry 101: 141–148.
3. FRITZ, C. E. & E. S. MARKS. 1954. The NORC studies of human behavior in disaster. J. Soc. Issues 10: 26–41.
4. SHEATSLEY, P. B. & J. FELDMAN. 1964. The assassination of President Kennedy: public reaction. Public Opinion Q. 28: 189–215.
5. DOHRENWEND, B. S. 1973. Life events as stressors: a methodological inquiry. J. Health Soc. Behav. 14: 167–175.
6. HOCHSTEIN, J. R. 1967. A critical comparison of three strategies of collecting data from households. Am. Stat. J. 62: 967–989.
7. MOONEY, W. H., B. POLLACK & L. CORSA. 1968. Use of telephone interviewing to study human reproduction. Public Health Rep. 83: 1049–1060.
8. COOMBS, L. & R. FREEDMAN. 1964. Use of the telephone interview in a longitudinal fertility study. Public Opinion Q. 28: 112–117.

9. LUCAS, W. & W. ADAMS. 1977. An Assessment of Telephone Survey Methods. The Rand Corporation. Santa Monica, Calif.
10. KERLINGER, F. N. 1973. Foundations for Behavioral Research. Holt, Rinehart and Winston. New York, N.Y.
11. DOHRENWEND, B. P., B. S. DOHRENWEND, M. S. GOULD, B. LINK, R. NEUGEBAUER & R. WUNSCH-HITZIG. 1980. Mental Illness in the United States: Epidemiologic Estimates of the Scope of the Problems. Praeger Publishers. New York, N.Y.
12. DOHRENWEND, B. P., P. SHROUT, G. EGRI & F. S. MENDELSON. 1980. Non-specific psychological distress and other dimensions of psychopathology. Arch. Gen. Psychiatry 37: 1229–1236.
13. DOHRENWEND, B. P., L. OKSENBERG, P. E. SHROUT, B. S. DOHRENWEND & D. COOK. What brief psychiatric screening scales measure. In Proceedings of the Third Biennial Conference on Health Survey Research Methods. S. Sudman, Ed. National Center for Health Statistics and National Center for Health Services Research. Washington, D.C. (In press.)
14. FRANK, J. D. 1973. Persuasion and Healing. Johns Hopkins University Press. Baltimore, Md.

APPENDIX

DESCRIPTION OF MEASURES OF MENTAL HEALTH AND BEHAVIORAL EFFECTS
USED IN THE NUCLEAR WORKERS STUDY

Upset

I am going to list some problems that people experience from time to time. Please tell me if any of them have bothered you at the times indicated.

	During the 6 Months before 3/28/79		Anytime during Crisis (3/28–4/11/79)		At the Present Time	
	Yes	No	Yes	No	Yes	No
Periods of Extreme Upset	1	2	1	2	1	2

Scoring note: Yes was scored 1 and negative response 0 for each period.

Reliability: No information. Each is treated as a single item index.

Interpretation: This item attempts to measure the subjectively experienced "upset" of workers during three time spans.

Demoralization

Typical questions from this 26-item scale are listed below:

1. How often since TMI have you had times when you couldn't help wondering if anything was worthwhile any more? (4 very often; 3 fairly often; 2 sometimes; 1 almost never; 0 never)

2. Since TMI, how often have you felt that nothing turns out for you the way you want it to, would you say (4 very often; 3 fairly often; 2 sometimes; 1 almost never; 0 never)

3. Since TMI, how often have you felt completely helpless? (4 very often; 3 fairly often; 2 sometimes; 1 almost never; 0 never)

4. Since TMI, how often have you felt completely hopeless about everything, would you say (4 very often; 3 fairly often; 2 sometimes; 1 almost never; 0 never)

Scoring note: All items are scored in the same direction on a five-point scale.

Reliability: The internal consistency reliability of this scale is 0.90 in TMI and 0.91 in Peach Bottom.

Interpretation: These 26 items are a sample from a larger set of items that have been developed in the Social Psychiatry Research Unit, Department of Psychiatry, Columbia University, to measure demoralization.[11-13] The 26 items correlate 0.98 with a composite scale formed from the larger set of demoralization scales.

"Demoralization" is the term used by Jerome Frank to describe the psychological symptoms and reactions a person is likely to develop " ... when he finds that he cannot meet the demands placed on him by his environment, and cannot extricate himself from his predicament."[14] Demoralization can coincide with diagnosable psychiatric disorders but may also occur in the absence of such disorders. The various sources of the intractable predicaments include, for example, situations of extreme environmental stress such as combat or natural disasters; physical illnesses, especially those that are chronic; and crippling psychiatric symptoms of, for example, the kinds associated with severe psychotic episodes. Hence, an elevated score in a scale measuring demoralization is something like elevated physical temperature; it tells us that there is something wrong; it does not in and of itself tell us what is wrong.

Threat to Physical Health

During the TMI incident (3/28/79–4/11/79) did your job expose you to:	If "yes", how much of a problem was this for you?			
	No Problem At All	Slight Problem	Sizeable Problem	Great Problem

1. Radiation?
   Yes
   No
2. Risk of catching diseases?
   Yes
   No

3. Even if your employer kept you fully informed (about any dangers or unhealthful conditions that you may have been exposed to on your job), do you feel that your health was *endangered more than usual* during the TMI incident due to hazards in the workplace?
   Yes
   No

4. Are you satisfied that you are now *safe* and not contaminated by radiation from the TMI incident?
   Yes
   Maybe
   Don't know
   No

Scoring notes: Question 1: No or No problem = 1, Slight problem = 2, Sizeable problem or Great problem = 3. Question 2: No or No problem = 1, Slight problem, Sizeable problem, or Great problem = 3. Question 3: No = 1, Yes = 3. Question 4: Yes = 1, Maybe or Don't know = 2, No = 3. Item scores were added and divided by four to obtain scale scores.

Internal consistency reliability: 0.53 in two samples combined; 0.48 in TMI and 0.54 in Peach Bottom sample.

Interpretation: This scale is intended to measure the extent to which the workers felt that their health was endangered by the TMI accident.

Certainty about Future Occupation

I would now like to ask you how you see the *future of your occupation.* For each of the following questions, please indicate how certain/uncertain you feel. Possible responses include:

Somewhat Uncertain	A Little Uncertain	Somewhat Certain	Fairly Certain	Very Certain
1	2	3	4	5

1. How certain are you about what your *future career* picture looks like? Are you ... (repeat response categories)?

2. How certain are you of the *opportunity for promotion* and advancement that will exist in the next few years? Are you ... (repeat response categories)?

3. How certain are you about whether your *job skills* will be of use and value five years from now? Are you ... (repeat response categories)?

4. How certain are you about what your *responsibilities* will be six months from now? Are you ... (repeat response categories)?

Scoring notes: All items are scored on a scale of 1 to 5 as indicated. Item scores were reversed so that 5 indicated most uncertain. Item scores were added and divided by four to obtain scale scores.

Internal consistency reliability: 0.65 in combined samples, 0.72 in TMI sample, and 0.36 in Peach Bottom. The low figure for Peach Bottom is probably due to the lack of variability in responses to these items in this sample, where the majority responded "very certain" to three of the four questions. This figure cannot therefore be interpreted as indicating that this scale is necessarily an unreliable measure in this sample.

Interpretation: These questions were designed to assess the workers' feelings of security or insecurity about their occupation without an indication of the basis for their feelings.

Perception of Hostility from Community

1. How do you think the performance of nuclear workers such as yourself was seen by *people in the community* during the TMI incident (3/28/79–4/11/79)? Please indicate on a scale of 1 to 10: 1 = made serious errors; 10 = performed very capably.

2. To what degree do you feel this view was *justified*? Please indicate on a scale of 1 to 10: 1 = completely unjustified; 10 = completely justified.

3. How much do you feel the general public *appreciated* the work of nuclear workers such as you during the TMI incident (3/28/79–4/11/79)? Please indicate on a scale of 1 to 10: 1 = very little appreciation; 10 = very great appreciation.

Scoring notes: As noted, each item is scored on a scale of 1 to 10.

Interpretation: Each item is treated as a separate index with no interpretation beyond the wording of each question.

Trust in Authorities

1. Do you feel that the information you were getting from state and federal officials during the TMI incident was truthful?

2. During the TMI incident (3/28/79–4/11/79), do you think your employer kept you fully informed about the dangers or unhealthful conditions that you may have been exposed to on your job?

Scoring Notes: Question 1: Yes = 1, Maybe or Don't know = 2, No = 3. Question 2: Yes = 1, No = 3.

Reliability: There was no relation between these two items. Therefore, they were not combined to make a scale, but were analyzed separately.

Interpretation: Each item is treated as a separate index with no interpretation beyond the wording of each question.

LOCAL PUBLIC OPINION*

Cynthia Bullock Flynn†

*Social Impact Research, Inc.
Seattle, Washington 98102*

Both survey data and interviews with people living close to Three Mile Island (TMI) indicate a substantial variation in response of individuals to the accident. At the extremes, we find some who were virtually oblivious to the potential gravity of the situation and others who were traumatized. This variation is, perhaps, one of the most unexpected results of the research that has been conducted in the local area. What accounts for the variation?

The attempt to measure and analyze the public response to the TMI accident has relied greatly on public opinion polls. The seven surveys reviewed for this paper were conducted by the Nuclear Regulatory Commission (NRC), the Pennsylvania Department of Health, Smith, Kraybill, Rutgers, Goldsteen, and Brunn. Although the methods used varied (see APPENDIX for brief descriptions of each survey), many of the results were similar.

The accident at Three Mile Island began at about 4:00 A.M. Wednesday, 28 March. The two-week period immediately following the accident was characterized by a gradual increase in concern on the part of officials and the general public through Monday, 2 April, followed by a gradual decrease in concern. Although the effects of the accident will continue to be felt in the area for some time, it seems appropriate to set apart the first two weeks for study, given the sense of urgency felt at that time.

EMERGENCY PERIOD BEHAVIOR

Individuals' Behavior

Generally, the public appears not to have been alarmed on Wednesday, 28 March. This was due partly to the fact that many people were not aware until the evening that an accident had occurred. Exceptions to the general lack of early concern included those who had close friends or relatives working at Three Mile Island. Since those reporting for the 7:00 A.M. shift were not allowed on the island, some indication of the seriousness of the accident was apparent to these people. The NRC survey shows that some evacuation occurred as early as Wednesday, but this was unusual.

---

*An earlier version of this paper appeared in Reference 3.

†On leave from the Department of Sociology, University of Kansas, Lawrence, Kans. 66045.

0077-8923/81/0365-0146 $01.75/2 © 1981, NYAS

By Thursday, media reports indicated that the situation at TMI was under control, and the public seems to have been reassured. Ron Drake, a local radio personality for over 20 years, joked about the accident on his Thursday morning show.[1] Again, a few people evacuated on Thursday, but the public generally remained calm.[2]

By Friday, 30 March, individuals began to react to the developments in vastly different ways. Those who appear to have been less affected continued in their normal activities. A Friday night card party at the Elks Club in Middletown was not canceled, and other social activities later that night also continued, even though a curfew was in effect after 9:00 P.M. Individuals who were less affected did not stay indoors or shut their windows; they shopped and went about their business as usual over the weekend. It did not occur to them to evacuate, and few of their friends evacuated. Some report being astonished to learn later how many had evacuated. Although by the weekend they were aware of a problem at TMI, the problem did not carry personal significance for them.[3]

Others in the area did not evacuate but seemed to be more aware of the possibility of the necessity for evacuating. Individuals in this group who remained behind usually made preparations for leaving, such as filling the gas tank and packing, but never did evacuate. The Rutgers study estimated that the percentage of people who made preparations but did not leave was 33% within 20 miles, and the Brunn study showed that two-thirds of those who did not evacuate considered doing so.[8,16]

In some cases, women and children were evacuated so that their safety would be insured. This was particularly true for the families of men with official responsibilities who did not want to have to be concerned about them if a general evacuation were ordered. The NRC survey showed that households in which some people evacuated and some did not were very sensitive to the danger of the situation (86% reported that the situation seemed dangerous). The primary reasons given for some persons remaining behind were that they were unable to leave their jobs or that they would have left only had they received an evacuation order. Many (45%) felt that whatever happened was in God's hands, and a third were concerned about looters.[2]

The households where no one evacuated exhibit a quite different pattern. The overriding reason given for staying was that they were waiting for an evacuation order; this reason was also mentioned most frequently in the Brunn study. This was followed by the feeling that whatever happened was in God's hands. The third reason for staying was that they saw no danger; this was mentioned two and a half times as frequently by households in which no one evacuated as compared to households where some members evacuated and others did not. Together, these three reasons suggest greater confidence in authority in the households where everyone stayed. Although the desire to remain for their jobs was something of a consideration for this group, it was not the overriding concern that it was for nonevacuees in households in which some persons evacuated.

Among those who did evacuate, there is variation in the response. It is clear from individual descriptions of behavior during the first days of the accident that the decision to evacuate was perceived as requiring individual choices. Individuals were left with the responsibility for deciding who would evacuate and when, where, and how they would evacuate. In some ways, the decision was more stressful for individuals whose children were in the elementary grades (but not preschool) or who lived just beyond the recommended five-mile limit because these individuals had more of the responsibility for the decisions themselves.[3] The decision about whether or how to evacuate appears to have been particularly difficult for housewives who were at home alone, separated from their children at school, and unable to reach their husbands because of jammed telephone lines. One resident, perhaps speaking for many who evacuated, reported:

> On Friday a very frightening thing occurred in our area. A state policeman went door-to-door telling residents to stay indoors, close all windows, and turn all air conditioners off. I was alone, as were many other homemakers, and my thoughts were focused on how long I would remain a prisoner in my own home and whether my husband would be able to come home after teaching school that day. Suddenly, I was scared, real scared. I decided to get out of there, while I could. I ran to the car not knowing if I should breathe the air or not, and I threw the suitcases in the trunk and was on my way within one hour. If anything dreadful happened, I thought that I'd at least be with my girls. Although it was very hot in the car, I didn't trust myself to turn the air conditioners on. It felt good as my tense muscles relaxed the farther I drove.[4]

Decisions had to be made about which, if any, of the normal day-to-day responsibilities would be met. For instance, one informant baked, decorated, and delivered a promised cake for a birthday party for Saturday on her way out of town. Decisions about the care of pets and livestock had to be made. It was not clear how urgent the threat was, which materially increased the stress of making all these decisions.

In a few households, the absence of a clear order for everyone to evacuate resulted in disagreement over whether to evacuate. About 12% of the respondents in the NRC survey said that members of their families disagreed somewhat or strongly over the decision.[2] Most of these families did not, in fact, evacuate; given the general level of tension in the area, the family members who favored evacuation were undoubtedly particularly upset.

Considering the limited nature of the governor's advisory, the extent of the evacuation was substantial. The advisory was just that; it was not an order to evacuate. Further, it only applied to pregnant women and preschool children within 5 miles of the station. Less than 6% of the NRC sample had family members who fell under these criteria specified by the governor. However, the surveys by the NRC and by the Pennsylvania Department of Health both indicate that 60% of those within 5 miles of TMI evacuated; this amounts to approximately 21,000 persons.[2,4] In the 5-

to 10-mile ring, 56,000 persons (44%) evacuated. In the 10- to 15-mile ring, which contains most of the Harrisburg metropolitan area, 67,000 persons (32%) evacuated. Thus, within 15 miles of TMI, it appears that a total of 144,000 persons, or about 39% of the total population, evacuated. Other estimates of the extent of the evacuation are summarized in TABLE 1. Given the differences in target population and methodology, one would not expect these estimates to be identical. Taking into account those differences, however, these data suggest that well over half of the population left from within the 5-mile area, and about a third left from the 5- to 15-mile area. These data imply that a significant number of persons made individual decisions to evacuate, although they had not been formally advised or ordered to do so. FIGURE 1 shows the distribution of evacuation by both distance and direction from the plant as estimated in the NRC survey.

Since the majority of persons who evacuated were not doing so because of the governor's order, why did they decide to leave? The main

TABLE 1

PERCENTAGE OF POPULATION EVACUATING BY VARIOUS DISTANCES FROM TMI

| Survey | Distance from TMI | | | Total |
	0–5 Miles	5–10 Miles	10–15 Miles	
NRC[2]	60	44	32	39
Penn. Dept. of Health[5]	60			
Smith[4]	50			
Kraybill[6]				42
Rutgers[16]	33 (0–10 miles)			
Goldsteen[15]				52
Brunn[8]	55	54	28	

reason given in five surveys was that the situation seemed dangerous.[2,5-8] In personal interviews, evacuees said they were frightened by the reports they received.[9-11] Another major reason for evacuating was the confusing information about the situation. Many assumed it was better to be safe than sorry, and in the absence of conclusive reassurance of the plant's safety, many chose to evacuate. A related reason for voluntarily evacuating was the desire to avoid the danger or confusion of a forced evacuation.

The surveys showed that some types of people were more likely than others to evacuate. The NRC survey showed that females were more likely than males to evacuate. Two-thirds of the children aged five and under were evacuated, and it appears that 90% of the pregnant women evacuated. In the NRC study, no systematic relationship was found between income, education, or occupation and evacuation behavior. However, according to the Kraybill study, the more highly educated were more likely to have evacuated.[6] Both of the surveys and the

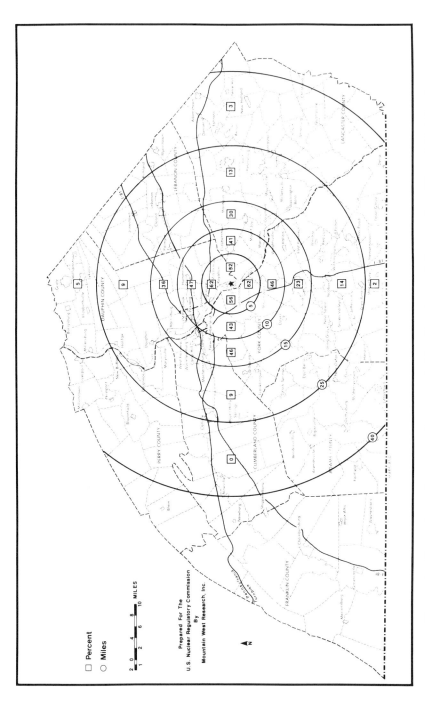

FIGURE 1.   Percentage of persons who evacuated by direction and distance from Three Mile Island.

personal interviews indicated that older persons were less likely to have evacuated. In part, this was because they were less likely to be included, directly or indirectly, in the governor's advisory.

The greatest number of those who did evacuate left on Friday, 30 March. Estimates of the percentage who left on Friday range from 55% to 72%.[2,7,8,11] It appears that most of the people who left then had not considered doing so prior to Friday. Although a few households stayed in motels and hotels, the overwhelming majority of the evacuees stayed with friends and relatives (estimates range from 74% to 90%). Because most people decided to leave on such short notice, their friends and relatives had not expected a visit. In some cases, facilities were less than ideal for unexpectedly accommodating entire families—complete with pets. Most evacuees went to friends and relatives in Pennsylvania (67 to 72%);[2,11] for those who evacuated a significant distance within the state, the most likely destinations were in and near Shamokin, Altoona, or Pittsburgh. Estimates of the median distance traveled are between 85 and 100 miles.[2,8]

The official evacuation center was located in Hershey, Pennsylvania. It appears that the maximum number of people at Hershey in one day was about 180, and that a total of as many as 800 may have stayed there for a short time. On at least one occasion, there were more reporters than evacuees at the center. Although their parents were undoubtedly concerned about the accident, the children at the Hershey evacuation center were entertained by clowns, given coloring books, and taken to the zoo. Generally, persons in the center remained calm.[10]

By the middle of the week following the accident, the perception of danger was considerably lessened. The median date of return to the area was Wednesday, 4 April.[2,8] However, the governor's advisory to pregnant women and preschool children was not lifted until 9 April, and schools within five miles of TMI did not open until the 11th. There was considerable variation in the amount of time spent out of the area, but there has been no systematic study of the decision-making process for returning to the area. Local informants cited the need to return to their jobs and a perception that the situation was under control as reasons for returning.[12,13]

During the two-week emergency period, the activities of at least half of the people in the area were disrupted.[2] During the week following 30 March, curfews were in effect over much of the area and evening meetings were canceled. Schools were closed, many of the children had evacuated, and, therefore, daytime activities involving children were canceled as well.[3] The main changes in day-to-day activities mentioned by NRC respondents were staying indoors, canceling plans, being on edge, and getting ready to leave. Other frequently mentioned responses were that someone was out of work, children were home from school, extra time was spent listening to the news, or someone worked more than usual.

Stress and Psychological Effects

The amount of stress experienced by people near TMI was both a function of the perceived amount of threat to physical safety and of the reliability of the information being used to ascertain the amount of threat. The perceived amount of threat varied considerably among individuals. For instance, respondents in the NRC study were asked about their perception of the seriousness of the threat at the time of the accident. Most respondents thought the threat was very serious (48%) or serious (19%), but more than a fifth (21%) thought it was only somewhat serious, and 12% thought the accident was no threat at all. Generally, those closer to the plant were more likely to perceive a serious threat than were those farther away. Conversely, those who thought it was no threat at all were located farther from TMI. Those who thought TMI was a serious threat at the time of the accident were younger, female, more highly educated, and of high income. Pregnant women were much more likely (64%) than average to view it as a very serious threat and much less likely to think it was no threat at all. Kraybill's study indicated that 76% felt that the threat was very serious and an additional 20% felt it was a little serious.[6] Those more likely to perceive the situation as serious included females, persons aged 25–34, the better educated, and those who evacuated (of those who had returned by 8 April).

Similar results were found when respondents in the NRC study were asked about their concerns over emissions from the plant. Sixty-one percent were very concerned with emissions at the time of the accident, and 26% were somewhat concerned, but 13% were not at all concerned. Those who did not evacuate were three times as likely (19% vs. 6%) to be unconcerned as were those who evacuated. Considering that the preaccident perceptions of TMI were either neutral or positive, these indicators of concern during the accident period represent a substantial change.

Kraybill's study indicated that nearly half (48%) did not feel they received sufficient information about emergency procedures during this time. People aged 25–32, the better educated, and those who evacuated were most likely to respond that they had not. Respondents in the NRC study found media such as local TV and radio most useful. Sources such as national network TV were less useful, and the print media ranked behind all radio and TV. Poor ratings for friends and relatives as information sources apparently resulted because they were perceived as having rumors rather than factual information.

The NRC study also included questions about the various sources of official information. The governor of Pennsylvania and the NRC were cited as the most helpful during the two-week period of the accident. Respondents perceived Metropolitan Edison as least helpful. The Rutgers study had similar results: 57% said the NRC was the most reliable source, followed by 19% who cited Governor Thornburgh.[16] Smith's study showed that Harold Denton of the NRC was viewed as the most legitimate source by 45%, followed by no one (30%), the governor (11%), and the media (11%).[4] Seventeen percent volunteered that Metropolitan Edison was *not* viewed as a legitimate source. However, both state and local officials fared poorly in the Brunn study, as did the utility.[8]

When asked, "Overall, how satisfied were you with the way you were given information during the emergency?" the median response for NRC respondents was in the middle of the four-point scale: half were very satisfied (12%) or mostly satisfied (37%), and half were very dissatisfied (22%) or mostly dissatisfied (29%). Generally, those farther from TMI were more likely to be satisfied with the information they received than were those closest to TMI. Those who were *least* likely to be satisfied were pregnant women (71%) and students (75%). There was a marked difference in overall satisfaction with information by evacuation status. Evacuees were much more likely to be dissatisfied (64%) than were nonevacuees (47%).[2]

The perceived lack of information was especially frustrating for those who had already evacuated. These persons were dependent on national media for information, and in some cases heard information that later proved to be inaccurate. Evacuees were unsure of whether they would ever be able to return to their homes and friends and were concerned because they had not thought to bring family photographs and important papers with them.[11,12] They were also concerned about the safety of their friends who were left behind.

Given the high degree of stress, it is not surprising that some of the people in the area reported experiencing psychosomatic symptoms because of the accident. Goldsteen's research indicated that persons in the area felt demoralized shortly after the accident and that students experienced an average of one physical symptom, such as stomachache, headache, or sleeping problems.[15] The NRC survey showed a higher level of stress symptoms for those persons living closer to TMI at the time of the accident for all 15 indicators: stomach trouble, headache, diarrhea, constipation, frequent urination, rash, abdominal pain, loss of appetite, overeating, trouble sleeping, sweating spells, feeling trembly and shaky, trouble thinking clearly, irritability, and extreme anger. Other indicators of stress among local residents included resumption of smoking during the emergency period, insomnia, short temper, and long-lasting indigestion.[4,14]

POSTACCIDENT EFFECTS

Most of the conspicuous signs of the emergency disappeared as suddenly as the emergency had appeared. There was no damage to public and private facilities (other than the nuclear generating plant itself), and by the second week in April, most evacuees had returned to their homes, businesses were open, schools and other institutional facilities had reopened, and daily activity appeared much as it had before the accident.

The presumption was made frequently by those at a distance from the plant site that real estate values would plummet, that tourism and agriculture would be adversely affected, and that the entire economic future of the area would be in question. Yet in the vicinity of the plant, real estate transactions continued to take place, dairy products were produced and sold, visitors came to have their pictures taken against the background of the Three Mile Island cooling towers, and industrial developments continued to move forward. A conspicuous characteristic of the postaccident environment was the discrepancy between the presumed severity of impact suggested by persons with little direct familiarity with conditions in the area and the absence of continuing effects alleged by many living in the area.

Continuing Stress and Psychological Effects

There is some evidence that stress persisted after the emergency period. Nearly a quarter of the respondents in the NRC study still perceived TMI as a very serious threat to their safety in the late summer of 1979.[2] Only 28% felt it was no threat at all. Even more respondents were still very concerned about emissions from TMI (41%), and somewhat fewer (25%) were not at all concerned. Of course, both perception of threat and concern with emissions had decreased by late July relative to their levels during the accident (61% very concerned). However, the fact that concern with emissions was considerably higher in July than it was before the accident (41% vs. 12% very concerned) showed that TMI had clearly become a substantially greater source of stress.

It appears that many of the psychosomatic indicators of stress have been reduced to their preaccident levels over time. Goldsteen's data indicate that feelings of demoralization increased sharply during the emergency period but these indicators of stress were short-lived.[15] Data from the NRC survey show a similar pattern for similarly measured indicators: overeating, loss of appetite, difficulty in sleeping, feeling trembly or shaky, trouble thinking clearly, irritability, and extreme anger.[2] However, the more somatic symptoms, such as rash, headache, stomach trouble, diarrhea, constipation, frequent urination, cramps, and sweating spells, continue to affect a small percentage of the population.

Attitudes toward TMI are another indicator of continuing psychological effects of the accident. The percentage of the population that feels that the disadvantages of TMI outweigh its advantages has changed from 27% before the accident to 50% since the accident. This is consistent with local estimates that at least a third of the people in the area are pronuclear and a third are antinuclear and that communities are probably split about 50–50.

There is continuing concern by local residents over the quality of existing evacuation plans. It is generally known that most areas did not have well-developed plans prior to the accident but that they had developed plans by Saturday afternoon (31 March) or Sunday morning. Since the accident, there has been additional work done on the plans, both by the county and municipal officials, with citizen participation in some cases. Some municipalities have already spent dozens of man-hours on revising their plans since the accident.

Another psychological effect is the persistence of rumors. This is perhaps more of a problem among the antinuclear people, who make a greater effort to keep continuously informed about developments. For instance, there are rumors that Metropolitan Edison is burning off the fuel remaining in TMI-1 at night so as not to alarm the populace. Interviews with local residents were commonly interspersed with requests for technical information about the plant and the accident. The ongoing discussion regarding Three Mile Island is still quite technical, and it is clear that in trying to understand what is currently occurring, many lay people are still confused.[3]

Movement from the Area

There continues to be some sensitivity to living near the nuclear plant. It is difficult to know the magnitude or extent of the concern without extensive interviewing, which has not, and probably will not, be done. However, we can examine the behavior of individuals as an indirect indication of the extent of their continuing stress and anxiety.

The most extreme behavioral response is to pack up and leave the area. Given the economic and psychological costs associated with a sudden move, this would certainly be an indication of extreme distress. The respondents in the NRC survey were asked whether anyone in the household had considered moving because of the accident. Nineteen percent indicated they had, and this response was given much more frequently by persons living nearest the station (30% within five miles).[2] This corresponds to 16% who said they had considered moving in the Brunn study.[6] The Pennsylvania Department of Health survey reported similar results, with 25% of the evacuees and 5% of the nonevacuees within five miles of TMI having considered moving.[5] Those who considered moving were younger and more highly educated than were respondents who reported that they had not considered moving. Evacuees were more than three times as likely to say that they had considered moving as compared to nonevacuees (33% vs. 9%).

Among the households that reported they had considered moving, 22% (25% in the Pennsylvania Department of Health survey) reported that they had definitely decided to move. This implies that as many as 5,100 households within 15 miles of the plant (approximately 4%) report that they intend to move. The number that will actually move remains to be determined, but it is significant that these responses (in the NRC survey) were recorded in late July and early August.

The census conducted by the State Department of Health of the population within five miles of the plant gives a preliminary indication of movement from the area.[5] As of 21 August, preliminary hand tabulations of data collected by the Department of Health indicate that 147 households within five miles of TMI were identified as having moved between 1 April 1979 and the end of July (about 1% of the estimated total number of households). Of the movers that had been contacted, 29% indicated that their move was motivated by the accident at TMI. If this percentage is applied to the total number of households that moved (147), an estimated 43 households may have moved due to the accident—less than 0.3% of households within the five-mile ring. Additional tabulations on movement from the area will be available in the future, but, to date, out-migration due to the accident appears modest.[5]

As an additional check on possible out-migration from the area immediately around the plant, elementary school enrollments since the 1974–75 school year were obtained from local officials. In no case is there clear evidence of an effect of the accident; therefore, even though many families living near the facility report stress and continuing threat due to the proximity of TMI, relatively few have been sufficiently concerned to relocate their homes because of the accident.[4]

CONCLUSIONS

During the emergency period, the perceived threat, the lack of good information, the evacuation experience itself, and the psychosomatic effects indicate that part of the population experienced considerable stress. On the other hand, a significant minority of the residents were not at all worried about emissions from TMI and did not feel at all threatened. The accident made very little practical difference to these individuals. On virtually every indicator examined, there was substantial variation in individuals' reactions. For this emergency period, then, it is oversimplifying reality to speak of "the" reaction of local residents. It is only in the postemergency period that we begin to see evidence of similar behavior among residents, and this behavior tends to indicate that in most respects, normal daily behavior has resumed.

REFERENCES

1. WISE, P. 1979. Personal communication from the president of the Middletown, Penn., Borough Council, 8 October.
2. FLYNN, C. B. 1979. Three Mile Island Telephone Survey: Preliminary Report on Procedures and Findings. U.S. Nuclear Regulatory Commission. Washington, D.C.
3. FLYNN, C. B. & J. A. CHALMERS. 1980. The Social and Economic Effects of the Accident at Three Mile Island: Findings to Date. U.S. Nuclear Regulatory Commission. Washington, D.C.

4. SMITH, P. 1979. Response received to previous TMI rebuttal. Trinity Parish Newsletter (Summer).
5. Pennsylvania Department of Health. 1979. Report on TMI census statistics questionnaires. Harrisburg, Penn. (Unpublished.)
6. KRAYBILL, D. B. 1979. Three Mile Island: Local Residents Speak Out. Social Research Center. Elizabethtown College. Elizabethtown, Penn. (Unpublished.)
7. SMITH, M. 1979. Preliminary tabulations. Franklin and Marshall College. Lancaster, Penn. (Unpublished.)
8. BRUNN, S. D., J. H. JOHNSON & D. J. ZIEGLER. 1979. Final Report on a Social Survey of Three Mile Island Area Residents. Department of Geography. Michigan State University. East Lansing, Mich.
9. LESNIAK, R. & M. A. LESNIAK. 1979. Personal communication from school board member, Central Dauphin School District, 21 September.
10. LIGHT, K. 1979. Personal communication from resident and member of Persons Against Nuclear Energy (PANE), 7 October.
11. KINNEY, P. 1979. Personal communication from PANE member, 20 September.
12. SIDES, S. 1979. Personal communication from the secretary of Middletown Borough, 13 July.
13. KELLEY, J. 1979. Personal communication from the director of evaluations, Pennsylvania Department of Welfare, 2 November.
14. Editor. 1979. Editorial. TMI Alert 1(2): 16.
15. DOHRENWEND, B. P., R. GOLDSTEEN, et al. 1979. Report of the Task Force on Behavioral Effects. President's Commission on Three Mile Island. U.S. Government Printing Office. Washington, D.C.
16. BARNES, K., et al. 1979. Human Responses by Impacted Populations to the Three Mile Island Nuclear Reactor Accident: An Initial Assessment. Department of Environmental Resources. Rutgers University. New Brunswick, N.J. (Unpublished.)

APPENDIX

The purpose of this section is to provide a brief description of the methodologies used for the surveys cited in the text. More complete discussions of the methodologies and findings are available from the published reports.

Brunn, S. D., et al., Reference 8.

A stratified sample of 178 addresses was chosen from the Harrisburg and York telephone directories, with a proportionately greater number chosen from communities nearer TMI. An additional 122 were randomly selected from the Carlisle, Duncannon, and Lancaster urban areas. One hundred fifty responses were received to the mailed survey. The measure of distance used was perceived distance from TMI.

Flynn, C. B., (NRC Survey), Reference 2.

A stratified sample of households was random-digit dialed. Households nearer TMI had a greater probability of selection. One thousand five hundred and four half-hour surveys were completed, and weights are available for estimating population totals within 15 miles of TMI. The measure of distance used in tabulations was the distance from the community to TMI. Perceived distance is also available.

Dohrenwend, B. P., R. Goldsteen, et al., Reference 15.

Data gathered by Raymond Goldsteen (so cited in the text) using a variety of methodologies. They indicate "strict probability sampling procedures . . . to select households at random" from the 20-mile radius of TMI and the Wilkes-Barre region (no further specification of methodology); place-stratified random sampling from telephone directories; birth listings in newspapers for mothers of young children; entire classrooms of students (selection procedures not specified); and a convenience sample of mental health clients. It appears that the data represent a combination of telephone, face-to-face, and mailed interviews.

Kraybill, D. B., Reference 6.

Respondents were selected by a multistage, simple random sample of residential telephone numbers from three directories: Middletown, Marietta, and Elizabethtown. All respondents live on the east side of the Susquehanna River within a 15-mile radius of Three Mile Island. Polling began on Monday evening (2 April 1979) after the mass media reported that the immediate crisis had abated. Interviewing continued through Sunday evening (8 April 1979) in order to include returning evacuees. The results are based on 375 completed interviews.

Barnes, K., et al. (Rutgers Study), Reference 16.

A questionnaire was mailed to a sample of 922 respondents selected from reverse telephone directories stratified by distance (five 5-mile zones up to 20+ miles) and direction (north, east, south, and west) from TMI. Equal sample sizes were selected from each of the twenty units. Three hundred sixty surveys were returned; more were returned from closer to TMI. Distance was determined by telephone exchange.

Pennsylvania Department of Health, Reference 5.

Exchanges within five miles of TMI were random-digit dialed in July 1979 to produce 690 respondents.

Smith, M., Reference 4.

One hundred thirty-five households with Middletown exchanges were randomly selected from the Harrisburg telephone directory. One hundred twenty-three schedules were completed.

STRESS IN THE COMMUNITY: A REPORT TO THE PRESIDENT'S COMMISSION ON THE ACCIDENT AT THREE MILE ISLAND*

Bruce P. Dohrenwend,† Barbara Snell Dohrenwend,‡
George J. Warheit,§ Glen S. Bartlett,¶
Raymond L. Goldsteen,‡ Karen Goldsteen,‖
and John L. Martin**

The charter for the President's Commission on the Accident at Three Mile Island (TMI) stated that, as part of its comprehensive study and investigation, it should include "an evaluation of the actual and potential impact of the events on the public health and safety" (Section 3 of the charter). The role of the Task Force on Behavioral Effects was outlined in the Table of Contents and Plan for Study for the Task Forces in the Area of Public Health and Health and Safety of the Workers dated June 12, 1979. According to this document, one objective of the task force was to examine effects on "the mental health of the public." In examining effects on mental health, a distinction was to be made between short-term and long-term effects.

The Task Group on Behavioral Effects was created on June 18th and met for the first time as a group on July 2-3. The accident at TMI took place between March 28 and April 10. Fortunately, during or shortly after the accident, on their own, several researchers from colleges and universities near the TMI site began sample surveys of the approximately 744,000 people living within 20 miles of TMI.[1] Most of these studies employed reliable measures of psychological effects with small, but carefully drawn, samples of the general population or of high-risk groups, such as mothers of preschool children within the general population. These studies provided the basis for identifying the immediate and short-term behavioral effects of the accident on the general population and several important groups within it.

METHODS

To be of use for purposes of the commission, the studies conducted by local researchers had to be suitably focused and expanded. The general

*Adapted from *Report of the Task Group on Behavioral Effects to the President's Commission on the Accident at Three Mile Island.*
†Department of Psychiatry, Columbia University, New York, N.Y. 10032.
‡Columbia University School of Public Health, New York, N.Y. 10032.
§University of Florida College of Medicine, Gainesville, Fla. 32601.
¶Pennsylvania State University, Hershey, Penn. 17033.
‖Capitol Area Health Research, Inc., Columbia, Md. 21045.
**Graduate Center, City University, New York, N.Y. 10036.

0077-8923/81/0365-0159 $01.75/2 © 1981, NYAS

TABLE 1

COMPLETED SAMPLE SIZES AND COMPLETION RATES ACCORDING TO TIME OF STUDY, PLACE OF STUDY, AND TYPE OF RESPONDENT*

Dates in 1979	General Population: Male and Female Heads of Household within 20-Mile Radius of TMI	Mothers of Preschool Children Sample from Birth Announcements in Harrisburg Newspapers	Mothers of Preschool Children Sample from Birth Announcements in Wilkes-Barre Newspapers	7th, 9th, and 11th Graders in Lower Dauphin	Clients of Community Mental Health Centers
Prior to 3/28			No studies in this period		
3/28 accident–4/10 reopening of schools			No studies in this period		
4/10–4/30	50 (0.67)†	—	—	—	—
5/1–5/31	54 (0.67)†	165 (0.79)‡	—	632 (0.91)	—
6/1–6/30	—	—	—	—	—
7/1–9/5	380 (0.65)	260 (0.79)‡	328 (0.66)	—	198 (sample of convenience)

*Percentages in parentheses indicate completion rates for each sample; usually about one-half to two-thirds of those not obtained were refusals.
†Overall completion rate for April and May combined.
‡Overall completion rate for May, July, and August combined.

strategy of the task force was to locate studies of high-risk groups in the general population and to seek control groups from whom comparable data could be collected. Each comparison was selected to provide strong clues as to the mental health and behavioral effects between the time of the accident in late March and early April and the time of last data collection in July and August.

Respondents

TABLE 1 presents a description of the people and places studied and the times of the various data-collection operations. The samples of male and female household heads from the general population and the special samples of mothers of preschool children were drawn at different times in April. They were also selected in such a way that the effects of distance from Three Mile Island could be analyzed. In the TMI area, the population within a 20-mile radius of TMI was sampled.

Strict probability sampling procedures were used in the study of the general population to select households at random (April and May samples) or by place-stratified random sampling from telephone directories (July–August sample). In Pennsylvania, a minimum of 90% of the population have telephones, so no marked bias should have been introduced by this procedure. Unfortunately, in the telephone sample, there was no prior designation of whether the male or the female head of the household was to be interviewed, and females are overrepresented in the resulting sample as a consequence.

The mothers of preschool children were also selected by strict probability sampling procedures. This time, however, the source was listings of birth announcements in the Harrisburg and Wilkes-Barre newspapers dating back to February 1977 and continuing through June 1979. The first sample of mothers from the TMI area was drawn in May, and they were interviewed by telephone. Later samples were similarly selected and interviewed during July and August in both the TMI and Wilkes-Barre areas.

The procedures for selecting respondents were different in the other studies. The study of teenagers involved pupils in the 7th, 9th, and 11th grades of the Lower Dauphin School District. All classrooms participated in the study.

The procedure for selecting the clients at community mental health centers was again different. Here, we focused on neither a whole population of patients nor a strict probability sample of such a population. Rather, the clients interviewed were a sample of convenience, consisting for the most part of persons with chronic mental disorders who were available, willing, and able to be interviewed by telephone or in person. They provide a criterion group whose responses indicate what constitutes a high degree of demoralization.

Household Heads in the General Population

The typical member of the large general population sample interviewed in July and August is female, married, and a high school graduate who did not finish a four-year college. Less than 15% of this sample did not finish high school, and only slightly over this percentage finished four years of college or more. About a third live within five miles of TMI and 14% have preschool children.

This July–August sample was stratified in such a way as to overrepresent less educated persons and persons living within a 5-mile radius of TMI. The smaller samples drawn in April and May were random samples of the population living within 20 miles of TMI, and males were systematically alternated with females in the households selected. The result is that males and females are almost equally represented. The educational level of these early samples is higher than that of the later sample; 49% of the April sample and 40% of the May sample are college graduates. Smaller proportions of the early samples, 16% in the April sample and 13% in the May sample, lived within 5 miles of TMI. Slightly older on the average than the July–August sample, these earlier samples of household heads included slightly smaller proportions with preschool children. In general, the April and May samples are highly similar to each other in demographic characteristics but differ as indicated from the larger sample of household heads interviewed in July–August.

Mothers of Preschool Children

The samples of mothers of preschool children are, of course, much younger than the samples of household heads from the general population. Their main difference from the Wilkes-Barre mothers, similarly self-evident, is that they are roughly 80 miles nearer to TMI. The TMI mothers of preschool children have somewhat higher proportions of graduates of four-year colleges than does the Wilkes-Barre sample. There appears to be little difference in the demographic characteristics of the TMI mothers of preschool children interviewed in May and those interviewed in July–August.

Seventh, Ninth, and Eleventh Grade Students

The sample of students ranges from 12 through 18 years of age. The sample of 632 students includes 27% 7th graders, 36% 9th graders, and 37% 11th graders. The sample is fairly well balanced for sex, with 56% females and 44% males responding to the classroom-administered questionnaire. Over half of these students come from households in which the father has completed high school. About a third have fathers who have not completed high school, and about 20% have fathers who have had one or more years of college. About one-third of the students live within

5 miles of the TMI plant, one-half live between 6 and 10 miles from the plant, while the remaining 20% live 11 to 30 miles from the plant.

Measures

Main Measures of Threat for the General Population and Mothers of Preschool Children

At 11 A.M. on Friday, March 30, Governor Thornburgh advised pregnant women and preschool-age children to leave the area within five miles of TMI. He reaffirmed this advice at a nationally televised press conference at 10 P.M. No comparably authoritative definition of the chief targets of the threat was made before or after this message. Accordingly, the two major measures of threat that we emphasize are (1) living within five miles of TMI, and (2) having one or more preschool children. In so doing, we are not implying that Governor Thornburgh created a threatening situation; rather, we are suggesting that his statement provided structure and focus in an ambiguous situation.

Note that we accepted the respondent's report of the distance of his or her home from TMI. A survey conducted for the Nuclear Regulatory Commission (NRC) found that some people who live more than five miles from TMI reported themselves as living within five miles.[2] If this error occurred in our survey, it could inflate relations between distance from TMI and mental health effects if, in addition, those who were most upset and otherwise affected were also most likely to underestimate the distance of their home from TMI. We do not have the information that would enable us to check whether there were errors in our respondents' estimates of the distance of their homes from TMI. At the same time, a consistency that we will discuss later between the NRC results and ours concerning the proportion of people living within five miles of TMI who left the area argues against assuming gross misreporting by our respondents.

Main Measures of Threat in the Study of Seventh, Ninth, and Eleventh Grade Students

Three main threat factors were identified as having potential for elevating psychological distress and physical symptoms in the teenagers. Two of these threat factors are the same as those identified in the previous samples: (1) living within five miles of the power plant, and (2) having one or more preschool children in the household. The third threat factor is whether or not they left the area during the accident. In the studies of the general population and mothers of preschool children, our approach was to examine the factors that influenced whether or not they left the area. However, leaving or staying in the area was a matter over which the teenaged subjects probably had little control. Therefore, the act of leaving or staying in the area is taken here as an additional characteristic of the TMI incident for these young people. The question being posed is whether or not temporarily leaving their homes served to increase or decrease the amount of stress these young people experienced. Conceivably, it could go either way.

Measures of Mental Health and Behavioral Effects
in the General Population and Mothers of Preschool Children

One of the most prominent behavioral effects was leaving the area. We were able to develop other measures as well from the interview and questionnaire material gathered in the studies of the general population and mothers of preschool children:

1. Recall of the personal "upsettingness" of the accident at the time it occurred.
2. Demoralization.
3. Perception of the threat to physical health.
4. Attitude toward continuing to live in the TMI area.
5. Attitude toward nuclear power in general and TMI in particular.
6. Attitude of trust or distrust toward authorities.

A full description of these scales is provided elsewhere.[3] This description includes the internal consistency reliabilities of the five multiquestion scales for which they could be calculated in the various samples from the general population, the mothers of preschool children, and the clients of community mental health centers. All reliabilities are at least adequate for research purposes.

We used the scores on the demoralization scale of clients of community mental health centers as an indicator of the points at which that scale indicates severe demoralization. Because of differences in the way men and women express their feelings, our procedure was to call scores above the mean for male mental health center clients an indication of severe demoralization in male respondents in general and scores above the mean for female clients an indication of severe demoralization in females in general.

Measures of Mental Health and Behavioral Effects in the Study
of Seventh, Ninth, and Eleventh Grade Students

Toward the end of the questionnaire, students were asked to rate each of the following on a five-point scale: (1) worry, (2) concern, (3) disturbed, (4) anxious. They made these ratings first for how they had felt at the time of the accident and, on the next page, how they felt since the accident. Thus, we have a self-perceived distress measure.

Students were also provided with a list of 10 physical symptoms, such as sore throat and sleep problems, and asked to check any ones they may have had during the time of the accident, March 28th through April 11th.

We summed the number of checks to arrive at a symptom score for each student. A full description of these scales is provided elsewhere.[3] The first is a measure of psychological distress, the second a measure of psychosomatic distress.

Strategy of Data Analysis

We were concerned with assessing mental health and behavioral effects as they varied with the threat factors at the time of the accident at the end of March and during the course of the five months that followed. We did not have an ideal situation for doing so in that we had no preaccident baseline on any of our measures of effect nor did we have perfectly matched control groups that were not exposed to the threat. Moreover, we do not have repeated post measures on the same respondents at various times during the months following the accident. Fortunately, however, we have been able to select meaningful comparison groups, such as the Wilkes-Barre mothers of preschool children, a place quite far away from TMI. And also fortunately, the investigations that we are relying on are based on probability samples interviewed at different times after the accident, so that we could piece together which effects were immediate, which were dissipated, and which remained strong.

To conduct the statistical analyses, we used a general linear model that allowed us to assess the effect of one factor while holding the other relevant factors constant. Thus, for example, when we report an effect due to distance of a person's home from TMI, we have controlled for having a preschool child in the family and for various characteristics of the person—such as age, sex, marital status, and level of education—that might have been confounded with distance of the person's home from TMI. The particular procedure has variously been called dummy variable multiple regression analysis and nonorthogonal fixed effects analysis of variance.

All of the effects reported were found to be statistically significant at the 0.05 level or better, using one-tailed tests. Because of the large number of tests that were conducted and the lack of independence of the behaviors and attitudes studies, the true probability of type 1 errors may be somewhat greater than 0.05.

RESULTS

Studies of the General Population Living around TMI and of Mothers of Preschool Children

How Upset Were People at the Time of the TMI Accident?

On the average, people living in the 20-mile area around TMI rated the accident fairly high on an 11-point scale from least to most upsetting at the time. The midpoint on this scale is 5, and the average rating by these respondents was 7.4. As we would expect if people were indeed rating the extent to which they were upset at the time of the accident, rather than their current level of upset about the accident, there was no change in this average between earlier and later interviews.

Women reported being more upset than did men, and people under 65 years of age were more upset than were older people. However, all groups averaged above the midpoint on the scale.

Over and above these differences related to personal characteristics, people with a preschool child were more upset than were others. Furthermore, mothers of preschool children who lived in the area around TMI were more upset than were mothers living at a greater distance, in Wilkes-Barre. In general, while people in the area found TMI a relatively upsetting event no matter what their circumstances, the most upset were those who could infer from advice given about evacuation and safety precautions that they were in danger by dint of living relatively close to TMI or having a child in the vulnerable age range.

Who Left the TMI Area at the Time of the Accident?

On the basis of our study of the general population, we estimate that 52% of the people living within 20 miles of TMI left the area at the time of the accident, the majority of them on Friday, March 30th. As shown in TABLE 2, the proportion who left differed between men and women and by marital status, age, and education.

TABLE 2 also shows that, over and above these differences related to personal characteristics, the decision to leave was influenced by the distance of the person's home from TMI. Although the basis for the estimation differs, our finding that 62% of those living within five miles

TABLE 2

ESTIMATES OF PROPORTIONS OF PERSONS IN THE POPULATION LIVING AROUND THREE MILE ISLAND WHO LEFT THE AREA AT THE TIME OF THE ACCIDENT

Type of Person	Percent Who Left
Men	41
Women	57
Married	57
Not married	38
Less than 65 years old	53
65 or older	42
Not a college graduate	59
College graduate	50
Condition Related to TMI	
Home 5 miles or less from TMI	62
Home more than 5 miles from TMI	48
Preschool child in family	77
No preschool child in family	48

TABLE 3

ESTIMATES OF PROPORTIONS OF GENERAL POPULATION SAMPLE AND OF SAMPLE
OF MOTHERS OF PRESCHOOL CHILDREN LIVING WITHIN 20 MILES OF TMI WHO LEFT
ON EACH DAY DURING THE ACCIDENT

Day	Percent of Leavers	
	General Population Sample (13% have preschool children)	Sample of Mothers of Preschool Children
3/28	2.4	2.0
3/29	2.4	6.2
3/30	59.5	65.8
3/31	17.0	14.0
4/1	10.5	7.8
4/2	4.9	3.6
4/3	3.2	0.7
Percent who left	51.8	72.4
Percent who stayed	48.2	27.6

of TMI left is consistent with the estimate in the study done for the NRC that 66% of households within a five-mile radius of TMI contained at least one evacuee; the same study found that the proportion of households in which some members evacuated and others did not was small.[2]

The decision to leave was influenced not only by distance of the person's home from TMI but also by whether there was a preschool child in the family, presumably as a consequence of Governor Thornburgh's advice on March 30th that preschool children should leave the area within five miles of TMI. Further evidence of the impact of this advice is shown in TABLE 3. Of those in the general population who left, less than 5% left before March 30th and the majority, 59.5%, left on that day. TABLE 3 also shows that among the 72% in the sample of mothers of preschool children who left the TMI area, almost two-thirds left on the 30th.

How Demoralized Were People in the TMI Area?

Demoralization is a common distress response when people find themselves in a serious predicament and can see no way out.[4] Sometimes this level of distress can approach that shown by persons suffering from mental disorders.[5] On our measures of demoralization, the overall mean is 28.3 for clients of community mental health centers, most of whom were suffering from chronic mental disorders. For the female clients in our sample, the mean was about 30; for the males, about 25.

FIGURE 1 shows that, on the average, demoralization in the community never reached the level of severity of that in the clients of the community mental health centers. It was, however, far higher on the average in the sample interviewed in April, closely following the accident, than in the samples interviewed in later months, as FIGURE 1 shows.

Moreover, 26% of those interviewed in April showed severe demoralization, as indicated by scores above 30.46 for females and above 25.56 for males. These scores, let us emphasize, are like the scores of the more demoralized clients in our sample from mental health centers. In view of the stringency of this definition of what constitutes severe demoralization, the estimate of the proportion in the general population who were demoralized in April should be regarded as conservative. In May and later months, in contrast to April, 15% or fewer in the general population scored above the means for our male and female mental health center

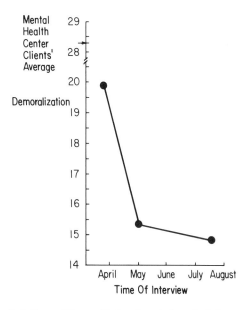

FIGURE 1. Relation of time of interview to level of demoralization.

clients. This difference between April and later months suggests that a substantial minority, perhaps 10%, experienced severe demoralization by the above definition at the time of and in the two or three weeks following the accident that was *directly attributable to the accident itself.*

Combining interviews across the entire period of the study, we find that the level of demoralization was higher among those living within 5 miles of TMI than among those living at a greater distance within the 20-mile area covered by the general population study. Almost a quarter, 22%, of those living within 5 miles of TMI scored above the mental health center clients' mean, whereas only 15% of persons living at a greater distance had demoralization scores this high.

Consistent with findings of previous studies, we also found that men and people who were currently married scored lower on the demoralization scale than did women and those not currently married.

Was the TMI Accident Perceived as a Threat to Physical Health?

On our multiquestion measure of perceived threat to physical health from the TMI accident and radiation, scores range from 1 to 3, with 2—the midpoint—indicating uncertainty. For the general population sample interviewed in April, shortly after the accident, the mean was 1.85. This level of perceived threat declined fairly steadily over the time during which interviews were being done, to 1.68. While some uncertainty remained, people were becoming more reassured.

Men and women differed, women perceiving more threat to their health than did men. People in different age groups also differed, the perception of threat generally declining with age. However, on the average, both women and younger people scored below the uncertainty point on the scale.

Over and above sex and age differences, those living within five miles of TMI were less certain that their physical health was not affected by the accident than were those living at a greater distance. This difference in opinions held by those living within five miles and those living farther away was found both in the general population and among mothers of young children living in the area of TMI. Realistically, mothers living still farther away, in Wilkes-Barre, felt less threatened on this count than did mothers living around TMI.

Attitude toward Continuing to Live in the TMI Area

In the general population, the average score on a measure of whether the individual devalued the area as a result of the TMI accident and would have liked to move away was just on the side of the uncertainty point, favoring remaining in the area. Scores on this multiquestion scale range from 1 to 3 with an uncertainty point of 2, and the average score in the general population was 1.90.

Men and women differed, with women holding more unfavorable attitudes, though still on the average favorable toward continuing to live in the TMI area. The attitudes of people in different age groups also differed. The youngest people, in their 20s, were the least favorable; the oldest, those 75 or older, the most favorable; and in between, there was a fairly regular increase in favorability with age. All but the youngest group, whose average was just above the uncertainty point, were favorable in general toward continuing to live in the TMI area.

People in the general population sample with a preschool child held more unfavorable attitudes toward continuing to live in the area than did those without a child in this age range. Their average score was near the uncertainty point rather than favorable. Reflecting the effect found in the general population of having a preschool child in the family, the mothers of preschool children in the TMI area who were sampled separately also had an average score near the uncertainty point, at 2.03.

Within the sample of mothers, those who did not graduate from college had a less favorable attitude, with an average score of 2.10, than did college graduates, whose attitude on the average was favorable. In addition, mothers living within five miles of TMI had a more unfavorable attitude than did those living farther from TMI. The average score in the latter group was at the uncertainty point, but the average in the former group was in the unfavorable range, at 2.37. In contrast with this difference, distance from TMI had a negligible influence on attitudes toward living in the area in the general population sample. Thus, the only people whose attitudes toward continuing to live in the TMI area tended to be negative were those who could infer from advice given at the time of the accident about evacuation and safety precautions that they were in danger on two counts, living relatively close to TMI and having a child in the vulnerable age range.

Attitude toward Nuclear Power Including TMI

In the general population living in the TMI area, the average score on the multiquestion measure of attitude toward nuclear power and restarting of TMI one and two was in the unfavorable range. Scores on this scale range from 1 to 3, with 2 being the uncertainty point; the average score in the general population sample was 2.23. Although comparisons from surveys using somewhat different questions can be hazardous, the results of a national poll summarized by Mitchell suggest that on the issue of nuclear power, people in the TMI area do not differ from the rest of the country in being uncertain and divided. In the national poll taken in May, 38% reported that they had not made up their minds, 36% described themselves as supporters, and 26% described themselves as opponents of nuclear power.[6]

Women in the TMI area reported more negative attitudes on the average than did men. Attitudes in the general population in the area also varied depending on whether the person had a preschool child, those with children in this age range having more negative attitudes on the average. Furthermore, in a sample of women with preschool children, those who had not graduated from college had more negative attitudes than did those who had graduated. Among the relatively favorable groups—men, people without preschool children, and college-graduate mothers of preschool children—only men had an average score indicating a leaning toward favorable rather than unfavorable attitudes toward nuclear power.

Trust in Authorities

Individuals' responses to our scale of trust in authorities, including federal and state officials and utility companies, covered the full range from complete trust (score of 1) to total distrust (score of 3). For the sample interviewed in April, the tendency, as FIGURE 2 shows, was to lean strongly toward distrust. This level of distrust appears to be higher than that found in national polls taken in April and early May.[6] Although the questions on this topic in these polls were not identical to ours, on most of them, somewhere between about half and a substantial majority of respondents gave the trusting rather than the distrustful response.

As FIGURE 2 shows, the level of distrust in the TMI area declined after April, but the decline was very gradual. The tendency as of July and August was still for opinions to lean, on the average, toward distrust.

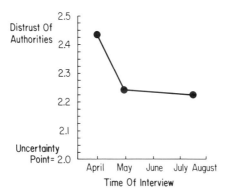

FIGURE 2. Change in distrust of authorities from April to August 1979 in general population in area of Three Mile Island.

Insofar as these results can be compared with the responses to somewhat different questions asked on the national polls, distrust in the TMI area seems to have remained above the national level. Elevation of distrust in authorities among people in the TMI area is also suggested by the finding that mothers of preschool children interviewed in the TMI area and in Wilkes-Barre differed, with the TMI-area mothers significantly higher on this measure.

Distrust was greater among women than among men. Comparing age groups, it was most elevated among people in their 30s, declining steadily with increasing age, but also lower among people under 30. However, both sexes and all age groups, on the average, scored above the uncertainty point on the measure, indicating a tendency to distrust authorities concerning nuclear power.

Study of Seventh, Ninth, and Eleventh Grade Students in the Lower Dauphin School District

The teenagers in the 7th, 9th, and 11th grades in Lower Dauphin were studied at the end of May, as TABLE 1 shows. No surveys were conducted with samples of these students either before or after that date. It is necessary, therefore, to rely on the students' recall of their distress and symptoms at the time of the accident by contrast with how they have felt since the accident more so than in the studies of adults in the general population. Nor do we have contrast groups as in the study of mothers of preschool children. Focus, therefore, is solely on contrasts in threat associated with living within five miles of TMI or living farther away, having preschool siblings or not, and being in a family that left the TMI area during the crisis by contrast with being in a family that stayed.

How Much Psychological Distress Did the Students Experience during the TMI Accident?

On the scale combining reports of worry, concern, disturbance, and anxiety, scores could range from 1 signifying psychological well-being to 5 indicating maximum psychological distress, with 3 the neutral midpoint. The students on the average reported a score of about 3.25 for the time of the accident. Moreover, a quarter of them scored 4 or more on this scale.

There was no difference between 7th, 9th, and 11th graders in level of psychological distress during the accident. However, somewhat higher levels of distress were reported by students living within five miles of the power plant. An even higher level of distress is found when we look at students who have a preschool sibling. They averaged about 3.75; those who do not have a preschooler in the home averaged around 3.12.

We also found a greater level of psychological distress for those students whose families left the area. They averaged around 3.50, while those who did not evacuate the area averaged around the neutral point, 3. We also found that females reported higher levels of concern in comparison to males, 3.50 and 2.75 respectively. Hence, during the accident, students in general tended to experience some psychological distress, and distress tended to be more pronounced for students in the more threatening circumstances.

How Distressed Did the Students Feel in the Period since TMI?

Students were also asked about their level of worry, concern, disturbance, and anxiety since the TMI accident. This second measure of

psychological distress, like the first, has a neutral point of 3 and ranges from 1 to 5.

Overall, the average level of distress as experienced since the accident was approximately 2. This value did not differ across the three grade levels. Furthermore, there was a sharp drop in the level of distress within two months of the occurrence of the accident for students in all three grades. The assurances that have come from authorities apparently helped to reduce these teenagers' psychological distress over the accident.

However, there are two groups of students for whom this dissipation of distress was not quite so clear. When we compare the group of students who have a preschool sibling with those who do not, we find that their level of distress had not decreased to the student average of 2. Instead, they scored just over 2.25. Similarly, we found that having left the area during the accident reduced the dissipation effect such that for those who left, the average concern since the accident was also at 2.25, compared with those who did not leave, whose average score was 1.75. It is interesting to note that those who stayed in a potentially hazardous area showed significantly lower levels of psychological distress both during the accident and in the two months following the accident, compared with those who left the area.

The sex differences noted earlier continued to emerge, with females scoring higher levels of distress since TMI, compared with males.

Was Distress Accompanied by Somatic Symptoms?

The students were asked to report the occurrence of any of 10 symptoms commonly associated with stress in youngsters. An additional category of "other" was included in case they had experienced a somatic problem not included in the list of 10. They were asked to report the occurrence of these symptoms during the time from March 28 to April 9. We simply summed the number of symptoms checked to compute a scale score for each student. Thus, the scale values could range from 0 through 11. In fact, no student had a somatic symptom score above 8.

The average number of symptoms reported by the entire sample was 1. This small average value is characteristic for all group comparisons we made. Our correlations do indicate, however, that as psychological distress increased, so did the number of somatic symptoms experienced, but that high levels of somatic symptomatology were relatively uncommon. We found small, but significant, elevations in the number of symptoms reported by two of the groups at high risk. Students living within five miles of the TMI plant and students who left the area reported more symptoms, compared with those who lived farther away and with those who did not leave the area. Somatic symptoms were especially prevalent in females and in youngsters in the lower grades.

CONCLUSIONS

We conclude, first, that the accident at TMI had substantial immediate psychological effects on the people living in the area, including both adults and teenagers. The majority of families living within 20 miles of the plant left their homes temporarily, and a substantial minority were extremely demoralized in the month or so after the accident. These responses were related to real threats involved in the accident, as defined for the public by the authorities. Among both adults and teenagers, demoralization or other symptoms of psychological distress diminished within about two months after the accident.

The second major conclusion is that the accident at TMI has had a lasting impact on the population of the area in terms of their distrust of authorities with respect to nuclear power. Their level of distrust five months after the accident exceeded that reported for national and regional samples immediately after the accident. Mothers of preschool children living within 20 miles were also more distrustful five months later than were comparable mothers living in Wilkes-Barre, about 80 miles from TMI.

Meanwhile, as the president's commission concluded in their report, there is a long term physical threat: "The process of recovery and cleanup presents additional sources of possible radiation exposure to . . . the general population."[7] Although we cannot predict exactly how the long-term psychological consequences of the accident will interact with this continuing threat, we must assume that responses of the population living in the area around TMI to the process of recovery and cleanup will be shaped by pervasive distrust of the authorities responsible for that process.

REFERENCES

1. Ad Hoc Population Dose Assessment Group. 1979. Population Dose and Health Impact of the Accident at the Three Mile Island Nuclear Station. (Preliminary Assessment for the Period March 28 through April 7, 1979.): Table 3-2. Food and Drug Administration Bureau of Radiological Health. Rockville, Md.
2. FLYNN, C. B. 1979. Three Mile Island Telephone Survey: Preliminary Report on Procedures and Findings. U.S. Nuclear Regulatory Commission. Washington, D.C.
3. Public Health and Safety Task Force. 1979. Report No. GPO 052500300732-1. U.S. Government Printing Office. Washington, D.C.
4. FRANK, J. D. 1973. Persuasion and Healing. Johns Hopkins University Press. Baltimore, Md.
5. DOHRENWEND, B. P., B. S. DOHRENWEND, M. S. GOULD, B. LINK, R. NEUGEBAUER & R. WUNSCH-HITZIG. 1980. Mental Illness in the United States: Epidemiologic Estimates of the Scope of the Problems. Praeger Publishers. New York, N.Y.
6. MITCHELL, R. CAMERON. Public response to a major failure of a controversial technology. In Accident at Three Mile Island: The Human Dimensions. D. L. Sills, C. P. Wolf & V. P. Shelanski, Eds. Westview Press. Boulder, Colo. (In press.)
7. KEMENY, J. G. et al. 1979. Report of the President's Commission on the Accident at Three Mile Island. U.S. Government Printing Office. Washington, D.C.

THREE MILE ISLAND: A RESIDENT'S PERSPECTIVE

Anne D. Trunk*

*President's Commission on the Accident
at Three Mile Island*

Edward V. Trunk

*Department of Mechanical Engineering
Pennsylvania State University
The Capitol Campus
Middletown, Pennsylvania 17057*

A resident of Middletown, Penn., can be classified either as being too close to the forest to see the trees, or as being close enough to an historical event to see details that others miss. This paper presents the perspective of a family that lived within the critical five-mile region of Three Mile Island (TMI). The family is unique perhaps in that it had lived in Middletown for 10 years (long enough to witness the construction phases of both Units 1 and 2). Also, the family's head is a mechanical engineering teacher who comprehends the terminology of the industry. Finally, the family has a preschool child but remained in the area until the crisis was officially over. Each of these factors played its role in developing a reliable source of news and in making possible the best interpretation under the circumstances. The six-month period of service on the presidential commission following the accident provided access to a wide range of information that the typical resident never will see. Therefore, the authors were afforded a unique insight into all events surrounding the accident.

HISTORICAL

In 1922, the Metropolitan Edison Company installed the first of three electric generator units in Middletown's new Crawford Station on the Susquehanna River. It was the first station in the United States to use pulverized coal and would eventually have a capacity of 116 MW. Middletown was receptive to this plant on its riverfront and engaged in a 99-year contract to buy the city's power needs at a set bulk price and distribute the power to its own residents. That system has provided Middletown with reliable cheap power for over 60 years. However, three events caused major changes in this relationship between utility and municipality.

Stack emissions became progressively more disturbing as both power output and fuel consumption increased. In 1946, two high-pressure boilers and a 3-MW topping turbine were added to the plant, increasing

---

*Address correspondence to 143 Race Street, Middletown, Penn. 17057.

0077-8923/81/0365-0175 $01.75/2 © 1981, NYAS

the capacity from an original 70 MW to 113.5 MW. The soot problem caused residents to file complaints. Cleaning exterior surfaces of cars, benches, and even sidewalks became a daily necessity. It was risky to leave laundry out to dry for extended periods. The complaints grew in number. Eventually petitions were circulated to clean up the stack exhaust. During the 50s, such a modification to the plant appeared technologically difficult and expensive. To the residents, it appeared that the utility was insensitive to their problem—a first sign of disharmony. Of course, it must be recognized that residential heating units were also predominantly coal fired at this time. The residents were partners in this problem.

The second event of significance occurred in 1970, when Met Ed converted the six low-pressure boilers to oil in response to both local and federal government pressure to clean emissions. The Department of Environmental Resources and the Environmental Protection Agency (EPA) had recommended the complete conversion to oil. Unfortunately, before the job was completed, the energy crunch hit. The government reversed its position and recommended leaving the remaining two high-pressure boilers on coal. The utility had already invested nearly $1 million for this irreversible change to eliminate the air pollution problem. It represents an epitaph to futility in government regulation. The conversion raised plant capacity to 116 MW. The generating station was now consuming 600 tons of coal and 168,000 gallons of oil each day. This was the time when oil prices slowly began to increase. For residents of Middletown, the soot problem ended. To Met Ed, the added costs meant that they were selling Middletown electricity at a loss.

The final historical event affecting the community-utility relationship was the construction of the TMI nuclear-fueled generating station. This gigantic project represented the equivalent of nearly 15 Crawford Stations. Numerous hearings were held each time one of the many necessary licenses was considered. Antinuke groups started to make their presence known at these hearings. Construction began on August 1, 1967. The utility did make an effort to sell the TMI concept to the public. On August 20, 1972, a picnic-style open house was held on the island, complete with guided tours, refreshments, and souvenirs. No one can truthfully say that they did not have an opportunity to learn about TMI and nuclear power prior to the accident. An observation tower and visitor's center were constructed directly across from the island to provide residents and sightseers with written and visual information. Tours were available to groups during the construction stage.

Unit 1 went on line on September 2, 1974, and Unit 2 followed suit on December 30, 1978. Both incurred cost overruns due to labor problems, inflation, etc. Public tours of the island ceased due to the Nuclear Regulatory Commission (NRC) regulations. To the utility, this investment would mean an initially higher electrical cost rate that would level out over the years as other fuels escalated in price. Met Ed would have been at the low end of the electric utility cost structure in Pennsylvania in the

1980s. To the people surrounding TMI, the utility appeared to be too sure of itself on safety and on having prepared for all possible occurrences. This "nothing can happen" attitude must have grown out of all the pressures from antinuke groups and from the licensing red tape. Residents labeled the attitude "cocky." They remembered this attitude and were later unforgiving when the utility could not satisfactorily explain the accident at TMI.

These historical events would later play a significant role in the relation between the utility and the public. Lack of candor, education, and information—all are essential ingredients of credibility and information gaps. The utility, public, NRC, and news media all fell short in the role they should have played to prevent the real accident of TMI—the psychological damage.

Meanwhile, the Crawford Station had become a liability. Oil became too expensive, and attempts to clean up the coal stack emission fell short of meeting the new standards. Met Ed was now incurring monthly fines paid into the EPA Clean Air Fund. The plant was retired in 1977 without public notice.

The Accident

The public is now well informed on the details of the accident that occurred at TMI Unit 2 on March 28, 1979. It is ironic that the real crisis existed on Wednesday only. The public did not react to the danger until Friday. Some of the main events of each day, and public reaction to them, are reviewed in this section.

March 28th—Wednesday

The accident occurred at 4 A.M., involving equipment failure, a core scram,† a heat-disposal problem, and incorrect operator management of the cooling system controls (later to be attributed to misinterpretation of gauge readings brought about by a sequence of events not previously encountered or trained for).

To the residents of the area, this day passed without concern, even though a general emergency had been declared by 7:24 A.M. The news arriving later in the day did not appear alarming, due to communication gaps and the early hour when the accident began. The public had not paid much attention to TMI happenings before March 28th. There had been some previous problems with Unit 2, but each time, the trouble had been rapidly attended to and the public reassured that things were under control.

---

†Scram means the rapid shutdown of a nuclear reactor by dropping control rods into the core to halt fission.

Few residents noticed on Wednesday morning that none of the four cooling towers were emitting the familiar steam plumes. Unit 1 was undergoing refueling and would soon be back on line. Only two towers had been operating before the accident. Unknown to the public, Unit 2 operators were frantically struggling with a scrammed reactor that, unknown to them, was losing primary cooling water through a stuck-open pressure-relief valve. The loss of this radioactive coolant ended at 6:22 A.M., when a backup block valve was finally closed. Through an unpredictable sequence of errors, the core was damaged and over one-third of the reactor cooling water was lost. The damage was aggravated through compounded mismanagement of the cooling system for another two hours. At about 10:30 A.M., the core was again fully covered and the damage stopped. By this time, Met Ed had already declared a general emergency at 7:24 A.M. due to high radiation levels in the containment building. State authorities had been alerted, and an NRC team was on site at 10 A.M.

This rapid sequence of early morning events underscores how much can happen close by a person without his knowing. Even when community leaders received word that morning, the townspeople were not alerted. Radio must be credited as the first news source to let the public in on the problem. Few caught it, and it did not sound serious to those who did. Radio would continue to be the best local news source. On March 28th and 29th, the utility followed their normal approach in handling communications—a faith that the plant was still in control and would soon be brought to cold shutdown. Jack Herbein, Met Ed's vice president of generation, became the main spokesman for TMI when he arrived from Philadelphia at 11 A.M.

As the news of an emergency at TMI trickled out, the news media started to tap many sources of information—not all knowledgeable of the details on what was occurring. Some conflicting reports were given. Odd lines of communication developed. Harrisburg's Mayor Doutrich learned about the emergency in a telephone call from a radio station in Boston. Middletown's Mayor Reid telephoned Met Ed's home office in Reading and was assured that no radioactive particles had escaped and that no one was injured. A few minutes later (11 A.M.), Mayor Reid heard a radio newscast stating that there had been radioactive particles released. At 4 P.M., Reading called back and told the mayor that there had been a slight radioactive release.

Perhaps the most dramatic news of the day was heard on the televised "CBS Evening News," when Walter Cronkite informed the nation that we had just taken "the first step of a nuclear nightmare." Just before this newscast, state and local officials had reassured the TMI community that there was no danger to public health. In his newscast, Cronkite mistakenly reported that the containment building's reinforced concrete wall was three-feet thick, when it was actually four. He then declared that a strong, deadly radiation was passing through this wall and being picked up a mile away. This technical mismanagement of the news was typical of the inaccuracy and sensationalism that the community would be subjected to in the days that followed. Distant news was always more depressing than was local. People's hopes would be successively raised and then dashed by conflicting news reports. TMI would later be labeled the most overrated and overplayed story in the history of American journalism.

The day ended for Mayor Reid as he walked the streets of Middletown with a Geiger counter to double check on his radiological team, which had found nothing earlier. Mayor Ken Myers of Goldsboro, located 1½ miles west of the plant across the Susquehanna River, ended his day by going door-to-door to talk with residents of his community about evacuation plans.

March 29th—Thursday

By Thursday, most of the residents around TMI assumed the crisis was over and were not expecting to hear any more alarming news. Talk shows carried interviews on TMI. Congressmen visited the plant for an update. The news media started assembling in larger numbers at press conferences. The typical reporter dispatched to Middletown was not prepared to interpret this type of accident, where no death or injury, no fire, water, or visible wreckage was involved. They took pictures emphasizing the cooling towers, and to this day, many still believe that radiation (rather than harmless water vapor) emanates from cooling towers. Science newswriters were not being utilized properly. The technical barriers that existed contributed to inaccurate and incomplete news reports.

Although some progress was being made in the plant, two additional events raised concern on Thursday. Another radioactive gas release within NRC limits occurred during the afternoon. Also in the afternoon, the utility began dumping about 400,000 gallons of wastewater, which they had been storing since the accident began and which were now close to overflowing storage tanks. This runoff water from showers, toilets, laundry, etc., contained a slight trace of radioactivity. The water was discharged with NRC knowledge. Both of these releases were not announced to the public beforehand. A political overtone became evident later when it was learned that the NRC stopped the wastewater discharge into the Susquehanna River at 6 P.M., passing the buck to the state for a resumption order. The order to resume dumping finally came down at midnight, after much debate about the wording of a press release.

On Thursday, Jack Herbein admitted that some damage to the core might have occurred. By day's end, residents were being told that the plant was stable and soon to be in a state of cold shutdown (Jack Herbein), that the plant atmosphere was one of "calm competence" (Lt. Governor Scranton), that there was "no cause for alarm nor any reason to disrupt your daily routine" (Governor Thornburgh).

March 30th—Friday

Friday was the day of crisis *off* the island. A guarded optimism about TMI was dashed unnecessarily by false rumors and some wrong decisions. It was the *only* day of panic, and none of what followed need have happened. There lies the tragedy.

The morning contained some wild rumors about an early venting. Radiation levels were not reported generally until Friday. Therefore, the 1,200-mR/hour helicopter reading was misinterpreted to the residents as a serious danger. No one compared this with the previous day's 3,000-mR/hour release concentration over the plant, which also had dispersed to a harmless dilution. These rumors were spread by the news media and by telephone. They gave the impression that the situation had deteriorated and that the plant was in trouble. The final trigger of panic had to be the poorly conceived evacuation order that originated at NRC headquarters in Bethesda, Md., and which was countermanded by Governor Thornburgh. The governor would not know later when they met that the NRC director who issued that order was Mr. Harold Denton. News of a possible evacuation leaked to the public and remained as a threat until the following week.

By 11 A.M., when Jack Herbein was having his tragic "loss of credibility" press conference, parents in Middletown were already pulling their children out of school. A blind panic had begun. The press was saying that Met Ed was not telling everything. Fears of the unknown were heightened. The demand to know "everything" would produce technical information that the press could not understand or interpret for the public.

The use of Jack Herbein as a TMI spokesman was an unfortunate choice. The utility lost the technical services of one of its sharpest nuclear engineers. He was not trained for the role of dealing with the press and had no patience for explaining details that appeared to him already to have been presented adequately. The press corps was treating him as a hostile witness. When they knew something about the 1,200-mR/hour reading about which he actually had not been informed, a communication gap arose. He was considered to be withholding information or lying. One year later, the press was asked to comment on the possibility that the media pressure on key people in a crisis might actually impede progress. The consensus was that "that is one of the prices we must pay for the public's right to know."‡ There is certainly a lesson in this for the future.

At noon, a TMI forum was held in the auditorium at the Pennsylvania State University, Capitol Campus, in Middletown. This was a response to requests for information from students and parents. The engineering and

‡Press panel response at a Pennsylvania State University symposium on TMI held on March 28, 1980, in Middletown, Penn.

technology faculty put the rumors of high radiation levels to the test (an engineer is trained in the scientific method). A radiation survey meter, even when pointed directly at TMI, indicated *no radiation* level. The rumors were proven false! By 12:30 P.M., Governor Thornburgh made his decision to advise pregnant women and preschool-age children to leave the area within a five-mile radius of TMI. He was also closing the schools in this area. The provost of Capitol Campus relayed the order that the campus would close until further notice, smiled, and shrugged his shoulders. The TMI forum ended, and students began leaving the campus. It was not clear why.

Citizens were weighing the governor's words "in the interest of taking every precaution" and that his advice was in the spirit of "excess of caution." He also reported that Friday's readings were no worse than Thursday's. His statement was actually reassuring to the majority of residents, helping them in their decision to stay. The Trunk family weighed the trauma involved in evacuating against the risk in staying. The decision was tempered by knowledge of the plant's design, accumulated over the years of living near TMI. The decision to stay eliminated any possibility of inflicting panic or psychological stress on the six children. It was learned later that psychological damage to children generally originated in the way parents conducted themselves during the crisis. The residents of this area had survived two floods, tropical storm Agnes in June 1972, and tropical storm Eloise in 1975. The lower half of Middletown had been evacuated during the Agnes flood. Although TMI was different, the majority of Middletown's residents demonstrated an ability to cope with the crisis. About two-thirds of the population in town concluded there was no need to leave.

That afternoon, Harold Denton, director of the Office of Nuclear Reactor Regulation, arrived in Middletown to become the chief spokesman for the TMI operation. People within a 10-mile radius of TMI had been advised to stay indoors, close windows, and turn off air conditioners. The public was unaware that a noncondensible bubble was slowing down the cooling rate of the core and that the words "core meltdown" had entered conversations.

It took the evening national newscasts to shock the community again. Walter Cronkite announced on the "CBS Evening News" that as many as 50,000 people could die as a result of a catastrophic meltdown. From this point on, many families began comparing local with national newscasts. It was reassuring to receive the Thornburgh-Denton statement that night that the 10-mile advisory had been lifted and that no immediate evacuation was necessary.

As residents ended their day, most surrounding towns had initiated evening curfews. The people reflected on the mixed signals they had received. Logic and careful weighing of the news helped to discount many false reports. A measure of insecurity would call for close monitoring of all news sources in the days ahead.

March 31st—Saturday

On Saturday, some residents initiated their own sources of information, not trusting normal news channels. Telephone lines became overloaded. Relatives calling in gave grim tales of what they were hearing. For example, in Florida, the word was that the community was being airlifted by helicopter. It was reported somewhere else that Middletown was completely wiped out. Two nursing homes in Middletown—Frey Village and the Odd Fellows Home—actually were evacuated, but because the remaining staff was too small to provide services, not because of any danger from TMI.

Most residents went about their routine of Saturday shopping and tending to household chores while trying to keep abreast of the news. Radio provided the best source of hourly news updates. The news media had invaded the area and made their presence known to the people. The emphasis was to focus in on anyone who displayed emotion or fear. Although many calm residents were interviewed, these reports were not aired. At the Hershey evacuation shelter, the press outnumbered the evacuees.

Although the press now was focusing on the human aspect, there were still conflicting reports about progress at the plant. The new concern on Saturday was what to do to eliminate the huge hydrogen bubble in the reactor. Once again there were speculative reports of a potential explosion, a meltdown, even a nuclear bomb situation, and the need for a mass evacuation.

At Herbein's last news conference on Saturday morning, he announced that the hydrogen bubble had been reduced to two-thirds its Friday dimensions and that the crisis had ended. He also announced that Harold Denton would henceforth be spokesman for plant status. An hour later, Denton took issue with Herbein. The difference between the two experts was played up in the news. Herbein was correct, but unfortunately Denton's views carried more weight at this time. Even as late as Monday, when Met Ed was saying that the bubble had gotten so small it might have disintegrated, the NRC insisted on a retraction of the story. Then at his 11:15 A.M. Monday news conference, Denton conservatively reported that the bubble had shown a dramatic decrease in size. It was not till Tuesday afternoon that Denton let the world in on what had been evident long before, "the bubble has been eliminated."

The news media were hearing two or more opinions on each issue and were unable to make accurate assessments for the public. As mentioned earlier, science reporters should have been used for this assignment. Knowledgeable engineering faculty at the local university were also passed up. One lived directly adjacent to the island. Residents picked up the growing caution in Denton's responses. The situation always turned out better than what he stated. Yet the overall effect of Denton's presence was one of increased confidence. The people looked up to him. In retrospect, Denton had as much trouble in communicating with the press as did Herbein. Denton's televised press conferences were mini-education sessions and shouting matches. He did not always relay the whole story, as illustrated by the hydrogen bubble scare episode.

April 1st—Sunday

The hydrogen bubble was still on people's minds when they attended Sunday church services. The threat of a mass evacuation seemed to linger on.

The community's spirit returned to normal when President Carter and his wife visited Middletown and toured Unit 2, together with Governor Thornburgh and Harold Denton. Most residents concluded that if the president was allowed to enter the island, the crisis indeed must be over.

Confidence in the utility's technical ability was still being questioned. The hydrogen bubble had caught everyone by surprise, and it appeared that the utility was not exploring for a way in which to eliminate it. The NRC had no plan. The utility called in an advisory team of experts, so it appeared that they did not know what to do. A recombiner system for removing the hydrogen was installed on a crash basis and was reported to be doing well by Monday. Even with all the experts at TMI, the public did not get a picture of competence. The hydrogen bubble episode and the initial mishandling of the accident, particularly as they were reported during the crisis, were severe blows to the industry's credibility. The antinuke movement would claim that nuclear power is not safe and that the public had been lied to. They would demand guarantees no industry can make. A later study would report that Met Ed's elimination of the hydrogen bubble had been systematic and professional. But who got that picture on Sunday?

The Following Week

Schools within the five-mile radius of TMI remained closed the entire week. Other schools reopened on Wednesday. Some residents started returning to check on their homes and possessions. Middletown's Mayor Reid issued a shoot-to-kill order for police regarding looters. It is interesting to note that there was no crime of any kind reported in this area during the TMI crisis. Once the hydrogen bubble scare ended, the plant seemed to be on a sure course to cold shutdown. By week's end, most of the ones who had left had returned. They had received more frightening news reports than had those that stayed. Many had hastily withdrawn bank savings before leaving and now returned with angry feelings. Some vented this anger on those that had stayed. Others had taken advantage of the opportunity and made a vacation out of it at the expense of the nuclear insurance. They thought it was great.

EPILOGUE

This paper has presented an historical perspective on the relation between the utility and community around TMI. The environment that existed in Middletown during the accident was reviewed on a day-by-day basis. On April 27th, cold shutdown was announced. Out of the accident, four poles of attitude have emerged. The pro-nuclear energy group concludes that the nation's worst nuclear accident has proven the soundness of the industry. No damage external to the plant had occurred. The first survey of the area made immediately after the accident by Elizabethtown College, Elizabethtown, Penn., showed that 62% of 1,386 respondents supported nuclear power. The anti-nuclear energy group has been the most vocal and emotional in this area. They are well organized. Recent abusive outbursts at meetings have awakened a reaction from an otherwise silent majority. A third group is uncommitted. They still see some problems to be solved but are not ready to scrap nuclear energy till something better comes along. The fourth group is unconcerned about nuclear issues. Their primary concern is their electric bill. They are vulnerable to persuasion if the arguments are economic ones.

Discussions on the subject have been strained by emotion. This conflict even permeates marriages where husbands and wives hold different views. An irony exists in that many intellectuals who ran during the crisis have now returned to become "instant experts" on all that had happened. One year later, the educational process is barely beginning.

This costly accident continues to provide much new knowledge to the industry. We are learning daily about what it takes to clean up a crippled nuclear plant. Residents of Middletown have a vested interest in this plant's survival. The majority of residents appear determined to see this history-making operation through to a satisfactory conclusion.

ADDED EXTEMPORANEOUS REMARKS

I am going to depart from my written material to tell you of a personal concern. The people in the TMI area are emotionally on edge. Emotions are getting in the way of clear thinking. The people are wide open to fear. The media still are feeding us material that provokes fear. The antinukes are telling us constantly that we have something to fear. Even some so-called experts are telling us to be afraid. If you tell anyone to be afraid often enough, he will become frightened!

The people need factual data, not speculation; the "what is," not the "what if." The infant mortality scare is a current example of the harm that improperly reported data can do. When properly presented, the infant mortality rate actually was shown to have decreased in the TMI area, although a few days earlier, we had been told that it had increased.

As a resident of this community, I demand a professional job from everyone involved. No one gets off scot-free in my book. They all had a role in TMI. They all have something to learn: the utility, the government, and, yes, the press too. Even the people, because they had a chance to learn about TMI before the accident and did not take advantage of it. Anyone who claims to come out with a clean slate on this one is being naive.

A PREGNANT PAUSE: THE EUROPEAN RESPONSE TO THE THREE MILE ISLAND ACCIDENT*

Dorothy Nelkin and Michael Pollak†

Program on Science, Technology, and Society
Cornell University
Ithaca, New York 14853

"Pennsylvania is everywhere." So chanted demonstrators throughout Western Europe after the Three Mile Island accident. European public opinion was deeply affected by this event. In densely populated countries, many nuclear plants are near populated areas, and identification with the people in Harrisburg gave immediate and powerful impetus to the antinuclear movement. However, the official response to the accident and to the public concern about nuclear safety differed considerably from country to country.

The accident intervened in an ongoing debate over nuclear power in Western Europe, but a debate that was far more intense and more immediate in some countries than in others.[1-3] The reaction to the accident in each country varied, depending on the stage of internal controversy over nuclear technology at the time. Interpretations of the problems revealed by the accident also varied depending on the political and administrative issues that were most salient to protagonists in the specific policy context.

This paper analyzes the influence of the Three Mile Island accident on the European nuclear debate, as it followed from the political and institutional factors that shaped national policies toward nuclear power and its opposition.

THE IMPACT OF THE THREE MILE ISLAND ACCIDENT

In Germany the future of nuclear power was an active and important item on the political agenda when the Three Mile Island accident occurred. German courts, backed by provisions in the Atom Law, had imposed a *de facto* moratorium on nuclear power until the waste disposal problem was resolved. A major international expert review was in progress to evaluate the government's proposed waste disposal and reprocessing facility at Gorleben. Its decisions were key to the future of

*This research was supported with grants from the German Marshall Fund and the National Science Foundation EVIST program. Material was drawn from a forthcoming book by D. Nelkin and M. Pollak, *The Atom Besieged: Extra-Parliamentary Dissent in France and Germany* (see Reference 1).

†Present affiliation: Centre National de la Recherche Scientifique, Paris, France.

0077-8923/81/0365-0186 $01.75/2 © 1981, NYAS

Germany's nuclear program. Hoping to influence these decisions, anti-nuclear groups had planned a demonstration in Hannover against the Gorleben project. The Three Mile Island accident was a timely event; aroused by the accident, over 50,000 people turned out, giving impetus to the antinuclear movement just at a time when internal tensions over priorities and tactics had weakened its influence. The size of the Hannover demonstration had a decisive effect on the decision to stop the Gorleben waste disposal project. The prime minister of Lower Saxony, Ernst Albrecht, declared that his government supported nuclear power but it could not proceed in the face of such massive public opposition.

Albrecht's ambivalence reflected a growing skepticism about nuclear power in Germany. The courts had defied the government policy in a series of antinuclear decisions. The ecologists, maintaining an antinuclear platform, had several electoral successes that threatened to shift the balance of power within Germany's three-party system. Several leading politicians had voiced their doubts. The Liberal minister of interior, Gerhard Baum, questioned whether the nuclear program was reasonable in view of its risks. In this context of ambivalence, Three Mile Island had an important impact. First of all, the accident attracted very wide public attention and evoked a very cautious public response. Public opinion polls found that 82% of the adult population claimed to have followed the accident closely and everyone had heard of it. Twenty-four percent wanted to stop the nuclear program and to close the existing plants; 37% favored continued use of plants already in operation but no further development. Only 30% favored continued expansion of the nuclear program.[4] The accident also reinforced the growing tendency to shift the burden of proof to the promoters rather than the critics of nuclear technology. After the accident, Chancellor Schmidt reminded nuclear experts at a European nuclear conference that the industry owed the public reliable information about risks, and he warned that unless public fears and concerns were taken seriously, the loss of confidence could be profound. "In our democratic system based on public participation nuclear energy cannot be implemented without broad consent."[5]

In Sweden, too, the Three Mile Island accident reinforced an existing conflict over nuclear policy. The debate in Sweden had dominated political life for years, and the reaction of Swedish politicians reflected the political turbulence associated with nuclear policy. Before 1970, Swedish nuclear policy had developed with a consensus of the five main parties and the major interest groups. However, in 1972, the Center party—backed by a growing ecology movement—challenged government plans for a major nuclear expansion. By breaking the establishment consensus, the Center party provided an official channel for popular dissent. Subsequently, in 1976, divergence among the parties over the nuclear issue turned it into an electoral theme that was in large part responsible for the electoral failure of the Social Democratic government after 44 years in power. The new center-right coalition government, strongly influenced by the antinuclear Center party, initially imposed a moratorium on all new power plant construction. But only several weeks after coming to power, the government approved a plan to build a plant in Barseback, a heavily populated area near the Danish border. This became a major target for antinuclear groups. By late 1978, disagreement over the future of nuclear power split the center-right coalition. With potential support by Conservatives and Social Democrats and with opinion polls indicating increasing public acceptance of nuclear power, the Liberal minority government proposed a policy for further expansion of nuclear power. However, the Three Mile Island accident once again hardened public opinion against the technology. Surveys indicated an increase in opposition to nuclear power from 41% before the accident to 53% after, and a decline in support from 37% to 26%. The accident also revived the interest of Sweden's powerful environmental lobby, and protest groups demanded that Ringhals II, using a reactor design similar to Three Mile Island, be abandoned. Thus, decisions were postponed once again. Anxious to keep the divisive issue out of the 1979 parliamentary elections, the Social Democrats called for a special national referendum to be held in the spring of 1980. The referendum will include three alternatives, each backed by political organizations: the Center party and antinuclear movement's alternative is no new construction and a phaseout of the existing reactors within 10 years; the Social Democrat/Liberal party choice is six additional reactors, but with state and municipal control and with laymen constituting the safety committees; the Conservative party choice is six reactors with no state control. The two pronuclear positions include an eventual phase-out of nuclear power consistent with maintaining employment and public welfare. The government has allocated $8.5 million for information activities: $4.25 million for the Center party, $2.36 million for the Socialist/Liberal position, and $1.88 million for the Conservatives. In addition, the Federation of Swedish Industries and the Employers Federation are spending an estimated $4 million on a pronuclear public relations campaign. While the referendum is not binding, all parties have agreed to abide by its outcome.[2,6]

For several years, Danish and Swedish antinuclear activists had worked together to oppose the construction of the reactor at Barseback, and a few days after the Three Mile Island accident, 25,000 Danes and Swedes demonstrated at the site and demanded increased studies of nuclear safety. Denmark is one of the few European countries not yet developing a civilian nuclear power program. But the political parties had been debating the issue, and the government had put its nuclear energy proposal on the parliamentary agenda. After the Three Mile Island accident, the Social Democratic prime minister, Anker Jorgensen, withdrew the proposal and—like his Swedish counterpart, Olaf Palme—proposed a referendum in order to prevent splits in his party over the issue. The Danish referendum will be held in 1981 after intensive public discussions.

Several weeks before Three Mile Island, the Swiss held a referendum to assess a proposal that would give authority to approve the siting of nuclear plants to people living within a 20-mile radius of each site. Antinuclear activists had proposed the scheme, hoping that local resistance would block the development of nuclear power. The voters, by a very narrow margin (51%), supported continued federal government authority, thereby allowing continuation of the nuclear program. However, conscious of the strength of the opposition, especially after Three Mile Island, the government proposed to tighten its licensing procedures by requiring utilities to provide evidence of the need of the power plant and guarantees of a safe disposal of nuclear waste. In addition, the government proposed required parliamentary approval of the licensing procedures. Although this theoretically would remove power from a federal bureaucracy, environmentalists opposed the proposal as a tactical means to creating a false sense of security while allowing the nuclear program to proceed under what in fact would be minimal restraint. Voters approved the government proposal, but the opposition used the Three Mile Island accident as an argument to demand a new referendum.[7,8]

The British reactor program has faced a long history of indecision. Great Britain is the only country in Europe that had opted for an advanced gas-cooled reactor system (AGR) instead of a pressurized water reactor (PWR) design. The accident was thus assumed to have little direct relevance, and in addition, public attention at the time was focused on the national election campaign. However, the accident did have a direct effect on an ongoing dispute about a possible shift to the pressurized water reactor system. Design problems and conflicting business interests had muddled the AGR program. The British Department of Energy was urging a change in the technology for commercial reasons, and, in fact, the Central Electricity Generating Board was evaluating the Babcock and Wilcox reactor design as one potential partner for a demonstration pressurized water reactor program. Labor party minister of energy, Anthony Benn, however, had firmly supported the British AGR system, emphasizing the safety record. "Safety must be a top priority ranking above every other factor including the economics of operation."[9] After Three Mile Island, Benn could say: "I told you so. No government anywhere in the world can now widely approve of nuclear systems of that type."[9] Benn ordered a stop to work on the PWR; but as soon as Margaret Thatcher won the election, she strongly advocated a PWR program that would double the number of reactors already in operation or under construction. This program was based on a reactor similar in design to the Three Mile Island reactor. The boldness of this proposal so soon after the accident revived the antinuclear movement, which had been relatively quiescent following a period of intense activity at the 100-day inquiry over developing the Windscale reprocessing facility. The movement's ability to mobilize public opinion during the forthcoming public inquiry will influence the implementation of the new British nuclear program.

The Three Mile Island accident had less of an impact in other countries, where, for a variety of reasons, the nuclear controversy has been less intense. The reaction in France reflected the political management of a conflict that had been under way for about seven years. French public opinion polls after Three Mile Island indicated that 16% of residents living near nuclear plants considered the accident catastrophic; 62%, serious. Of those interviewed, 75% felt that despite the differences in the technologies, a similar accident could occur in power plants in their communities, but 80% believed that if such an accident did take place in France, the people would not be informed of the truth. Despite the striking lack of confidence in official information, the accident affected overall public opinion towards nuclear power in France less than in any other country. Forty-five percent of the population favored continuing the nuclear program, and 37% opposed it, a slightly more positive response than pre-Three Mile Island polls had indicated.[10,11] The accident, however, abetted the ongoing debate about the adequacy of public information and provided material for the ecology press, which reported the events in technical detail.

Ecologists suggested that commitment to nuclear power in France precluded consideration of safety. Responding to French government reassurance, Le Canard Enchaîné reminded its readers that officials had reassured the public after the New York City blackout, stating that this could not happen in France. Fifteen months later it had.[12] A French ecology magazine printed a mock interview:

Question: If the Three Mile Island reactor had killed three million people and contaminated ten thousand square kilometers of American soil instead of just being a source of worry, what would [Prime Minister] Raymond Barre have thought?

Answer: He would have thought it's very troublesome. So poorly informed are the French they would still assume that nuclear plants are dangerous.

Question: What would he say?

Answer: He would say there is no question of stopping the French programs. The lack of public understanding about nuclear power is a problem. If the people were correctly informed they would know that nuclear power presents absolutely no danger.[13]

Despite their cynicism, French ecologists could not mobilize around this event. The internal schisms after clashes with the police during the 1977 demonstrations, the consistent inability to influence nuclear policy in the courts or through administrative channels, and the electoral victory of the presidential majority in 1978 had left the antinuclear movement in a slump. This inability to mobilize allowed the government to ignore the scattered protests and to minimize the importance of the accident and its relevance to the French context.

In April 1979, immediately after the accident, French government officials announced that plans for rapid expansion of the nuclear program would proceed on schedule. A commission was sent to Pennsylvania to investigate the cause of the accident, and Giscard promised to use its findings to improve French nuclear safety procedures. Meanwhile, Prime Minister Raymond Barre announced on television that France used a different pressurized water reactor model than did the Americans and that in any case, "the event was more significant in its psychological impact than in the technical reality. . . . The scenarios in the United States could not happen in France. We have safety systems which take into account such accidents."[14] Similarly, officials from the Commissariat à l'Energie Atomique and Electricité de France argued that the standardization of reactor design, the concentration of efforts on a single technique, and centralized coordination in France all helped to avoid the problems of control so evident in the American system. And Minister of Industry Giraud saw "no reason to upset our program. We must remember that there were no victims despite the fact that the accident was the most serious that could ever happen."[15,16] The main problem in France, claimed officials, was one of public information; and indeed, beyond several technological changes in the instrumentation systems to improve reactor safety, the most important effect of Three Mile Island was a decision to publish the annual reports of the Radiation Protection Service and to progressively release evacuation plans.

The response in Italy was minimal because of the existing state of the nuclear program, described by an industry observer as a "pregnant pause" reflecting the "years of frenetic drift that have plagued all Italian institutions."[17] Italy's four small nuclear plants were over 15 years old. In 1977, the Parliament had approved a long-range energy plan calling for 12 light water reactors to be completed by 1990. As construction proceeded on 2 of them, local opposition provoked a general dispute over the power of the central administrative authority to control siting decisions. One of the new 860-MW plants was ending a testing phase when Three Mile Island occurred, and it was pulled out of operation for inspection for an indefinite period. Meanwhile a consultative commission including pro- and antinuclear representatives has been appointed to debate the future of the nuclear program.

Nuclear policy in Holland also has remained in a state of nondecision. Besieged by grass roots activism, the Dutch program to develop three 1,000-MW nuclear power plants was first delayed until the completion of safety studies in 1975, then until the end of the elections of 1977, and then until the result of a major national debate planned for 1980 and 1981. In this context, Three Mile Island had little policy impact.[2,18]

Similarly, the accident aroused only minimal political interest in Austria where, in a referendum in November 1978, the public had rejected the government plan to begin operation of its first nuclear power plant at Zwentendorf, near Vienna.[19] During the electoral campaign of spring 1979, the conservative opposition exploited the Three Mile Island accident to argue that the socialist government had been irresponsible in supporting nuclear energy. The government ignored this and started diplomatic efforts to reach bilateral agreements concerning the regulation of power plants planned close to the Austrian border.

ANALYSIS

Differences in the response to the Three Mile Island accident ranged from a *de facto* moratorium in Germany to an expansion of the nuclear program in France. The variation reflected the character of the internal debate in these countries. The accident, merely an intervening factor in a long-lasting controversy, simply served to reinforce existing trends. In those countries where the controversy over nuclear power was already very intense, the accident helped the critics of this technology to further dramatize their concerns. In Germany, the Gorleben international review was under way; in Denmark, Parliament was discussing the introduction of nuclear energy; in Sweden, the coalition had just split over the nuclear issue; and in Switzerland, a referendum had just taken place. In Great Britain, the accident influenced ongoing government debates over a possible shift to the pressurized water reactor technology. The accident had less impact in countries where nuclear policy for different reasons seemed to be immutable: in Holland, the decision to go nuclear had already been postponed; in Austria, people had decided against this form of energy; and in France, few people expect to be able to influence a policy determined by the centralized administrative structure.

The Three Mile Island accident only served to mediate a controversy that was already shaped by the differences in the political and economic contexts in which it had evolved. The state of the nuclear controversy has reflected several factors. The degree of energy dependence is clearly important: those countries relatively rich in primary energy resources— such as Holland with its natural gas or the North Sea countries with their oil—planned only small nuclear programs and were receptive to public concerns. Countries poor in resources have organized major nuclear programs and are understandably reluctant to cut back their initial plans. Yet in Germany and France, both heavily dependent on imported energy resources (Germany 58%; France 76%), the nuclear debate has evolved in quite different ways.

While the availability of resources has influenced nuclear power expansion, the character of the nuclear debate rather reflects two political factors: the structural relationships between governments and their nuclear industries, and the available channels through which critics are able to influence policy decisions. Public ownership characterizes the energy sector in all European countries; even where there are private utilities, public authorities, through shareholding, have a decisive stake in the promotion of nuclear power. The very size of past investment in the nuclear option tends to reduce government autonomy in this policy area. The pressure of plant suppliers on government policy is especially important in those countries with national nuclear firms. France, Germany, and Sweden have their own national nuclear industries, and their governments have taken direct economic responsibility for the nuclear sector through support of research and development and maintenance of long-term contracts with nuclear supply firms. The high capital costs of developing a national nuclear program brought close collaboration between these governments and their industries even before the decisions were made to commercialize nuclear technology on a large scale. Collaboration grew after 1974, when the sharp price increase of oil brought pressure to expand alternative energy sources. These three governments soon became the most active promoters of nuclear energy, and in each case, the nuclear program became a source of intense conflict. As criticism of nuclear policy developed throughout Europe, only a few countries—Austria, Denmark, and Norway—still had the option to renounce this technological choice. In these cases, the lack of commitment to a national nuclear industry and the relatively limited economic ties between government and nuclear plant supply firms allowed governments to maintain greater regulatory independence and responsiveness to public concerns.

More important in determining the intensity of the nuclear debate in different countries were differences in their political traditions and administrative arrangements: the official channels for conflict resolution (e.g., the court system); the possibilities for citizens to participate in administrative decisions at different government levels (e.g., the *Länder*, or states); and the openness of large political organizations to dissident popular demands.

In Germany, Switzerland, and Sweden, the existence of official channels through which the public could influence substantive policy decisions gave the opposition the means to delay the expansion of nuclear power and to sustain a level of conflict over time. In Germany, the ecologists had created a political climate that was relatively sympathetic to their cause. Then, backed by provisions in the Atom Law concerning the disposal of nuclear waste, they had been able to stop several projects through the courts. In addition, the dispersal of power in a federal system provided the possibility of influencing policy through the *Land* governments.

In Switzerland, with its constitutional elements of direct democracy, active citizen groups have had considerable impact on the nuclear debate.

Sweden is one of the few countries where the debate over nuclear power has taken place within the traditional political process. Here, years of conflict among the political parties over nuclear power had turned this issue into a major and persistent public controversy.

Political and legal structures in Germany and Sweden provided channels for public influence and points of tension through which activists could create divisions within the political establishment. Similar channels exist in Holland and in Austria, but there the antinuclear activists had won their case by the time the Three Mile Island accident occurred, and this, as we have seen, limited the salience of the event for these countries.

In France, it is the lack of channels for public influence that has allowed the government to continue its nuclear program essentially unchanged. For many years, public protest was as widespread and intense in France as in other countries; however, neither the courts nor the parties nor the local governments could channel protest to the policy level. The accident simply reinforced the administrative centralization of decisions in this policy area.

A PREGNANT PAUSE

The Three Mile Island accident added critical economic and political ramifications to European nuclear planning. In the face of widespread skepticism about nuclear power after the accident, the International Atomic Energy Agency (IAEA) sought to prevent a major slowdown in the expansion of civilian nuclear power programs. IAEA—as well as the International Energy Agency (IEA) and the Nuclear Energy Agency (NEA), both at the Organization for Economic Cooperation and Development—reinforced its programs of information exchange on nuclear accidents and launched efforts to improve safety research. Despite such efforts, a European Economic Community (EEC) commissioner predicted that the psychological and political impact of the accident would delay development of European nuclear power in every country except France for at least a year. In 1973, the EEC had developed a long-term energy plan that included major increases in the share of nuclear energy in electricity production. By 1979, examining the actual implementation of national nuclear programs, the EEC revised its nuclear forecasts downward. Whereas in 1973 it had projected there would be 160,000 MW in production in 1985, in 1979 it expected only 71,000 MW. Meanwhile, the economic benefits of nuclear power relative to other sources of energy are increasingly in question.

The new safety measures called for by the lessons of Three Mile Island will only compound the growing economic problems of the industry. After the publication of the *Report of the President's Commission on the Accident at Three Mile Island*,[20] the European press remarked that only two or three countries (Germany, France, and Great Britain) could ever afford to introduce the safety and personnel-training measures that were proposed in the Kemeny report. To insist on such conditions in smaller European countries or in the Third World would preclude their use of nuclear power.

Despite a growing ambivalence about the technology, soon after Three Mile Island, many European governments began concerted efforts to convince the public that the risks of nuclear energy are acceptable and that nuclear energy is more important than ever, given the increasingly turbulent politics of oil. The German government is trying to lift the quasi moratorium imposed by the courts by changing the legal conditions for defining "safe disposal of nuclear wastes." Experts argue that intermediate storage facilities, sufficient for a few decades, can in fact satisfy the meaning of the law. The new British Tory government has announced plans to introduce the pressurized water reactor line and to expand its nuclear program. France continues to develop its nuclear program unchanged. The outcomes of the Swedish and Danish referenda remain to be seen.

While Three Mile Island has been a subject of endless discussion and a source of prolonged delay, as a single dramatic event, it has only reinforced the political logic of the ongoing debates. These debates, however, were as much about political and administrative relationships—the structure of regulation, the role of experts, the authority of central governments over local environments, the influence of public opinion and dissent, the public availability of information—as about the technology itself.

REFERENCES

1. NELKIN, D. & M. POLLAK. 1981. The Atom Besieged: Extra-Parliamentary Dissent in France and Germany. MIT Press. Cambridge, Mass.
2. NELKIN, D. & M. POLLAK. 1977. The politics of participation in the nuclear debate in Sweden, the Netherlands, and Austria. Public Policy 25(3): 333-357.
3. 1979. Nuclear power around the world. IEEE Spectrum (November): 97-109.
4. Institut für Demoskopie Allensbach. 1979. Allensbach Report. Report No. EO3. Bonn, Federal Republic of Germany.
5. Frankfurter Allegemeine Zeitung. 1979. May 2.
6. LARSSON, J. 1979. Swedish energy policy in the shadow of Harrisburg. Swedish Information Service (June): 3.
7. New York Times. 1979. May 18.
8. New York Times. 1979. May 21.
9. Manchester Guardian. 1979. April 5.
10. Groupe Expansion. 1979. La Lettre de l'Expansion (No. 462, 7 Mai).
11. Louis Harris Poll. 1979. Les Français et l'Energie Nucléaire après l'Incident d'Harrisburg. For Electricité de France. Paris, France.
12. Le Canard Enchainé. 1979. 4 Avril (3047).
13. Charlie Hebdo. 1979. April 5: 438.
14. Le Monde. 1979. April 4.
15. La Recherche. 1979. Interviews. (July): 102.
16. La Gazette Nucléaire. 1979. Press reviews. 26/27(Mai-Juin).
17. Nucleonics Week. 1979. November 8.
18. VAN DER HUEVEN, E. 1979. Dutch Energy Policy Leaves Hot Issues Undecided. Amsterdam, The Netherlands. (Unpublished paper.)
19. HIRSCH, H. & H. NOWOTNY. 1978. Information and opposition in Austria's nuclear energy policy. Minerva 15(3-4): 314-334.
20. Kemeny, J. G., et al. 1979. Report of the President's Commission on the Accident at Three Mile Island. U.S. Government Printing Office. Washington, D.C.

PUBLIC REACTIONS: PANEL DISCUSSION

Moderator: David L. Sills
Panel Members: Rupert F. Chisholm,
Barbara S. Dohrenwend, Cynthia B. Flynn,
Robert C. Mitchell, Dorothy Nelkin, and
Anne D. Trunk

C. B. FLYNN (*Social Impact Research, Inc., Seattle, Wash.*): I just want to mention to Dr. Mitchell that we asked the question in our survey, "How many miles do you think a nuclear plant should be located from the nearest community? Our survey was taken in the Three Mile Island [TMI] area, and the median answer was 30 miles. For what that's worth, toss it in: half the people felt less than 30, and half more.

R. C. MITCHELL* (*Resources for the Future, Washington, D.C.*): But of course, most of those people actually live closer to a plant.

C. B. FLYNN: And that was the general point: as far as I know, surveys conducted near the nuclear plants show more acceptance for that sort of thing than do surveys conducted farther away.

B. S. DOHRENWEND (*Columbia University School of Public Health, New York, N.Y.*): Let me just comment on what seemed to me as perhaps a difference between the results that Dr. Flynn was reporting and the kind of results that I was reporting with respect to trust or distrust. I think there's a time difference. That is, you were talking, Dr. Flynn, about opinions that people reported at the time of the accident. And one of the things that has struck me—both in terms of your description of the results you were getting and also in Dr. Mitchell's results—is the lack of crystallization, the lack of firm opinions in this area. I think what may have been happening in the TMI area is that as an effect of their intensive exposure, there has been more crystallization of attitudes. We have information on other attitudes as well that shows this. I would suspect that in the TMI area, people have a clearer idea of what they think. I believe that distrust is one of the consequences of the exposure there, and this differentiates them, as I mentioned, from other parts of the country.

C. B. FLYNN: I might just add that the question wording was significantly different. As Dr. Mitchell pointed out, that does make a difference. You were asking about trust, and we were asking about what was the most useful source of information during the accident. You would expect different results.

I might just add, too, that a survey was done two weeks ago, asking the same people one year later how they felt about how government officials handled the situation. Whereas in April of 1979, 72% of the people approved of how the government had handled the situation, the survey

*Dr. Mitchell spoke at the conference, but did not provide a paper for publication.

0077-8923/81/0365-0196 $01.75/2 © 1981, NYAS

showed that now only 32% approve—a very significant drop, about 40 points. So I think you rightly point out that time differences will make a significant difference in this local area.

M. FIREBAUGH (*Institute for Energy Analysis, Oak Ridge, Tenn.*): In response to Robert Mitchell's comments on Alvin Weinberg's comments on the Roper poll, I just wanted to clarify what the Roper poll actually asked. Roper mailed that question to me along with the results. What Roper said was this: In light of the fact that any fix will cost money, which of the following would you propose to essentially fix up nuclear power? And then he went down a list. One was stationing an NRC [Nuclear Regulatory Commission] man full-time in every plant. The second or third question down was about remote siting, making sure that the plant was at least 50 miles from where the respondent lived. The list went all the way on down to shutting down all plants. The surprising result to us was that the remote siting won out. It had something like a 69% favorable response; they want the plant at least 50 miles away. Since this result buttressed Dr. Weinberg's remote-siting position, he quoted that number. Incidentally, the last option—closing down all nuclear plants—got I think a 20% favorable response.

R. O'CONNOR (*Journalist*): I'd like to address a question to Mrs. Trunk. At various times during the Kemeny commission, I believe that 8 of the 12 commissioners supported a moratorium vote on nuclear power. However, a moratorium vote was never carried by the commission, and I understand that this was due to a procedural rule change instituted by the chairman. I'd like to know at what point and under what circumstances this rule change was initiated. Also, how can Mrs. Trunk explain her abstention from the moratorium vote that did manage to get six as opposed to the seven requisite votes at that time?

A. D. TRUNK (*President's Commission on the Accident at Three Mile Island, Middletown, Penn.*): At the beginning of the last session, if there was no change, Dr. Kemeny sent a letter around saying that it would take seven votes to pass a resolution, and we all voted on this. It was unanimous, and that's how it stood.

To get something talked about would just take six votes; but to get it voted on would take seven. All the commissioners voted on that, so it really was never changed.

I abstained from the first moratorium vote because it said that no new construction should be done until the NRC and the government had a chance to read our report. That didn't mean anything to me. It didn't say that the plants on line would be safe, and I just couldn't see voting for something that was so ineffectual. I want the plants that *are* running to be safe, and that was my concern.

There is a moratorium. I felt the utilities were not putting in their applications for new plants. So it created a moratorium in itself. That was the least of my worries. I was more worried about having the plants on line working well.

P. STRUDLER (*National Institute of Occupational Safety and Health,*

Rockville, Md.): I have two questions about tactics. The first is to Dr. Chisholm, and the second is to Dr. Flynn and Mrs. Trunk. My question is, How did you get in to work with the population that you wanted to reach? Did you come in under the mantle of the Kemeny commission? Did you come in under Metropolitan Edison? Did you have to deal with the unions? Did you have to develop union credibility?

R. F. CHISHOLM (*Pennsylvania State University, Middletown, Penn.*): We came in under the umbrella of the president's commission. And I must say that the utility companies involved were very very supportive of our efforts. They really bent over backwards to open the doors and make it easy for us to approach the people we wanted to approach in terms of giving lists of employees, etc.

One plant is unionized, the other is not. TMI is unionized, Peach Bottom happens not to be. Obviously the union played a big role here, and we worked from the top down all the way from the AFL-CIO headquarters in Washington through the district director and local union president finally. We naturally sought to get their support of the study, which we did get. We sent a letter to all people who were going to be contacted, indicating the nature of the survey, the reasons behind it, and that it did have support of the local union, with their agreement of course. So we did have a broadly based sanction. We tried as much as possible under the time constraints to give people information and a chance to ask questions; in the letter, we gave phone numbers where they could call and ask questions of myself or other individuals involved in the study.

P. STRUDLER: Dr. Flynn and Mrs. Trunk, my experience with doing community organizing as a response to a disaster in a small, defined community is that if you're an outside agent, you really have to integrate yourself into the community before you can expect them to listen to you at all. Thinking of that awful news clip from the "NBC Nightly News" several weeks ago—where Dr. Vollmer from NRC was not even allowed to speak in an auditorium of 500 to 700 angry people—do you feel that the NRC has adequately integrated itself into the community, and do you feel that some of the animosity might not exist if there had been a greater attempt to integrate in?

C. B. FLYNN: I guess the point of my talk was that the answer to your question depends on who you talk to. Some people in the community find the NRC's office in Middletown a great resource. They use it; they trust it. Others do not. What else can you say? There seems to be quite a difference of opinion on that subject. I don't consider myself an employee of NRC, although I'm under contract to them and I present myself that way. I do maintain a residence in Middletown, and I feel I talk pretty well—

P. STRUDLER: Do you go to church? Do you participate in social functions?

C. B. FLYNN: I go to church. I get my car fixed at the gas station. I shop at Fox's. I get my toothpaste at the market. And just like everybody else, I read the Middletown press, if that's what you're asking me.

UNIDENTIFIED SPEAKER: Dr. Mitchell, you mentioned the importance of the potential for a catastrophic accident in a nuclear power plant. Yet there are certainly other technologies, which are routinely accepted, that also present the possibility of catastrophic accidents, not least among them being alternative energy technologies. How do you account for the tremendous concern among the public and in the popular press about nuclear technology?

R. C. MITCHELL: Could I ask you about how alternative technologies present a similar potential for catastrophe?

UNIDENTIFIED SPEAKER: There is a University of California study that pointed to at least one dam in California that, if it burst, could wipe out on the order of 100,000 people. Another study said that there's the possibility in Boston Harbor, I believe, for liquified natural gas operations to cause an explosion equivalent to a one-megaton hydrogen bomb. Those are two examples that come to mind.

R. C. MITCHELL: They're very good examples indeed, especially the liquid gas. With nuclear energy, I believe we have those kinds of hazards extrapolated to many more installations. It's a technology that was inaugurated by several big explosions that took place in another nation. I think the invisibility of radiation also plays a role, and—as we've heard numerous times during our symposium—this is a technology that the news covered very badly and that the experts claim they have great difficulty in explaining to the public. It's all those things. It's likely that the liquid gas problem would become a real concern to people near those places, but even so, it's a much more localized kind of experience than a large-scale radiation release would be, as I understand it, given the wind in a particular direction, etc. The magnitude of a nuclear disaster, however small the probability, is quite striking.

E. MARMORSTEIN (*New York, N.Y.*): I'd like to direct my comments to Mrs. Trunk. I can sympathize with the feelings of people in the Harrisburg area, and I can certainly understand a leaning towards wanting to blame the public for fearing without cause or overreacting, let's say. However, I really can't believe that you would blame the public for the lack of learning before the accident. I think that if we're going to blame anyone for lack of learning before the accident, we should look closely at where that lack of learning came from. I think there's a very strong case against the utilities and the government. A case in point is the hydrogen bubble. The hydrogen bubble effect has been known about since the early 50s. It's a simple zirconium reaction. I'm not even a scientist, and I can understand that. It was not a surprise. The surprise was that such an accident could happen in the first place. The Rasmussen report said it couldn't happen. Now, where was the public supposed to get their learning from when the Rasmussen report and the Brookhaven report said that this couldn't happen? I don't see how we can blame the public when the experts who are supposed to be educating us didn't believe these things themselves.

About the hydrogen bubble, this was known and yet there was no facility built into the plant to vent the hydrogen gas. Why didn't the utility do their homework before they built the plant? That's one thing I'd like to know.

The other thing is that the gauges in the plant were designed to handle only up to a class 8 accident. During the accident, the gauges all went off beyond measurement capabilities. Why weren't they designed to handle this kind of accident?

As far as the government is concerned, the utility had only 20 monitors in the area around the plant for the first three days of the accident. Why didn't the government have independent monitors there all along? They didn't do their homework either.

As far as learning goes, I think it's not too late for the public to learn. And your attitude now seems to be that the accident has happened, it's too late for us to learn. One thing that's going on now, which is extremely upsetting, is that the assumption is still being made that the greatest exposure was near the plant. However, someone in the first session this morning admitted that the plume was above the houses in the five-mile radius and was blown away from the plant. Therefore, the highest radiation exposure—according to NRC figures that were stated by Tom C. Kepford in the intervention hearings—was not nearest the plant. It was 10 to 12 miles away in the northwest sector. You can study the wind factors. There was a temperature inversion for six days. That's where the greatest radiation was. That's not reaching the media, and I don't hear utilities talking about it. When the cancer studies come out, you can bet that that's going to be the control group. Because the model that people have in their minds is still wrong. So when you talk about learning, please don't blame the public. That's all I'd like to say, and I'd like to hear any comments.

D. L. SILLS (*Social Science Research Council, New York, N.Y.*): Thank you. Those were reasonable questions but not really addressed to any of the expertise here.

A. D. TRUNK: Could I just answer that. I am the public. I belong on an advisory board at an elementary school. It was my job—it should be my job—to see that my children are educated. I don't have to have the utility teach anything. I don't have to have the government come in and teach them. There are teachers in school teaching science classes who should be giving them this education. I went around in Middletown and instructed these teachers. Dickinson College is now starting a nuclear energy course. Elizabethtown is starting in their high schools for adult education. It's not too late, but I feel that our children should get started now. It's up to us to see that it gets through, and it should have been there before the accident also.

G. KUENSTLER (*In These Times, Chicago, Ill.*): I have a three-part question, especially for Drs. Dohrenwend and Flynn. Do you have data on suicides? Do you have a control for education? And thirdly, in the light of the public mistrust that you've found and given that the NRC's plan is based on the possibility of evacuating a 10-mile radius within four hours in a class 9 accident, would you comment on the implications of these high levels of mistrust for speedy and permanent evacuation of a population within the 10-mile radius?

B. P. DOHRENWEND: The data on suicides and other data, such as use of medical facilities, use of prescription drugs, and so forth, exist. But so far as I know, these have not been tapped, partly because there have been other resources. Somebody could, of course, analyze these records anytime they wanted to.

You asked about a control for education. I'm not exactly sure what you mean. Let me just say that in general, what we've found is that the attitudes and feelings were generally negatively related to education. That is, better-educated people were less antinuclear, less—I don't know about distrustful. Not that there were large differences, but that was the general trend. I mentioned that in connection with distrust, but it was not a picture of a small college-educated group who formed the anti-nuclear activist organization, but rather that there was pervasive distrust by people of less education in the area. As for the effect on evacuation, I don't think one could predict for the distrust directly to that because there would be so many other factors, including the organization, involved. As I understand it, evacuation plans were just totally unrealistic as they stood.

C. B. FLYNN: I don't know that I have a lot to add, except that the evacuation planning at this point as I understand it is for 20 miles, not 10. That expansion essentially occurred during the time period of the accident.

R. S. BARRETT (*Hastings-on-Hudson, N.Y.*): I'm an industrial psychologist who consults with the utilities on select issues. My question is directed primarily toward Dr. Chisholm but also to anyone else who's doing surveys. In my interviewing of nuclear power plant operators and supervisors, I have encountered a great deal of defensiveness on their part about nuclear power. I wonder if you've encountered this in your survey and, if so, how you took into account this attitude in your interpretation of the results.

R. F. CHISHOLM: We have encountered it in response to a number of open-ended questions that were included generally towards the tail end of our interviews. We have not yet analyzed those data to the point where I can say that there is a basic pattern. In terms of how we took it into account, I guess I would like to ask, How would you suggest that one take it into account? I don't know of any way of doing that. But I would like to say that when we began the survey, I talked with the president of the local electrical workers union. He said in effect: Hey, I don't think you're going to find anything. We support your survey, and we'll be interested in the results. But I don't think you're going to find anything because in my contacts with nuclear workers, they seem to be telling me that they feel much safer when they're at work at TMI in the plant than when they're back home with their families or talking with neighbors. But I don't really know how you can take into account the distrustful attitudes, and I'm not sure you should.

R. S. BARRETT: Well, I was struck by the response that the people at Three Mile Island thought the accident was less severe than did other nuclear power plant operators of a distant plant. The first thing that came to my mind was that their attitudes might well have been colored by defensiveness and did not really reflect their true opinions.

R. F. CHISHOLM: Okay, in response to several questions, we wonder whether or not that defensiveness was showing up. For example, there is a very clear-cut difference between those at TMI who felt their health was in danger via radiation during the incident and those at Peach Bottom—a very big difference. On the other hand, more Peach Bottom people felt that during the accident, they were exposed to—how was the question worded now—to other factors that would cause disease, not radiation but other factors that would cause disease. More people by a large margin at Peach Bottom than at TMI cited that. Now we wonder whether or not that is a defensive reaction. It's a very interesting result, and we have not come up with any final interpretations. It just surfaced within the last couple of weeks. But that's an incident along the lines you're talking about.

K. C. LESKAWA (University of Michigan, Ann Arbor, Mich.): At the time of this event, I was the only locally raised person pursuing a doctorate degree at the nearby Hershey Medical Center. About three to four months after this accident, the person in charge of the radiation safety group at the Hershey Medical Center—a man named Ken Miller—published an article in the Journal of the American Hospital Association in which he suggested that if this sort of thing should ever happen again, a total information blackout should be imposed within a 50-mile radius of the site. Basically, the response of the faculty and the staff at the medical center was one of absolute terror. But I feel as though it does pose a somewhat interesting question. Do you think that this was a viable ethical alternative? How do you feel the psychological effects would have differed if this sort of thing was in effect?

D. L. SILLS: To whom are you addressing the question? Would anybody like to reply to the ethicality of a 50-mile-radius blackout? David Rubin, do you want to comment on this?

D. M. RUBIN (New York University, New York, N.Y.): Well I find that a sort of Soviet response to the problem. It's not the way we handle information in this country. It's antithetical to at least the spirit, if not the rule, of the First Amendment. I think you'd find reporters going absolutely nuts. If you think that there was a panic—an information panic—among reporters having difficulty at the site without that, imagine what would be going on at the perimeter of the 50 miles and the rumors that would be flying then, if there were such an information blackout. I believe it would be foolish—worse than foolish. There are better ways to deal with the problem.

THE RESPONSE OF THE NUCLEAR REGULATORY
COMMISSION TO THE ACCIDENT
AT THREE MILE ISLAND

Robert J. Budnitz*

*Office of Nuclear Regulatory Research
U.S. Nuclear Regulatory Commission
Washington, D.C. 20555*

The topic that I'm going to discuss briefly today has to do with the institutional response of my agency, the Nuclear Regulatory Commission (NRC), during and after the accident. In order to talk about the response, I'll present a list of the flaws that the accident has revealed.

What I mean by a flaw is something that was imperfect either technically or institutionally. Now I think it's fair to say before I go through the list that there's hardly anything on it that wasn't known to many people before, in some cases for many years. But for a variety of reasons, none of these flaws were addressed before the accident.

I am going to discuss several of these flaws—by no means an exhaustive list—and tell you what has been done since the accident, within our institution and others, to address them. Now, many of these flaws are not things that can be cured in a short time. Rome was not built in a day, after all. Some things that we have already begun to do will not be fully accomplished for some time—perhaps another year, perhaps several years.

TECHNICAL FLAWS

Safety Design

The first flaw is the fact that the safety design or, more important, the safety philosophy of design, operation, and regulation did not adequately treat some important kinds of accidents. This is something that we are now remedying by looking carefully at some important types of accidents that have never been analyzed thoroughly before. The small loss-of-coolant accidents are one example. Certainly accidents in which operators play a major role, rather than the previously analyzed types in which operators didn't play a major role, are getting extra attention; and they deserve it.

Human Factors

The next technical flaw is the neglect of human factors. This runs through a number of different issues, including questions about the way

*Present affiliation: Teknekron, Inc., 2118 Milvia St., Berkeley, Calif. 94704.

0077-8923/81/0365-0203 $01.75/2 © 1981, NYAS

people in the control room are selected and trained, the way control rooms are designed, the way operating procedures for both routine and emergency operations are set down and operators trained for them, and so on. All of a sudden, not only NRC but the industry is interested—in fact, vitally interested—in addressing this and solving it as quickly as possible.

The industry's response has been to establish the new Institute for Nuclear Power Operations (INPO). Our response has been, in part, a major research program and also a major effort to incorporate human considerations into our regulations and the agency's review process.

Emphasis on Hardware

The next flaw is actually the converse of the last one. The NRC's regulatory review process and the way we thought about reactors had so great an emphasis on hardware that the human factors were neglected. Now I don't decry the emphasis on hardware. Thank God it was there. The fact that so much tremendous effort had been put into design review, care in construction, and operation and maintenance is in substantive measure the reason that the reactor rode through the accident as well as it did. It was because people thought carefully about those things. So that emphasis on hardware should not be replaced by an emphasis on human factors. It has to be there as well, and then the human-factors emphasis placed on top of it. The idea that the emphasis on hardware was a mistake is something that I don't agree with, and I know that both the industry and our agency don't agree with it either.

Siting Policy

There is now a substantial element of the community that is questioning the long-standing policy under which the NRC, and the Atomic Energy Commission before it, sited the present reactors. There is going to be a large and difficult rule-making proceeding in the next couple of years to look at this carefully. It is now generally understood that the way in which the criteria were written—the criteria that had to be addressed when one picked the site for a reactor—was inadequate because the safety design and the whole siting question didn't think about some accidents that are now considered to be quite important. Whether these accidents really are important enough to change the siting policy is going to emerge only after this very extensive rule making has been undertaken and completed. But certainly the whole question of siting in this country is being reevaluated, and this is also going on overseas.

Instrumentation and Controls Inadequacies

The accident revealed to all of us some serious flaws in the way operators—during an accident or during normal operation—received and processed information and the way it was presented to them, the way the operators were able to utilize the information and to respond properly during an accident, and so on. A lot of instruments that we now wish were in all reactors, which weren't there before, are being put there. Instruments that enable direct reading and in many cases redundant ways of indicating critical parameters, which many operators and many plants didn't have before, are being installed.

Also, the accident at Three Mile Island and the incident at Crystal River—which was stopped so promptly that it never became an accident (that was the incident in Florida a little more than a month ago now)—told us that there are some very serious problems with the way automatic control systems are installed and run in some of these reactors. We're going to be looking at them.

These, of course, comprise only a partial list; I could go on and on. But they're good examples of things that were never addressed, even though they were known to exist. There was a general feeling in the community that despite these flaws, reactors were safe enough. Which brings us to the list of institutional flaws.

INSTITUTIONAL FLAWS

Complacency

You see, there was a very large segment of the regulatory community, in the NRC and in the industry as well, who felt that the safety regulations and the safety practices that were in place before the accident were adequate. In fact, there was a view on the part of some people, maybe not such a small number, that the safety regulations and practices were not only adequate but a vast overkill. They felt, therefore, that the job of the industry and of the NRC was not to think so much about improvements as it was to assure that the existing regulations were met.

Concern with Regulation Rather than Safety

So the general notion in the NRC was that if the regulations were obeyed to the letter, the plant was going to be safe enough. Now we know that for a whole variety of reasons that is not enough.

Complacency explains why a number of other flaws are on this list. People who thought that improvements should be made had a burden of proof placed on them; they had to show why the present situation was not safe enough. Now, thank God, the whole institution of nuclear power has turned around. And yes, I'm sure that that is so. I am sure that NRC and the industry have modified their attitude very thoroughly. Now, I'm not sure that that will be true in 10 years. In fact, one of the biggest jobs that we have is to vigorously keep up the vigilance, so that in 5 or 10 years, this present attitude remains.

But I am personally convinced that there has been a tremendous change in everybody's attitude—that the institutional flaws I'm mentioning here are not only being corrected, but will remain institutionally important over the years.

Of course, complacency can return for a number of reasons. You remember what it was like before there had never been an accident. In fact, there are people who will tell you today that there still has not been an accident because Three Mile Island didn't result in any prompt fatalities. But the complacency was more a result of a kind of choir singing to itself than anything else; everybody kind of agreed with each other. Those few who didn't, really had a hard time getting heard. This has changed, and I think 100% for the better.

Let me go through some of these other institutional questions for you briefly and show you what I mean.

Poor Feedback of Operating Experience

The Nuclear Regulatory Commission, until the accident, had no systematic way of taking operating experience from reactors and feeding it back into the safety review process or safety operations or safety design.

I want you to imagine a situation in which an airplane has a problem that doesn't result in a crash but does result in a serious event in the air, though the pilot is able to land safely. Then that aircraft company and that pilot write a report, go back and fix their airplane—that particular airplane—and stick the report in a drawer and nothing ever happens to it. In the eyes of the flying public, that would be a completely unacceptable situation. Yet that's the way a large number of events were coped with by the nuclear industry and by the Nuclear Regulatory Commission until this year.

Now that's astounding to say, but it's true. Some events were analyzed very carefully, but a number of them were not. We are reacting to this now. We now have a special office in our agency—and the industry is setting up similar offices, not only in specific utilities, but also as a group—to ensure that feedback gets back. This is a major change, and I am sure that it is going to result in improved safety. It cannot fail to do so.

Again, that doesn't mean that the institutions we've just set up are going to be adequate. We don't really know. It may take more resources, or it may be more difficult than we think, but that institution is there and I think for the good.

Lack of Self-Policing

Although the NRC had inspectors that went around to the plants, the industry didn't police itself. The parts of the industry that were confident about their operations, that is, the utilities that were very good, didn't care as much as they should have about those parts of the industry that they knew were not so good. They didn't see, as they now do, that their fate was linked with these less safe reactors. So the industry was not self-policing. This has changed, and it's a very significant modification. It means that the NRC's burden will be shared in an important way with the industry. The industry can and will be doing some of the policing itself—finding some problems, fixing them, and reporting them.

Disincentives to Safety Improvements

I spoke about the NRC's concern with regulation. Our agency was, I think, so concerned with regulation that there were a lot of disincentives to safety improvements. I'll give an example to explain what I mean by disincentives.

Suppose that a utility has built a reactor at a particular site. The reactor had been run through the regulatory review, licensed, and built. Two years later, the utility decides that they want to build a second reactor—unit 2. Let's suppose that they want to buy an identical reactor from the same vendor. However, the manufacturer or the architect/ engineer knows of a few safety improvements that could be built in.

There was an important financial disincentive to building those safety improvements into the second unit. It was so much easier to get the old reactor design licensed from the NRC by running it through the same regulatory review process than it was to review all these new add-ons that typically the improvements were not made. It would take much longer for the NRC to review the improvements, and time was money— still is money.

That's just unacceptable. That's just one typical disincentive to safety improvements, and there is a long list of these. There were financial disincentives having to do with the way the state allowed or, in some cases, did not allow safety improvements into the rate base unless the NRC required them. So if NRC said, Thou shalt install that safety gadget, the public utility commission would put it in the rate base. But if NRC said, You can do it if you want, but we don't require it, some public utility commissions wouldn't let that go on the rate base. That's crazy.

The disincentives question is now being addressed. Although the problem isn't licked, the fact that all of the institutions are sensitive to it is certainly going to be a big step in the next few years towards overcoming it.

Inadequate Emergency Response

I won't say too much about this last item on the list, but everybody realizes that NRC and the industry were not prepared to respond to that accident. The state was not prepared. The Department of Energy and the

Environmental Protection Agency were not ready for their assisting roles. Nobody was prepared; and that's being fixed.

Summary

So there were some very serious, fundamental institutional flaws, each of which had been recognized before but none of which had been fixed effectively before the accident.

HOW SAFE IS SAFE?

I'm not saying that all of the fixes that we are working on here are going to be enough. There's a lot of good faith, and there's a lot of money, effort, and good hard blood, sweat, and tears going into these. I can assure you of this, because I know people—friends and colleagues of mine—who are working day and night, and have been for a year, on these things. I am hopeful, and I think most of us in the NRC are hopeful, that the things we and the industry are doing will be a major change and will be enough in the end.

But I'm not sure they're going to be enough, and there are two reasons. The first reason is that we might lose our momentum or there might be a turning away of motivation. The second reason has to do with the most fundamental institutional flaw of all. You see, *nobody knows exactly how safe is safe enough.* The Nuclear Regulatory Commission doesn't have a policy that can define this. Because they don't know how safe a reactor has to be in order to get a license or retain a license or not have a certain upgrade, decisions are made in a kind of random, haphazard way within the agency and within the industry. It would be nice if the NRC could find a way of enunciating a policy like this, but there are problems. It would be nice to know that certain reactors are considered safe enough, that is, that they do not require certain additional safety features because the design is already adequate.

Now, you can understand why that's such a difficult problem. It involves not only a public perception of risk, but a *changing* public perception of risk. Certain things that were thought to be safe enough two years ago are no longer thought to be safe enough because of the accident. Otherwise we certainly would not be fussing about the release of a few kilocuries of krypton while failing to prevent, for example, much more serious core-melt accidents in other plants. The inordinate attention given to low-risk activities is a result of the confusion on this question.

Until the NRC—with political and public information and feedback—has addressed and decided this question, we are going to continue in the dark. We're not going to know whether the site for a new reactor that's under construction is really going to be thought safe enough in five years. We are finding that sites where reactors have been located for many years—Indian Point right up here is an example—were said to be safe enough in the 60s and are no longer considered safe enough.

The agency has committed itself to answering this question, the commissioners have said they will. But they're not even clear about how to ask or answer the question, because it's a multidimensional one. It isn't just a question of how many prompt fatalities will result after an accident; there's a lot more to it.

I think it's going to be a long time before the question is answered. I just hope it's not a long time before the question is at least addressed. Otherwise we're going to continue in a quagmire—moving in what we think is the right direction but without any guideposts as to how fast is fast enough and how slow is unacceptable.

THE REACTION OF CONGRESS TO THE ACCIDENT
AT THREE MILE ISLAND

Paul Leventhal

*Subcommittee on Nuclear Regulation**
United States Senate
Washington, D.C. 20515

I will speak briefly about the congressional reaction to Three Mile Island (TMI), hopefully putting the role of Congress into some historical perspective.

I think the choice of the topic of institutional reaction, rather than response, is probably a wise one insofar as Congress is concerned because Congress at this point has reacted but has not yet responded to the TMI accident. It is now more than a year since the accident. Yet there has been no legislation passed in Congress dealing with TMI-related problems, such as the need for improved emergency planning, better siting policy, and stiffer civil and criminal penalties relating to safety issues. In a moment I'll go into the reasons for that, but first I would like to make some historical observations.

Until the end of 1973, the energy issue did not exist. Energy as such was not an issue before Congress or, for that matter, before the public. It wasn't until the Arab oil embargo that "energy" became an issue. Congress has been dealing with it, therefore, for somewhat less than 10 years.

Prior to the time of the Arab oil embargo, the only consideration of energy as such was in the context of atomic energy. The only energy policy that this nation had was its atomic energy policy, which began at the end of World War II. In the Congress, consideration of that policy was monopolized by a single committee—the Joint Committee on Atomic Energy—and its basic orientation was almost exclusively promotional and developmental. Regulation was not an area of major concern before the joint committee. The committee was preoccupied by the oversight of the nuclear weapons program and the development of the nuclear power program. In the context of the nuclear power program, it concentrated on the development of improved reactor designs and ultimately of the greater reactor. So regulation as we understand the term today was not a principal focus of the joint committee. There was a parallel situation in the executive side of the government. There was a single agency responsible for all things nuclear. That was the Atomic Energy Commission (AEC). It too was preoccupied with the weapons program and with the development of the nuclear power program.

With the Arab oil embargo—and a sudden awakening by the general

*Dr. Leventhal is science advisor to this committee, not a member of Congress.

0077-8923/81/0365-0210 $01.75/2 © 1981, NYAS

public and Congress to the energy crunch we found ourselves in—there was, for the first time in the history of the Congress, a really close scrutiny of the atomic energy program by the Congress at large. Prior to that time, it had been left to the joint committee. Those matters that reached the floor generally were disposed of in a fairly routine fashion. There wasn't much consciousness or awareness by members of Congress generally to atomic energy matters.

In 1974, the Atomic Energy Commission was abolished by a reorganization bill, which established the following year the Nuclear Regulatory Commission (NRC) to take over the AEC's regulatory responsibilities, such as they were—a relatively small portion of the Atomic Energy Commission.

That bill also established the Energy Research and Development Administration (ERDA), which subsequently was replaced by the Department of Energy and which inherited the promotional regulatory weapon side of the AEC.

Within two years, the Joint Committee on Atomic Energy was abolished by Congress and replaced by a relative multitude of committees and subcommittees looking at various aspects of the atomic energy question. So what had been a monopolistic situation on the Hill suddenly became a highly diverse and competitive one.

In the Senate, the jurisdiction of the joint committee was basically split between two committees. The Energy Committee, newly established at that time, took over promotional and developmental jurisdiction from the joint committee. The Committee on Environmental Public Works, of which the subcommittee that I work for is a part, took over the regulatory jurisdiction. We oversee the NRC and authorize its budget.

Let me also mention that the Foreign Relations Committee and the Governmental Affairs Committee of the Senate share jurisdiction over the nuclear proliferation/nuclear export question as a result of the abolishment of the joint committee.

On the House side, there was a similar, but slightly more complex, division of jurisdiction resulting from the abolishment of the joint committee. You have the Interior Committee and the Commerce Committee essentially sharing jurisdiction of nuclear regulation and of the NRC in particular. You have the Science and Technology Committee taking over the developmental aspects of the joint committee's jurisdiction. You also have the House Government Operations Committee, with general oversight and responsibility, and there is a rather assertive committee under the chairmanship of Congressman Moffat that is presently an active player in the area of nuclear safety regulation.

There is a similar diversity now in the executive branch. You have the Department of Energy, the NRC, and the State Department each with a role in nuclear matters; the State Department being involved in the licensing process of nuclear exports and the question of nuclear proliferation. The National Security Council also is involved in that. You have the Environmental Protection Agency with a role in radiation standards, and you have the CEQ on behalf of the White House looking at radiation and nuclear-related problems.

So what you have today in the wake of TMI, and even before TMI, is a lot of diversity, a lot of activity. Basically, the situation is a reflection of the growing awareness and confusion of the public in the whole area of nuclear energy. There exists in Congress, as it does out in the public at large, what I perceive as a dialogue of the deaf. The staunch nuclear advocates are on one side, and the staunch nuclear opponents are on the other side—talking at each other, not listening to everything that's being said by the other, using terms and jargon that tend to confuse the public. Prior to TMI, the vast majority of the public tuned out the nuclear debate. That all changed with Three Mile Island. Suddenly, the public wanted to tune in on the nuclear debate, and I think the situation that exists now in Congress—one of delay and hesitation in terms of what types of legislation to enact—reflects the confusion of the constituency and of the country at large.

Now, to talk a little bit about the congressional reaction to TMI specifically. What is actually going on? There are some things going on; there are some important things going on. They all take time. I'm involved in a special Senate investigation of the Three Mile Island accident. It's really the third principal investigation established as such to look into the Three Mile Island accident—the first being the Kemeny commission; the second being the NRC's special inquiry, the so-called Rogovin investigation. In the Senate, there was a small authorization of one-half million dollars for a one-year effort both to investigate the accident and to conduct a series of policy studies. We had hoped to have our investigation report out by the end of the year. However, the report is being revised and reviewed so that it can be issued as a committee report rather than as a staff document. Since it is a collegial process, it simply takes some time.

We do anticipate the report coming out within the next month. The policy study should be out by the end of the statutory life of our study, which is the end of June. Since we're a smaller investigation, we're trying to focus on particular elements of the accident rather than to take the broad retrospective view that was used by both the Kemeny and the Rogovin investigations.

We are looking at three specific areas and anticipate making a substantial contribution in each of them.

The first area is one that the other two investigations did not look at in detail: the recovery from the accident, the ongoing recovery and cleanup operations. We are looking at it from both the health and safety standpoints, as well as from the financial implications standpoint and in terms of legal and procedural delay. We're also looking at the preaccident contributors to the accident, most particularly at the evolution of Three Mile Island from a plant that was originally to be built at Oyster Creek, New Jersey, to see what impacts that transfer of site had on the plant itself. The third principal element of our investigation is to provide what we anticipate will be the most detailed narrative of the first day of the accident, written in a fashion that can be understood by the lay reader. Both Rogovin and Kemeny went into a lot of detail on the first day of the accident. They did it largely in the technical appendices of their report. Our coverage of the accident in terms of everything of significance that happened during the first day—particularly the accumulation and the communication of information within the plant site and to NRC and state officials—will be covered in the narrative portion of the report.

One issue we're looking at, which was raised earlier by Henry Myers from the House Interior Committee, is the possibility that information about the severity of the accident was known to utility officials but, for one reason or another, was not communicated. We're looking at the question of whether or not there was a cover-up; if there was not a cover-up, whether it was simply a matter of lack of competence on the part of the plant personnel; and if it was the latter, at what the underlying reasons were for operator error.

It should be obvious that if there was not a cover-up, then there was something severely wrong with the reaction of the plant personnel and, for that matter, of the NRC inspectors who were on the plant site. We're trying to analyze in detail what some of those problems were, and in our policy studies, we'll be looking at what needs to be done to correct them.

Another principal focus of our study is what the NRC and the industry are doing in the following areas: human factors, particularly the training of operators; the design of control rooms; and the preparation of emergency operating procedures. What is being done after TMI with respect to operating experience?, that is, collection of the operating experience of various reactors, analysis of it, and dissemination to other sites so that problems can be anticipated and avoided. The last principal focus of our policy studies is emergency response.

As I mentioned before, legislation has not been passed yet in Congress. The legislation that is pending, however, is the fiscal year 1980 authorization bill for the NRC, which has in it provisions that were passed by the Senate and the House relating to Three Mile Island. Among the principal provisions is one that my subcommittee chairman, Senator Hart, cosponsored with Senator Simpson, the ranking minority member. This provision would require that before a new plant could get a license to operate, there must be an approved emergency evacuation plan for that site. This particular proposal has run into a lot of opposition from the nuclear industry. It's been subject to attack as a moratorium proposal. Moratorium is something of a cold word and one that members of Congress are extremely sensitive to. The provision is undergoing some revision to provide suitable assurance that it is not a moratorium provision. But it is not a moratorium provision since it does not apply to existing plants. The NRC would have the option of writing regulations through which it could take action against existing plants in the absence of an upgraded and improved emergency plan, but this would be at the commission's discretion. So it is not a moratorium provision per se. It is not an antinuclear provision, but it is being criticized as such by representatives of the industry.

I think it's essential that Congress does require that a new plant not get an operating license if there is not an approved emergency plan for that site. I also think it essential that emergency plan requirements be upgraded in the form of new regulations.

The pending legislation also seeks to require that the NRC upgrade siting policy regulations so that emergency planning zones and ingestion zones surrounding nuclear plants can be established and enlarged. It also provides that siting policy be based on demographic considerations, with compensatory design factors not taken into consideration in terms of the siting of plants.

The bill provides for increased civil penalties and for the establishment of criminal penalties, the latter for willful violation of certain provisions of the Atomic Energy Act. It also would require prior notification when there is to be shipment of hazardous waste—this arose as an issue when TMI wastes were shipped through states without the governors being notified. There is also a House-passed provision that would require the NRC to make a presentation to Congress comparing the specifications and regulations that apply to new nuclear power plants with those that apply to older plants, in order that Congress can assess the adequacy of the regulations on older plants.

I should add that there have been two investigations on the House side, not of a separate special nature as was the Senate investigation but as part of the committee staffs. The Interior Committee under Chairman Udall did an investigation of Three Mile Island. This has resulted in Udall's offer of an omnibus nuclear safety bill, which is now pending in the House. There was also an investigation by Congressman McCormack of the House Science and Technology Committee. The two investigations differ somewhat in their findings. Udall raises specific safety concerns. Congressman McCormack's basic conclusion is that while the accident was serious, there is not a serious safety problem resulting from Three Mile Island.

To summarize, I'd say that the overall congressional reaction to Three Mile Island is one of caution, hesitation, and continued sensitivity to code words, particularly moratorium and antinuclear. In the wake of Three Mile Island, Congress is by no means antinuclear, but it is surely less pronuclear—or at least complacent nuclear—than it had been prior to the accident and during the period of the Joint Committee on Atomic Energy. There is growing interest in conservation and in the leading energy alternatives, specifically solar energy and synthetic fuels. There is a need for Congress to oversee closely the ongoing recovery operation. There are a lot of potential problems there that do bear close scrutiny. Congress needs to focus on some of the safety issues that I've already discussed. At the same time, it has not devoted much attention of late to what I would consider the bottom-line problem relating to nuclear power, that of nuclear safeguards and nuclear proliferation—the spread of atomic bomb capability through commercial nuclear exports combined with the pretty much acknowledged inadequacy of international safeguards to detect and give early warning of a diversion of nuclear materials. The Congress was deeply involved in that in 1975-77, leading to passage of the Nuclear Nonproliferation Act. However, other elements of foreign affairs and the Three Mile Island accident threaten to divert Congress' attention from this important issue.

THREE MILE ISLAND AND THE SCIENTIFIC COMMUNITY

Allan Mazur

Department of Sociology
Syracuse University
Syracuse, New York 13210

The immediate response to Three Mile Island (TMI) among the general public was one of increased awareness of the problems of nuclear power and increased opposition to it.[1,2] The response of scientists and engineers might be expected to differ from that of the general public. Technical people are more likely than are laymen to understand the technical issues involved in the accident, and they probably were more interested in the general topic of energy before the accident. The problems of nuclear power have been discussed fairly well in the technical community for some time, and opinions for or against atomic energy seem to have been set before the accident so there may not have been a great shift in opinion afterward. On the other hand, the magnitude of the accident and of the media coverage given to it may have been sufficient to alter all but the most firmly set opinions.

In order to explore this topic with the limited resources at hand, I wrote to several prominent scientists who had taken public positions for or against nuclear power well before TMI. I will use their responses to answer three questions. First, has their alignment in favor of or against nuclear power changed? Second, of those who changed, to what did they attribute their change—to TMI or to some other cause? Third, whether or not these scientists changed, how do they interpret TMI in the context of their current beliefs about nuclear power?

These men are not typical of the rank and file of the scientific community, but they do serve frequently as spokesmen and opinion leaders for the community. One would like a survey of the broader scientific community as well, but that was beyond my resources.

In the period 1975–76, there was a battle of petitions for and against nuclear power,[3] which provided a convenient pool of prominent scientists who had taken public positions on the issue well before TMI. One petition was circulated by the Union of Concerned Scientists (UCS), an antinuclear group based in Cambridge, Mass. It said, in part:

> ... the country must recognize that it now appears imprudent to move forward with a rapidly expanding nuclear power plant construction program. The risks of doing so are altogether too great. We, therefore, urge a drastic reduction in new nuclear power plant construction starts before major progress is achieved in the required research and in resolving present controversies about safety, waste disposal, and plutonium safeguards. For similar reasons, we urge the nation to suspend its program of exporting nuclear plants to other countries pending resolution of the national security

216

0077-8923/81/0365-0216 $01.75/2 © 1981, NYAS

questions associated with the use by these countries of the by-product plutonium from United States nuclear reactors.

There were over 2,000 signers from the technical community; however, most publicity was given to a smaller list of 50 particularly prominent signers, which I will refer to as the "anti list."

The Nobel physicist Hans Bethe from Cornell University, a frequent spokesman for nuclear power, circulated a counterpetition that said, in part:

> Uranium power, the culmination of basic discoveries in physics, is an engineered reality generating electricity today. Nuclear power has its critics, but we believe they lack perspective as to the feasibility of non-nuclear power sources and the gravity of the fuel crisis.
>
> All energy release involves risks, and nuclear power is certainly no exception.... Contrary to the scare publicity given to some mistakes that have occurred, no appreciable amount of radioactive material has escaped from any commercial U.S. power reactor. We have confidence that technical ingenuity and care in operation can continue to improve the safety in all phases of the nuclear power program, including the difficult areas of transportation and nuclear waste disposal. The separation of the AEC into ERDA and the NRC provides added reassurance for realistic management of potential risks and benefits. On any scale the benefits of a clean, inexpensive and inexhaustible domestic fuel far outweigh the possible risks.
>
> We can see no reasonable alternative to an increased use of nuclear power to satisfy our energy needs.

This was signed by 33 prominent scientists, and I will refer to it as the "pro list."

Both petitions avoided extreme statements in order to enlist signers with moderate views, and a wide range of opinions were represented on each side, but there is no doubt that in the political context of the time, the petitions validly sorted signers into those who were relatively pro and con on the issue.

I wrote personal letters to 24 names on the pro list and 18 names on the anti list (chosen primarily for their prominence and the availability of mailing addresses), asking each if there had been any changes in his views since the petition, particularly after TMI, and if so, what they were. I received responses to 63% of my pro letters (including three notifications that the addressee was deceased or otherwise unavailable) and to 78% of my anti letters (including one notice of unavailability). I also had post-TMI publications by 2 pro names and 2 anti names who did not provide personal replies, bringing the number of addressees for whom I had relevant information to 14 pros and 15 antis.

Responses came in a variety of forms, from brief statements to lengthy and detailed letters, and in many cases relevant articles or speeches by the addressees were enclosed. In some cases respondents stated clearly that they were or were not a proponent or opponent of nuclear power. In other cases I had to infer an alignment, and I used the following guide. I regarded a respondent to be pronuclear if he advocated the continued construction of light water reactors, whether or not he favored the breeder reactor. I regarded a respondent to be antinuclear if he opposed the construction of new power reactors, whether or not he wanted to close down those already in operation or near completion. In two cases I could not discern the respondent's current alignment and labeled it "ambiguous."

The first question I want to address is, What changes in alignment have occurred since the petitions were signed? Of 14 people who signed the pro list, 13 now support nuclear power and 1 response was ambiguous. Of 15 who signed the anti list, only 9 now oppose nuclear power, 1 gave an ambiguous response, and 5 now favor some development of nuclear power, though usually in a cautious way.

What accounts for the changes in alignments, whether TMI or some other cause? In the first place, some of the changes may be more artifactual than real. Three signers of the anti list, now classified as pronuclear, wrote that they had not been particularly antinuclear at the time of the signing. One wrote:

> A careful reading of the [petition] . . . shows that it was merely a statement of caution [and] not a statement of commitment to a specific position. . . . I personally do not find my position changed: I still believe in cautious progress.

Another wrote:

> I agreed to sponsor the UCS declaration of '75 after it was modified by eliminating a proposal for a moratorium on power reactors. . . . The lessons I draw [from TMI] are . . . [that] we should accept the risks of nuclear accidents and *cautiously* expand the light water fission reactor installations, as well as to plan for the introduction sometime in the 21st century of breeder reactors.

A third wrote:

> I was never happy about that declaration and I have increasingly regretted that my name was on it. Indeed, I asked them to remove my name from their propaganda material that they do send around. I am not an opponent of nuclear power. I am only an opponent of the way the nuclear industry has constructed reactors with many shortcuts in order to save money and without enough consideration for safety.

Thus, of all of those who appeared to change alignments, only two remain as clear cases of change, both moving from an anti to a pro posture:

> My views on nuclear power have indeed changed appreciably in the last three years. In spite of Three Mile Island, I am less anti-nuclear than I used to be. Several factors are involved: (1) I now believe the hazards of coal are distinctly more serious than those of nuclear power; (2) the risks of our dependence on foreign oil are very great indeed; (3) I still favor moving to solar and other forms of renewable energy as rapidly as possible, with vigorous support for such developments, but I do not think the transition can

be made rapidly enough to obviate the need for some additional nuclear power in the next twenty years.

My views have indeed changed since that time and about a year ago I asked the UCS to remove my name from future mailings of the declaration. At the time I signed the declaration I had four major concerns—reactor design, nuclear waste disposal, proliferation, and operator training and reactor management. I have become convinced in the intervening time that the basic reactor design is sound and that nuclear waste disposal is a political and not a technical problem. Proliferation remains a problem and operator training and management is clearly a problem. After balancing risks to the environment, risk to the individual, and risk to the economy, I now believe we should build more nuclear power plants.

Thus, in neither case of clear change was TMI a factor in the realignment, and both changes in position appear to precede the accident. Both of these realigned respondents report that major problems with nuclear power no longer seem as serious to them as before, but it is not clear if this rethinking is the cause of the change or only another indication of changed attitudes. In any case, TMI was not the change agent in either case.

How do the scientists perceive and interpret the accident? Even though the accident did not cause any basic realignment toward nuclear power, it still could have shaped attitudes considerably. However, typical responses were that "Three Mile Island did not change my thinking," or "only strengthened" or "confirmed" preexisting beliefs. A majority of the respondents explicitly said that their views had not changed, or at least not very much; and it made no difference whether their prior views were for or against nuclear power. There were exceptions, but few.

In general, the scientists had no trouble interpreting TMI in a manner that is consistent with their preexisting views, a phenomenon that has become familiar in the year since the accident. Those who oppose nuclear power see TMI as being very close to the catastrophe they had warned of:

> The near disaster at Three Mile Island emphasized the emptiness of the nuclear establishment's statements on reactor safety.

Proponents of nuclear power emphasize the successful operation of the safety system in spite of numerous mishaps:

> Even under such unfavorable conditions, nobody was killed or even injured.

There was general agreement, on both sides, with the findings of the Kemeny commission: the accident was not caused by important equipment failures but rather by various human failings, which encompass the operators, the corporations that manufacture and run reactors, and the federal agencies that regulate them. However, within this broad area of agreement, proponents and opponents placed different emphases.

Proponents often wrote in terms of improving operator selection and training, redesigning reactor control panels and procedures, and the various responses of the nuclear industry to improve safety. Many proponents—not all—were optimistic that improvements could and would be instituted in a straightforward manner:

With better training of the operators and simple additional instrumentation for their guidance, a repetition can and almost certainly will be avoided.

There is every reason to believe that appropriate [corrective] steps are now being taken.

A few proponents even regarded the accident as beneficial:

I think it will have a healthy effect in improving safety standards and quality of design.

I think that the Three Mile Accident will turn out to be a blessing in disguise.

Opponents came at these issues differently, focusing less on surface features of operators and control panels than on basic problems with the nuclear industry and government agencies:

What most disturbed me about the accident was the clear revelation of the incompetency and greed of the private power companies involved. I had not previously realized the extent to which their attempts to save money and time have enhanced the dangers which would exist even under the best of circumstances.

The Three Mile Island accident is, to my view, simply another demonstration of their [i.e., industry and the regulatory agencies'] lack of concern. I am convinced that whatever interest they manifest in making the use of nuclear safe is less the result of conviction than it is the desire to find a sop for the public; that they are convinced that the dangers are negligible.

Some proponents also expressed concern about the basic inability of the corporate-regulatory apparatus to handle nuclear power adequately, but this was primarily an issue for the opponents.

The most basic issue raised by proponents, and it was widespread, was American dependence on foreign oil. Opponents of nuclear power were much less likely to mention the oil problem, and they certainly did not discuss it with the urgency of many proponents, such as these:

My interest in seeing us move ahead with the nuclear alternative grows rather than diminishes as the links to our overseas sources of petroleum are placed under greater and greater hazard because of the ever increasing complication of international affairs.

I believe that any new reactor put in action in the United States reduces slightly a chance of a world war for the Mid-East oil. Such a world war would be so terrible that even a slight reduction saves more lives than would be lost by an accident.

In summary, none of the prominent scientists discussed here realigned themselves on the nuclear issue because of TMI; and in fact, TMI seems to have had little impact on most of them, according to their self-reports. Most respondents interpreted the accident to fit into their preconceptions about nuclear power. Thus, opponents tended to see it as a near catastrophe, symptomatic of the inability of private corporations and government regulators to manage reactors in a safe manner. Proponents emphasized that no one was killed; that the radiation release was relatively small, and therefore the safety system worked; that safety could and would be improved; and that the risks of nuclear power are minor compared to the risks of our dependency on foreign oil. The one clear effect of the accident was that it emphasized the role of human failure in reactor accidents, whether at the level of the operator in the control room, the corporate boardroom, or the federal agency.

If TMI had a minimal effect on these prominent scientists, compared to its effect on the general public, what of its effect on the overall technical community? Information on this point is fragmentary. A poll of members of the Institute of Electrical and Electronics Engineers, taken by the journal Spectrum two months after TMI, found that 83% of respondents favored further development of nuclear power.[4] Pre-TMI attitudes were not measured, but it is difficult to believe that engineers' support for nuclear power before TMI could have been much higher than it was after the accident.

I recently surveyed engineers, physicists, biologists, and sociologists via the campus mail at my own university and found that neither the alignments nor the substantive views of engineers and natural scientists were much affected by TMI, according to their self-reports. As with the prominent scientists, the accident typically "confirmed" or "reinforced" preexisting views, whether for or against nuclear power. Only among my sociologist colleagues did I find several respondents who regarded TMI as an important determinant of their present views, and the effect here was always some degree of lost faith in nuclear power, from mild to severe. The sociologists, compared to the engineers and natural scientists, were less informed about nuclear power or energy in general, and in this regard were more like the nonscientific public. Perhaps their views on nuclear power were less set at the time of the accident and therefore more susceptible to being influenced by the event.

REFERENCES

1. MITCHELL, R. 1979. The Public Response to Three Mile Island. Discussion Paper D-58. Resources for the Future. Washington, D.C.
2. MAZUR, A. The Dynamics of Technical Controversy. Communications Press. Washington, D.C. (In press.)
3. BOFFEY, P. 1976. Nuclear power debate: signing up the pros and cons. Science 192: 120–122.
4. CHRISTIANSEN, D. 1979. Opinions on nuclear power. IEEE Spectrum 16(November): 29.

THE REACTION OF THE NUCLEAR INDUSTRY TO THE THREE MILE ISLAND ACCIDENT

Robert A. Szalay

Atomic Industrial Forum, Inc.
Washington, D.C. 20014

INTRODUCTION

The accident at Three Mile Island (TMI) firmly established a working premise for the future—the ultimate responsibility for the safety of nuclear power plants belongs to their operators and managers. With the full support and resources of the nuclear industry behind them, utility owner/operators have recognized and accepted that responsibility. This enhanced management awareness, though not directly quantifiable, may have been the accident's most positive contribution.

Correspondingly, industry and government studies resulting from TMI have reached similar conclusions: the design of nuclear plants is fundamentally sound, but changes should be made in the organization, procedures, and attitudes of the institutions responsible for nuclear safety.[1,2,6,8] These studies showed no inherent problems in nuclear plants that make them unacceptable from a safety standpoint, but did demonstrate weaknesses in information processing, operator training and procedures, and certain aspects of the safety review process. Over the last year, the industry has worked to correct these problems.

Individual company actions show a willingness to alter practices. New organizations have been formed to deal specifically with the lessons learned. Industry resources have been concentrated on those technical/regulatory issues that could produce the most safety-effective measures. Together, these individual and collective efforts have resulted in safety gains that will continue to be realized.

THE INDUSTRY'S IMMEDIATE RESPONSE

The initial surprise caused by the TMI event gave way quickly to concerted industry action. Each utility operating or constructing a nuclear plant established a full-time task force to review every aspect of plant design, procedures, and training. Within days of the accident, reactor operators were being retrained and new instructions were issued in the form of how to recognize and respond to a similar event. As a result, the direct problem of the operator's confusion in responding to a specific small loss-of-coolant accident was corrected, for the most part, almost immediately.

At the same time, a massive response by hundreds in the nuclear industry was initiated to help at TMI and stabilize the situation. Over

0077-8923/81/0365-0222 $01.75 © 1981, NYAS

one-thousand individuals from industry, government, and the scientific community became involved in the recovery operation and investigation. Many of the lessons subsequently to be followed were spawned by the experience gained by the industry leaders who participated in the postaccident recovery.

While actions of the individual utilities and the response to bulletins released by the Nuclear Regulatory Commission (NRC) resulted in immediate upgrading of individual plant safety, the industry realized that corrective measures were required by the industry as a whole. Within two weeks of the accident, a new senior policy task force was created—the TMI Ad Hoc Nuclear Oversight Committee. Under the chairmanship of Floyd Lewis, chairman and chief executive officer of Middle South Utilities, the committee was charged with coordinating and overseeing the overall response to the accident. The Nuclear Oversight Committee consists of eight top utility executives representing investor-owned companies, public power systems, and rural cooperatives. The committee has functioned from the start with the support and resources of existing industry organizations.

As a result of this combined industry and oversight committee effort, three new and independent organizations were created to help solve the problems reflected at TMI:

● The Nuclear Safety Analysis Center, for detailed safety assessment;

● The Institute of Nuclear Power Operations, for improved operations and training;

● The Nuclear Electric Insurance Limited, for financial protection due to extended plant outages.

Additionally, the Atomic Industrial Forum (AIF) formed its Policy Committee on Follow-Up to the Three Mile Island Accident, chaired by Byron Lee, Jr., of Commonwealth Edison Company, to establish an official industry liaison with NRC. The Atomic Industrial Forum's policy committee reviewed the major lessons learned, coordinated the various industry technical efforts, and with the help of engineers from all over the country, gave guidance to the entire industry.[6]

THE NUCLEAR SAFETY ANALYSIS CENTER

On May 3, 1979, the Nuclear Safety Analysis Center (NSAC) was officially formed. The prime objective of NSAC is to provide the utility industry with the best possible technical information on questions of nuclear safety.

NSAC is headed by Dr. Edwin L. Zebroski, former director of the Electric Power Research Institute (EPRI) Nuclear Systems and Materials Department. NSAC has a staff of 50 technical experts drawn from EPRI, the national laboratories, reactor manufacturers, architect-engineers, and the electric utilities.

The basic charter of NSAC is to study in great depth what happened at TMI, what might have happened, the best solutions to the safety concerns identified there and at other nuclear power plants, and to provide a continuous source of accurate information.

To date, this has resulted in:

• Development of the most authentic sequence of events of the TMI accident;[3]

• Institution of a system to analyze every nonnormal occurrence at any nuclear plant, based on information from licensee event reports, in order to identify any needed corrective action and immediately distribute such information to nuclear utilities to assure the appropriate attention and action;

• Establishment and completion of special studies, which indicated that the safety margins at TMI during and following the accident were substantial, and can be improved generally;[4]

• Completion of a joint NSAC/INPO investigation of the Crystal River nuclear incident, and publication of a report.[5]

INSTITUTE OF NUCLEAR POWER OPERATIONS

After an early, intensive review of the many issues raised by the TMI event, improving the quality of nuclear operations and training was singled out by the operations subcommittee of the AIF policy committee as having overriding importance.[6] The Institute of Nuclear Power Operations (INPO) was proposed as a means to assure this improvement. Following accelerated senior management attention, the intent to establish INPO was announced on June 28, 1979.

The immediate focus of the institute is to establish "benchmarks of excellence" in the management and operations at nuclear power plants. These criteria are being evolved by experienced operating personnel and professional consultants. A major effort is under way to conduct independent and periodic evaluations of operating practices at nuclear plants in order to assist utilities to meet the established benchmarks. Other areas to be addressed are training of plant personnel, increased use of simulators, review and analysis of nuclear power operating experience, emergency preparedness, and human factors in plant design.

In November 1979, Atlanta, Ga., was selected as INPO's headquarters. In January 1980, Eugene P. Wilkinson was confirmed as the first president of INPO. Mr. Wilkinson had been the first captain of the first nuclear submarine, the USS Nautilus.

Task forces are now drafting the standards for excellence in operation, training, management, and emergency planning. In addition to selecting full-time staff, INPO is developing a close working relationship with the individual utilities and other industry organizations. INPO has a 1980 budget of $11 million, contributed by member utilities depending upon the size of their nuclear programs, and will have a staff of 200 professionals.

The philosophy of INPO is to encourage excellence in every aspect of nuclear power plant operations and training and to see that it is achieved. It recognizes that a problem at any nuclear plant is a potential problem for every nuclear plant. Mutual self-interest is thus a major factor in its motivation.

NUCLEAR ELECTRIC INSURANCE LIMITED

Nuclear reactor operators are currently insured for public liability and property damage through various private and government insurance programs. But the tremendous financial burden of a reactor accident and the high cost of a prolonged nuclear plant outage necessitate an additional form of coverage. Nuclear Electric Insurance Limited (NEIL) is a mutual assistance insurance program that deals with a utility's financial risk from a nuclear accident as distinguished from liability or property damage.

This mutual insurance concept (the utilities will be insuring themselves) would cover a portion of the replacement power costs incurred in the event of an extended reactor shutdown caused by accidental physical damage.

Based on commitments now in hand, and the premiums to cover these commitments, it is anticipated that insurance coverage will be available by June 1980.

AIF POLICY COMMITTEE ON TMI

The AIF policy committee was established by the Board of Directors of the Atomic Industrial Forum shortly after the Three Mile Island event. It is made up of top-level executives from all four nuclear steam supply vendors, from the major architect-engineers involved in nuclear power, plus a broad representation from utilities owning nuclear power plants.

The objective of the policy committee is, simply, to help consolidate the industry's approach to all the various TMI technical response efforts. The insights developed through the wide involvement of both people and organizations helped individual utilities pursue a relatively consistent course in response to the lessons of TMI.

Seven subcommittees specifically addressed the critical issues of emergency response planning, operations, systems and equipment, postaccident recovery, safety analysis considerations, control room design, and unresolved generic safety issues. Reports were issued in August through October 1979 and distributed to individual companies for use in the development of their internal safety programs.[6] Interactions between industry and NRC staff efforts in the process of developing these reports were healthy and productive. An example of the work produced is a model "emergency response plan," which utilities can adapt to their own plants.

The policy committee is currently working actively on the NRC staff-proposed action plan.[9] This interaction is aimed at assuring that the constructive safety efforts now in motion are not diluted by the large number of requirements of lesser value proposed in the action plan. The policy committee has proposed a constructive process for setting priorities that can lead to an orderly and positive increase in overall safety. It has also pointed out the tremendous impacts that could result without further rationalization of the proposed action plan requirements.

A LOOK BACK, AND AHEAD

From the industry's vantage point, the immediate lessons of TMI were applied soon after the accident. Of the other lessons of TMI, the most important were taken care of through the initiatives of utilities in augmenting their safety practices, in responding to NRC bulletins and orders, and in implementing programs to address the NRC short-term lessons learned.[7] Collectively, the industry established INPO and NSAC, both of which are aimed at raising the general level of safety awareness.

Two principal insights gained from the accident were (1) the incomplete coverage in procedures and training and in regulation of the small-break loss-of-coolant accident and (2) the contribution of human error in delaying correct response of the plant systems, which were otherwise capable of preventing or terminating damage at any point. The actions taken thus far by industry and NRC on small-break loss-of-coolant accidents have reduced very substantially the probability of repeating this event and of human error contributing to such an accident progression.

Additionally, new mechanisms have been set up by the industry to enhance the cumulative learning process:

• Each utility is devoting increased in-house effort towards evaluation, understanding, and implementation of actions in response to various kinds of malfunctions;

• The Nuclear Safety Analysis Center (NSAC) is providing a focus for analysis of issues that have generic elements affecting a number of plants of similar or related designs;

• The Institute of Nuclear Power Operations (INPO) is setting industry standards for training and certification of operators and supervisory personnel and for training of the managerial chain that oversees safety practices; and equally important, it is evaluating utility performance against these standards.

These activities suggest a fundamental change in the attitude and operating styles of utilities.

Many specific changes have already been invoked by the NRC and are implemented, or are in the process of implementation, by utilities. About 30 changes in procedures and in some design features are called for in the recommendations of the NRC Lessons Learned Task Force.[7] For example, all reactors now have a "saturation meter," which gives the operator one independent and clear indication on the margin he has to unwanted bulk boiling conditions in the reactor coolant. Direct indication of power-operated relief valve position is another resultant addition. Combined with improved operator training and emergency procedures and the addition of shift safety advisors, changes already made or well under way have resulted in additional safety.

The question of the margins available during an accident more serious than TMI in terms of environmental effects has been given much attention. Studies performed by the Nuclear Safety Analysis Center conclude that there was no imminent challenge to the integrity of containment at any time, and that ample means for termination of damage or escape of radioactivity from the reactor core were available by design.[3,4] Such means, according to the NSAC analysis, could terminate the progression of the accident even if the operator's perception of the nature and seriousness of the accident had been further delayed by many hours. The assertions that "a disaster was narrowly averted" have no valid substance.

Moreover, improvements have already been made that lead to the reduction of the likelihood of lapses of the kind that contributed to the TMI accident. The human error at TMI was in failing to perceive the actual progression of damage to the reactor for several hours. This arose from the constraints of conflicting procedures, regulations, and training, all of which contributed to the TMI accident. All operators now have had extensive training in the conditions that can lead to core uncovering, and have greatly improved procedures and technical backups to assure that correct actions are taken to prevent a minor malfunction from progressing to serious damage to a reactor core.

This enhanced capability to prevent serious degradation of the core and containment, and thus significant environmental releases, adds further credibility to the adequacy of current plant designs. The single failure criterion and the "design basis accident" philosophy in place still provide conservative and appropriate design and licensing bases. However, event tree/fault tree methodology, which draws on the best available realistic information, will be used to a successively greater extent to judge any further changes in the existing design bases.

At the operations management level, which includes operator training and emergency procedures, best estimate analysis, including selected multiple failures, is receiving greater emphasis. This combination of conservative design bases and enhanced operational capability offers the best approach to continuing improvement in safety margins.

CONCLUSION

The nuclear industry has responded to the Three Mile Island accident vigorously and forthrightly, and the result has been a substantial improvement in nuclear power plant safety. It is taking pride in its individual and collective safety programs, and for good reason.

Problems remain to be faced in the regulatory and political arenas. The inertial tendencies of the regulatory process were having a paralyzing effect on the nuclear industry before Three Mile Island; these tendencies have been magnified since then and could have critical impacts on the industry's ability to supply needed energy now and in the future. This is evident in the "licensing pause" that has held construction permits and operating licenses hostage. It is also apparent in the NRC's slowness in providing a rational program for dealing with the continuum of proposed changes in licensing requirements.

Decisions need to be made at the highest level of government on the nation's overall energy strategy and nuclear power's role in it. Such a statement of national energy policy should form the preamble for NRC's articulation of nuclear power plant regulatory and safety goals. These goals must be set in recognition of the competing risks from other electric generation alternatives.

Without these goals and a regulatory process designed and managed to achieve them, uncertainty will prevail. In the current financial environment, few things could be more damaging, particularly when our energy needs are taken into account.

Throughout the year following the Three Mile Island accident, the nuclear industry has clearly shown its determination and capability to produce safe nuclear energy, and in greater quantity. It is now up to the administration and Congress, with the support of the public, to create a regulatory and political climate that will allow the benefits of nuclear power to be fully realized.

REFERENCES

1. KEMENY, J. G., et al. 1979. Report of the President's Commission on the Accident at Three Mile Island. U.S. Government Printing Office. Washington, D.C.
2. Nuclear Regulatory Commission Special Inquiry Group. 1980. Three Mile Island—A Report to the Commissioners and to the Public. Nuclear Regulatory Commission. Washington, D.C. (The Rogovin report.)
3. NSAC. 1980. Analysis of the Three Mile Island-Unit 2 Accident—NSAC-1 Revised. Report No. NSAC-80-1. Nuclear Safety Analysis Center. Palo Alto, Calif.
4. NSAC. 1980. Mitigation of Small-Break LOCAs in Pressurized Water Reactor Systems. Report No. NSAC-80-2. Nuclear Safety Analysis Center. Palo Alto, Calif.
5. NSAC/INPO. 1980. Analysis and Evaluation of Crystal River-Unit 3 Incident. Report No. NSAC-3/INPO-1. Nuclear Safety Analysis Center/Institute of Nuclear Power Operations. Palo Alto, Calif.
6. AIF Policy Committee on Follow-Up to the Three Mile Island Accident. 1979. Reports of subcommittees on: Emergency Response Planning, Operations, Systems and Equipment, Post-Accident Recovery, Safety Analysis Considerations, and Control

Room Design, August/September/October. Atomic Industrial Forum. Washington, D.C.

7. Office of Nuclear Reactor Regulation. 1979. TMI Lessons Learned Task Force Status Report and Short-Term Recommendations. Report No. NUREG-0578. U.S. Nuclear Regulatory Commission. Washington, D.C.

8. Office of Nuclear Reactor Regulation. 1979. TMI-2 Lessons Learned Task Force Final Report. Report No. NUREG-0585. U.S. Nuclear Regulatory Commission. Washington, D.C.

9. NRC. 1980. Action Plans Developed as a Result of the TMI-2 Accident. Report No. NUREG-0660, Draft 2. U.S. Nuclear Regulatory Commission. Washington, D.C.

10. TMI Ad Hoc Nuclear Oversight Committee. 1980. A Report to the President and the American People: One Year after Three Mile Island. Washington, D.C.

REASSESSING THE NUCLEAR OPTION: THREE MILE ISLAND FROM A STATE PERSPECTIVE

Emilio E. Varanini III

California Energy Commission
Sacramento, California 95825

INTRODUCTION

In the year since the accident at Three Mile Island (TMI), an enormous amount of analytical work has been performed and published, hundreds of articles have been written, and conferences nationwide have discussed the post-TMI prognosis for nuclear power. Buried within this mass of material is almost any conclusion one may wish to draw about the causes of the accident, its consequences, and its implications. Some have called the TMI accident a disaster mitigable only by an immediate retreat from the use of nuclear power. Others take a sanguine view of the accident, seeing it as a simple iteration in the overall learning process for nuclear power.

Although the bulk of reaction to TMI lies between these two extremes, the common views that emerge have serious implications for nuclear power. TMI has shaken the confidence that almost all facets of society have had in our ability to deal with nuclear technology. The credibility and competence of the nuclear industry, utilities, and the federal government have been questioned as a result of TMI. Most important of all is the general loss of public confidence in nuclear power and its regulatory institutions. Recent news clips showing intense local opposition to the utility's plans to vent radioactive gases into the atmosphere provide empirical evidence of the public's reaction to TMI. A notable characteristic of this reaction was the citizens' total lack of confidence in the integrity of either the utility or the Nuclear Regulatory Commission (NRC). The citizens simply did not believe statements that gas ventings would be harmless. The significance of the public confidence issue is summed up in a quote from the President's Commission on Three Mile Island:

> We are convinced that, unless portions of the industry and its regulatory agency undergo fundamental changes, they will over time totally destroy public confidence and, hence, they will be responsible for the elimination of nuclear power as a viable source of energy.[1]

These circumstances imply an expanded role for state government in voicing regional public opinion about nuclear power specifically, about regional energy policies in general, and in developing an alternative regulatory framework to that provided by the federal government.

0077-8923/81/0365-0230 $01.75/2 © 1981, NYAS

STATE INTERESTS

The implications of TMI for state government are manifest; state roles will never be the same again. States can be expected to expand their jurisdiction in all areas of energy planning, especially those areas that directly affect the health and safety of citizens.

TMI can be expected to accelerate the trend toward the enhancement of state capabilities in forward or proactive planning. Apart from traditional state land use authority, until quite recently state regulation of power plants was confined to the financial and economic aspects of utilities' proposals. This regulatory authority was exercised by state public utilities commissions and focused on providing reasonable rates to customers. States rarely became involved in decisions on alternative technologies or fuels.

Several states, including California, have adopted broad planning programs directed toward optimizing energy supply, environmental quality, and economic considerations. For example, California's Warren-Alquist Act instituted an "open planning" process.

The principal characteristics of the "open planning" process are (1) that it provides a public forum in which the major decisions about how much to build, what to build, and where to build are very much at issue, rather than untouchable "givens," as they are under traditional licensing schemes; (2) that the utility or other project proponent is expected to demonstrate the overall validity of the choices it has made in proposing the project; (3) that it provides a genuine opportunity to challenge and debunk, if possible, major planning mistakes that characterize traditional utility supply planning and regulation; and (4) that the process eliminates any project proposal that cannot withstand the test and substitutes an alternative that does.

In addition, states can be expected to develop an independent capability to understand nuclear technology and to clarify responsibility at all levels of government for emergency preparedness and response.

The following discussion covers six general issues: (1) need for power, (2) supply planning, (3) economics, (4) alternatives, (5) land use policy and power plant siting, and (6) emergency planning. Together, these issues provide a comprehensive discussion of state interests resulting from the accident at TMI. Additional sections on state authority and potential problems in state-federal relationships discuss issues that may become important in the near future.

Need for Power

Increasingly sophisticated state forecasting capabilities allow a more accurate determination of electricity demand. Utilities have traditionally used macroeconomic forecasting models that rely on only a few demographic and economic variables, including gross state product, personal income, retail sales, population, energy prices, and past electricity sales. Macroeconomic models establish historical relationships between these variables and electricity consumption, and then extrapolate future demand.

The flaw of macroeconomic forecasting is its insensitivity to changes in historical correlation between measured variables and electricity demand. For example, growth in personal income does not by itself dictate growth in demand. As consumers choose to spend their extra income on goods that do not conserve electricity—a change that has been occurring—the correlation on which macroeconomic models are based breaks down.

Microeconomic models based on end use of electricity provide a more accurate forecasting capability. Such models allow an almost empirical determination of how specific energy-conserving activities contribute to electrical demand. For example, the forecast of demand in the residential sector in a particular service area would be based on the product of numbers of households, the fraction of households having a particular appliance, and the average electricity consumption of that appliance (i.e., end use). This procedure is followed for all end uses, and the resulting sum is the forecast.

End-use forecasting also results in a more accurate peak demand forecast, since energy demand is calculated at the point of consumption rather than at the point of generation. Since transmission losses can amount to 10% or more over long distances, adding generation requirements into the forecast to meet needs on a distant point gives an inflated demand figure. The same demand could be met by generating power close to the consumption point.[2]

Microeconomic models are more sensitive to price elasticity than are the macromodels. With a micromodel, the forecaster can simulate more accurately the effects on electricity consumption of increasing oil prices, inflation, and market saturation. In addition, the micromodel can simulate the effects of alternative policies, such as conservation in the residential and commercial sectors, thereby providing a basis for demand-reducing policies.

The deteriorating situation in the Middle East injects another variable—an extremely unstable variable—into the determination of need for new power plants: the necessity of displacing oil consumption. Although nuclear power does not displace overall a great deal of foreign oil, some areas are far more dependent on nuclear power than are other areas. As oil displacement strategies increase the need for alternative supply options, the federal government can be expected to support expanded nuclear deployment. At the same time, TMI is likely to place nuclear power lower in the hierarchy of possible state supply options. If states are to meet national oil displacement goals, the importance of integrated energy planning becomes more apparent. The following sections discuss several components of integrated planning.

Supply Planning

As stated above, states are looking at demand for electricity and methods for meeting demand (i.e., supply planning) in an integrated or unified manner. The key feature of unified supply planning is consideration of a variety of state interests. For example, in addition to planning for peak demand growth, reserve margins, facility retirements, and oil displacement, such factors as public health and safety, preservation of environmental quality, maintaining a sound economy, and conservation of energy resources will receive more consideration. These general principles imply a variety of specific concerns, such as preserving agricultural land and fresh water supplies, preserving coastal ecology and esthetic value, protecting citizens from harmful effluents and accidents, preserving air quality, assuring a supply of capital for nonenergy economic growth and jobs, and providing fair rates for electricity. Other important considerations include diversity of supply to enhance system reliability, costs to rate payers and society overall of overbuilding, and the diminishing returns that can defeat the purpose of economy of scale. States are starting to consider all these factors in establishing policies that will result in an optional supply mix.

Within this framework, TMI has special implications for nuclear development. TMI associates a certain level of technical uncertainty with the nuclear option. This uncertainty has broad implications, affecting nearly all of the factors above that are important to supply planning. For example, unresolved questions—whether real or perceived—about nuclear technology factor into both utility plans to rely on nuclear power to meet future demand growth and state interests in balancing a variety of social variables.

Regulatory ratcheting from state and federal government entities is also a factor in determining the placement of nuclear power in the supply mix. New regulatory measures add another level of uncertainty about power availability in grids with a large nuclear component. Frequent shutdowns for technical backfitting and general upgrading of operations procedures are certain to reduce system reliability and increase costs.

Although expanded state supply planning functions are typically lodged in new agencies (the California Energy Commission, for example), the rate-making bodies that have long been part of state government also have a role to play. Sophisticated rate-making methodology is being developed and deployed to shift rate-making authority from its historical role as a means of counteracting railroad rate monopolies and price fixing to a role that supports state efforts to achieve a balance of public and private sector goals. Allowing utilities a rate of return on research and development investments in preferred technologies is one example of enlightened rate-making strategy. These economic issues are discussed in more detail in the next section.

Economics

Although the full range of TMI impacts is yet to be determined, the economic issues are certain to be among the most significant. Moreover, the economic issues have the most direct bearing on state government, since rate-making decisions on all costs must eventually be made by state authorities, generally the public utilities commission (PUC) or its counterpart.

The economic implications of TMI fall into two categories: (1) costs of the accident itself, and (2) economic factors associated with comprehensive state energy planning. These two areas are discussed in turn below.

Costs of the Accident

The magnitude of the costs of the TMI accident and the consequent direct and indirect burdens on rate payers and the utility will provide experience through which states can develop criteria and formulate policy for the economic acceptability of the nuclear option. The purpose of this paper is not, however, to calculate estimates of TMI-related costs. This exercise is better left to economists and to the Pennsylvania PUC. The important point is that the costs will be substantial. It is more appropriate for me to focus on the different categories of costs, on which entity will bear the different costs, and to outline the implications for state decision making.

In testimony before the California legislature, the California Energy Commission placed the TMI economic impacts into three categories:

1. Public liability.
2. Costs to the utility (and its insurers, stockholders, and rate payers).
3. Indirect costs that have longer term "ripple" effects on the economy.

Public Liability

Third-party claims for TMI are as yet undetermined. It should be noted, however, that pending lawsuits include claims of several billion dollars.

Under the Price-Anderson Act, public liability is limited to $560 million per incident. This amount is covered by a three-part system. The act requires first that a utility operating a nuclear plant carry private liability insurance up to the maximum amount available ($140 million). Next, each nuclear utility can be assessed up to $5 million to cover additional claims. This amounts to about $350 million in coverage. The balance of coverage up to a total $560 million is provided by the federal government; simple arithmetic places this amount at $70 million.

Although the total of TMI claims will probably fall within the

Price-Anderson Act limit, a question still arises as to whether $560 million is sufficient. This limit was established in 1957 and has not changed to keep pace with inflation. According to the *Report of the* [Pennsylvania] *Governor's Commission on Three Mile Island,* the equivalent limit in 1979 dollars should be almost $1.4 billion.[3] The only role for state government, under existing law, is to assess the adequacy of Price-Anderson liability and attempt to influence Congress on altering the current liability limits if these are determined to be inadequate.

Utility Costs: Property Damage

Not all of the property damage caused by the accident is covered by insurance. The currently available insurance limit is $300 million. Estimates of costs to either repair ($576 million to $1.02 billion) or replace ($1.6 billion) the TMI facility exceed the limit considerably.[3] Although additional coverage may be obtained through lawsuits against suppliers of faulty equipment that contributed to the accident, the TMI accident will probably result in substantial uninsured costs to the utility.

Other Economic Effects

The California Energy Commission's testimony before the California legislature also identified a number of additional issues that may affect states' rate payers or produce other economic consequences that have public policy implications:

- Modification to other nuclear power plants.
- Improvement in emergency response plans.
- Loss of business to nuclear vendors and constructors.
- Loss of capital value of nuclear utilities.
- Increase in nuclear power plant insurance premiums.
- Increase in future costs of raising nuclear utility capital, both equity and debt.
- Possible shutdown of existing nuclear power plants, or abandonment of plants under construction.
- Loss in property values near existing nuclear plants.

Economic Aspects of Comprehensive Energy Planning

These are hard times for utility companies. With many utility stocks selling currently at below book value and with utility bond ratings at a low point, capital is difficult to raise. With nuclear costs escalating at a rapid rate—a problem that is likely to become worse as a result of TMI—the ability of utilities to finance nuclear power plants is questionable.

Part of states' increased forward energy planning capabilities is a more sophisticated approach to developing "preferred" alternatives for meeting electricity needs. The emphasis of this increased sophistication is on utility investment strategies in such alternatives as conservation, geothermal, cogeneration, clean synthetics, wind, biomass, and solar. State policies should offer financial incentives for investing in those alternatives that are preferable on a regional basis. In addition, utilities can be directed to participate in customer loan programs for conservation retrofits and solar energy. As these loan programs evolve, utilities can be directed more toward total energy service corporations providing a variety of services rather than just expanding electricity supply through capital-intensive new power plants.

Alternatives

Recent years have seen an intense debate between advocates of conventional energy technologies—oil, natural gas, coal, nuclear, and now synthetic fuels—and advocates of the so-called soft path energy alternatives, such as solar, conservation, biomass, wind, etc. The alternative energy future recently gained considerable credibility by virtue of the Harvard Business School study, which indicates that conservation and solar energy are the best investments for matching energy supply with demand.[4]

As indicated in previous sections, alternative energy technologies will be advanced as comprehensive state energy planning is developed and states move to maintain the flexibility in supply mix that allows a regional approach to meeting demand. Some alternatives, such as solar energy, that are realistic in the Southwest have less potential in other parts of the country. Moreover, alternative technologies are more consistent with the broad range of societal goals on which state energy planning is based.

In general, alternatives are being identified as investment targets by utilities, governments, and private sector businesses, including lending institutions. As alternatives become preferential choices, nuclear energy will be moved toward the last resort category in utility supply plans.

Land Use

Although states traditionally have had exclusive authority over land use, it is only recently that states have developed the capability to site power plants. Pennsylvania, for example, does not have an authority that oversees the placement of new power plants, whether nuclear or fossil fueled. One suggestion of the Pennsylvania legislature's Commission on Three Mile Island is to explore the possibility of establishing such a state authority.[5]

TMI demonstrated the potential problems with siting nuclear plants near populated areas. As a result, states with siting authority may move toward remote siting of new nuclear plants. In addition, states can be expected to take action to limit population growth through land use controls around existing plants that are remotely sited.

Much of the post-TMI literature discusses the need to site nuclear power plants in remote areas. As a result, it is likely that NRC will adopt a remote siting strategy as part of its licensing authority. Nevertheless, concurrent state siting jurisdiction has payoffs for state officials in terms of learning experience and contact with federal and utility officials. California, for example, had access to firsthand information during the progress of the TMI accident due to prior work in nuclear power plant siting.

Remote siting of nuclear plants raises difficult questions about preservation of open space and agricultural lands as well as questions concerning rural areas assuming the burden of a power plant site for the benefit of city dwellers who use the power. These "not in my backyard" issues already have arisen around the country and are likely to become even more problematical as a result of TMI. Avoiding these issues gives utilities further incentive to look away from new investments in nuclear power.

Emergency Planning

One of the most disturbing aspects of TMI from a state perspective is the dichotomy between states' plenary responsibility for public health and safety and states' limited responsibility for controlling or interpreting the flow of information between a utility and the federal regulatory authorities. The utter confusion that characterized communications during the course of the accident and that culminated in near hysteria over the possibility (nonexistent, as it turns out) of a hydrogen explosion is an indication of just how important information flow can be.

In general, state response to TMI is to upgrade emergency planning, personnel, and equipment. Post-TMI literature has produced an exhaustive list of advisable actions in the event of a radiologic emergency, ranging from extensive evacuations to provisions for housing livestock. Some of the more important issues for the state are to make sure that (1) the duties and authority of various entities are specified clearly; (2) the utility and state/local authorities share information on the progress of an emergency and necessary actions; (3) the public is informed of the potential hazards of radiation; (4) local health authorities are provided with adequate information; (5) a reliable communications network exists between state, local, and federal authorities, including NRC; (6) "on-the-shelf" evacuation plans exist; and (7) continuous off-site radiation monitoring is performed.

Another prospective action states could take is to limit population growth in the area of nuclear power plants. Of course, some states do not have the luxury of implementing this option since nuclear power plants are already sited in areas of high population density, for example, at Indian Point in New York.

Who is responsible for preparing emergency plans? Probably the most comprehensive treatment of the issue is contained in a report for NRC by the National Academy of Public Administration, entitled *Major Alternatives for Government Policies, Organizational Structures, and Actions in Civilian Nuclear Reactor Emergency Management in the United States.*[6]

This report describes the general responsibilities of varous entities as follows:

> NRC is responsible for assuring that plants are built and operated in a way that does not threaten public health and safety. Operating through the NRC licensing process, the builder of the plant has primary responsibility for its safe design. Operating under NRC regulations, the utility company has primary responsibility for safe operations onsite, warning governmental authority of any dangers offsite, and cooperating with governments in emergency planning and response. State and local governments have primary responsibility for protection of public health and safety as these may be threatened by releases of radiation or otherwise; e.g., fires. Federal agencies are responsible for helping state and local governments, industry, and the public in emergency planning and response. As discussed above, the responsibilities of the NRC may include direct command action at the site.[6]

Proposed NRC regulations tie new reactor licenses as well as contin- uation of existing licenses to NRC approval of a local emergency plan. The regulation would give the licensee the incentive to take responsibil- ity for preparing the plan. It is important for state and local government to become involved in this first to assure that it is consistent with the fulfillment of states' responsibilities over public health and safety and second to prevent the development of more than one plan (one by the utilities and one by local government) in any area.

A number of options for financing emergency planning have been suggested, including federal grants to state and local government and levying a tax on licenses. Again, the best discussion of this issue is in the major alternatives report.[6] California has taken the tax levy approach, with legislation requiring each nuclear power plant to contribute to a fund for emergency planning. This regulation places a limit of $2 million on utility assessments for emergency planning around four nuclear plants.

ADDITIONAL STATE CONCERNS

State Inspection and Audits

The state of Illinois made the following recommendation for state action:

It is recommended that the State of Illinois conduct Independent Safety Audits, as necessary, covering major nuclear facilities operating and under construction within its boundaries, these independent audits should also include a review of the performance of relevant Regulatory Agencies. It [is] further recommended that these audits be carried out under the responsibility of the Illinois Commission on Atomic Energy, and/or other appropriate State Agencies.[7]

Similarly, the Pennsylvania House of Representatives Select Committee on Three Mile Island suggested studying the feasibility of placing a state inspector at each nuclear power plant.[5]

Both of these suggestions would increase the state's knowledge of the safety status of nuclear plants. The success of such a state policy is largely dependent, however, on the cooperation of the operating utility. Furthermore, the new NRC policy of placing federal inspectors at each plant may obviate the need for such state actions as long as adequate communications exist between state officials and federal inspectors.

Reviewing the TMI Recovery Process

The aftermath of the TMI accident provides an opportunity for states to learn about the environmental, public health, and economic aspects of a serious nuclear accident. For example, venting radioactive gases from a containment vessel is a problem over which states arguably have regulatory authority under the 1977 Clean Air Act amendments.

The economic consequences of the accident are among the most serious. This issue is discussed in more detail in an earlier section. States now have the opportunity to observe the actions of the Pennsylvania Public Utilities Commission and financial problems of Metropolitan Edison as a case study for establishing their own policies in preparation for possible nuclear accidents in the future.

Finally, states should observe carefully the reaction of citizens around TMI. This would aid in planning emergency response and in obtaining empirical data on the public's perception of risk from radiation.

Participation in Federal Proceedings

States may wish to become intervenors in federal proceedings regarding the safety of nuclear plants. For example, the California Energy Commission elected to participate in the Atomic Safety Licensing Board's proceeding on the Rancho Seco nuclear plant and has succeeded in raising a number of technical issues regarding the safe operation of the plant. Another possibility for state actions includes the study of particular reactor safety issues, both technical and institutional, to attempt to influence federal decisions regarding these issues.

STATE AUTHORITY

Comprehensive state energy planning is not without challenge. Two lawsuits filed against the California Energy Commission will provide further definition of state authority to expand jurisdiction into areas where utilities formerly acted unilaterally:

1. *Pacific Legal Foundation v. California Energy Commission.*[8]
2. *Pacific Gas and Electric v. California Energy Commission.*[9]

The first suit challenged three California laws that tie future reactor deployment to (1) completion of a feasibility study on placing reactors underground; (2) the existence of a reprocessing technology; and (3) the existence of a high-level waste-disposal technology. A federal district court found that the undergrounding and reprocessing issues were moot, but that the waste-disposal law was unconstitutional.[8]

The second suit was decided in April 1980 in favor of the plaintiff.[9] The state authority issues in this suit are much broader. The decision struck down state regulation of:

1. Design, siting, construction, or location of a nuclear power plant.
2. The effect of a nuclear power plant on environmental quality.
3. Transport, handling, and storage of wastes and fuels.
4. Special design features to account for seismic or other potential hazards.
5. Type of fuel or source of fuel.
6. "Other factors" (unspecified) relating to the safety and reliability of a nuclear plant.
7. Monitoring the construction and operation of a nuclear power plant.

It could also:

1. Prohibit state authorities from even requiring the submission of information from utilities regarding the above matters.
2. Prohibit state authorities from regulating, including even requiring submission of any information about, sites other than those sites proposed to the NRC.

These cases have been consolidated for appeal in the Ninth Circuit Court of Appeals.[10]

PROBLEM AREAS

Increasing capabilities for states in comprehensive energy planning may create conflicts with the federal government over energy policy goals and their implementation. While the issues discussed above imply a need to retain state or regional flexibility in meeting national energy policy goals, the federal government appears to be focusing on energy technologies that fit more easily into a framework of centralized direction and financing. These are the "conventional" technologies—oil, natural gas, coal, synthetics, and nuclear energy. Alternative technologies to which states are likely to give more consideration within their increased planning capabilities—conservation, solar, wind, geothermal, cogeneration, biomass—are not emphasized in national energy policy because their success depends to a large extent on local initiative.

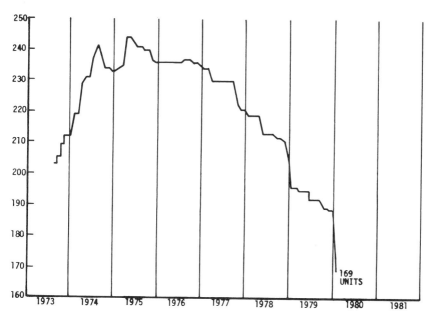

FIGURE 1. Profile of nuclear reactors built, being built, or planned since 1973. Data as of January 31, 1980 (from Reference 11). Announced or ordered units for which applications were not under review by the NRC were not counted in the statistics prior to September 1973 or after December 1979. In January, in addition to 4 cancellations, the following were dropped from the statistics: 3 nonoperating plants, 9 plants indefinitely deferred, 1 plant denied state certification, and 6 ordered or announced plants.

It is not difficult to understand federal motivation in taking control of energy decision making. The unstable situation in the Middle East, the uncertainties of continued foreign oil supplies, and the economic burden of importing oil can be expected to result in action by the federal government for reasons of national security and political survival for the current administration. Unfortunately, federal plans to expand coal production, increase the number of nuclear power plants (ostensibly to offset oil imports), and develop a synthetic fuels industry could ride roughshod over states' interests in developing more sophisticated methods of dealing with overall energy planning. The proposed (and nearly established) Energy Mobilization Board (EMB) is evidence of federal intentions to exercise more authority in energy decision making. Although nuclear power is not within the scope of EMB's authority as currently proposed, it is quite possible that nuclear power will be included within the near future. Under such a scenario, nuclear power may become a federal enterprise or nuclear plants may be sited on federal land using taxpayers' and rate payers' money.

Another method of expanding nuclear power is to continue to build public confidence through announced regulatory reforms. This method, however, would require a concomitant effort to build the confidence of utilities. New orders for light water reactors in the United States have been declining for several years. The NRC's *Program Summary Report* for February 1980 shows a linear decline since the end of 1976 in reactors being built or planned (see FIGURE 1).[11] A number of factors account for this decline, including rapidly escalating costs for plant construction, increasing regulatory constraints on manufacturers and utilities as public awareness of potential hazard grows, the need to "overbuild" nuclear facilities to meet increased demand for safety and the consequent loss of economics of scale, and a growing public awareness of the problems associated with the entire nuclear fuel cycle, from uranium mining to waste disposal. These factors have made nuclear facilities the first to be abandoned in utility supply plans. TMI can only exacerbate these problems. New designs are bound to increase capital costs of nuclear plants and increase the construction time requirements. New regulatory requirements are bound to increase the overall costs of nuclear plants and to make utilities even more wary of the nuclear option. Thus, attempts to build confidence may defeat their own purpose unless the utilities are assured that new regulations will not be stifling. State governments may wish to look critically at the federal government's dual role of providing the necessary assurances to both the public and the utilities.

REFERENCES

1. KEMENY, J. G., et al. 1979. Report of the President's Commission on the Accident at Three Mile Island: 25. U.S. Government Printing Office. Washington, D.C.
2. California Energy Commission. 1979. Demand Forecast Testimony of Robert K. Weatherwax of the Energy Commission Staff. Sacramento, Calif.
3. Commission on Three Mile Island. 1980. Report of the Governor's Commission on Three Mile Island: 50, 147. Office of the Governor of Pennsylvania. Harrisburg, Penn.
4. STOBAUGH, R. & D. YERGIN, Eds. 1979. Energy Future: Report of the Energy Project at the Harvard Business School. Random House. New York, N.Y.
5. State Government Committee. 1980. Report of the Select Committee—Three Mile Island. Pennsylvania House of Representatives. Harrisburg, Penn. (Summary of hearing suggestions.)

6. National Academy of Public Administration. 1980. Major Alternatives for Government Policies, Organizational Structures, and Actions in Civil Nuclear Reactor Emergency Management in the United States. Report No. NUREG/CR-1225: 42. Prepared for the Nuclear Regulatory Commission. Washington, D.C.

7. Ad-Hoc Nuclear Power Reactor Safety Review Committee. 1979. Nuclear Power Reactor Safety in Illinois: A Report to the Honorable James Thompson, Governor of the State of Illinois: 12. Illinois Commission on Atomic Energy. Chicago, Ill.

8. Pacific Legal Foundation v. State Energy Resources Conservation and Development Commission, 472 F. Supp. 191 (1979).

9. Pacific Gas and Electric Company v. State Energy Resources Conservation and Development Commission, 489 F. Supp. 699 (1979).

10. Pacific Gas and Electric Company et al. v. State Energy Resources Conservation and Development Commission et al., U.S. Court of Appeals, Ninth Circuit, Nos. 80-4265 and 80-4273.

11. NRC. 1980. Program Summary Report. Report No. NUREG-0380 4(2). U.S. Nuclear Regulatory Commission. Washington, D.C.

PUBLIC INTEREST GROUPS

C. P. Wolf*

*Department of Social Sciences
Polytechnic Institute of New York
Brooklyn, New York 11201*

The topic of public interest groups raises some awkward, but essential, questions not only for nuclear development and regulation but for democratic governance in technological society as a whole. What public concerns were aroused by the accident at Three Mile Island (TMI)? What is the legitimacy of such concerns? How can they be usefully addressed and expressed? The awkwardness comes in trying to identify the "public interest" and in structuring "meaningful" public involvement in relation to public concerns. Yet these issues can scarcely be avoided so long as we strive to maintain responsible and responsive public institutions. While these questions are not unique to the situation at hand, the accident at TMI does pose them in especially acute form.

The topic also underscores the polarization and politicization of the nuclear debate, and the partisan stance of the parties at interest. The terms of the debate often have been polemical; the tone, shrill; the outcome, indefinite. These factors weigh against the value of public discussion and decision as viewed by those in institutional control. Yet the ordeal of TMI suggests that the institutions, as much as the technology, were *not* in effective control.[1] As the event unfolded, institutional adequacy—both of safe nuclear operations and of emergency preparedness and response—became increasingly tenuous. Formation of the President's Commission on the Accident at Three Mile Island (Kemeny commission) is symbolic of the seriousness of the accident and of the evident deficiencies in its management. In the fallout from TMI, new opportunities were opened for involvement of the public in general and of public interest groups in particular.

THE "PUBLIC INTEREST"

It would seem reasonable to begin the identification of "public interest groups" by a definition of the "public interest." It is a subject political philosophers have been pondering for decades, if not centuries, and the end is nowhere in sight.[2] Surely we have learned enough by now to realize that there isn't going to be a definitive definition, so long as the political process remains open and public choices are freely taken. In our own political culture, moreover, we experience great difficulty in formu-

---

*Please address all correspondence to P.O. Box 587, Canal Street Station, New York, N.Y. 10013.

0077-8923/81/0365-0244 $01.75/2 © 1981, NYAS

lating and operating with collective concepts such as this. The reduction-
ist tendency is to identify the public interest as a summation of special
interests, and to convert the notion of "public goods" (e.g., environmental
quality) to private (market) goods. This makes for grave complications
when trying to practice a "politics of consensus," e.g., in devising a
national energy plan.

Although the supersummative nature of the public interest is elusive,
at least we can distinguish between it and "public opinion." The majori-
tarian definition does not satisfy our understanding of its deeper meaning; individuals can speak in the public interest, and "the public" can
voice its collective feelings from very private motives. This is the
distinction between "self-" and "society-regarding" attitudes that political scientist Edward Banfield introduced and others have elaborated.
Similarly, a long tradition of survey research has revealed a sharp
disjunction between personal and social or societal futures.[3]

Clearly there is no one "public"; we should properly speak of publics
in the plural, not in the singular. In acknowledging this pluralism, the
concept of representation is called into question.[4] For example, a Mobil
ad, "Who's the Public in 'Public Interest' Politics?" (The New York
Times, 1 November 1979), asserts that most "public interest" groups
"don't represent any broader interest than that of their own members."
Moreover, "the interests advanced by many of the public interest lobbies
are actually 'special' interests—limited interests, often held by extremely
small groups who are in obvious disagreement with the American
majority." Their minority interests are held to be "anti-growth, antibusiness, anti-energy, and dedicated to an elitist, big-government view of
America." Conversely, the identification of special business interests
with the public interest, e.g., "What's good for General Motors is good for
America," has received its share of ridicule.

One evasion of the conceptual problem is to emphasize the preemi-
nence of local or regional interests over the national interest. Thus, TMI
area residents are now claiming a psychological exemption from further
nuclear operations.[5] Public acceptance regularly exhibits an inverse
relation to distance from nuclear power generating facilities. The "anywhere but here" syndrome encountered in any sort of nuisance facility
siting cannot provide a general prescription for siting decisions.
Nevertheless, there is a proportionality between the right of interest and
the intensity of impact. This finds expression in the legal concept of
"standing," but again that principle is weakened by the admissibility of
class action suits, which convey the sense of disinterested concern.

Not only is it arguable what constitutes public interest, but who the
public themselves are remains a murky question. A generation ago,
Walter Lippmann declared the public to be a phantom—an insubstantial,
incorporeal, and inconvenient fiction. What has happened since then is
completely the opposite: the institutionalization of public involvement in
public decision making, e.g., as directed by revised Council on Environmental Quality guidelines for administering the National Environmental

Policy Act of 1969 (NEPA). The vital distinction here is between the public and the electorate. Previously, the public interest was operationally defined as the will of elected officials in representative assemblies. This paradigm of "normal politics" has now been supplemented by agency mandates to mount public involvement programs. While there is lingering uncertainty over the role of public participation in decision making, the intent is clear. In essence, meaningful participation means power sharing. Naturally this complicates the decision process, but it is held to be a necessary complication.

Another usual distinction is that between publics and experts. Its translation in decision and action turns on the further separation of technical and political issues. On purely technical questions, expert judgment is generally upheld. The difficulty comes in the circumstance that few (if any) *public* issues are of this character. More often, they are of a kind Alvin Weinberg calls "trans-scientific,"[6] namely, questions asked of science that cannot be answered by science—e.g., those attended by great uncertainty, such as low-level radiation effects, or having strong political overtones, such as radioactive waste management. This order of question was recognized in the conference prospectus:

> One of the clearest lessons of the last few years is that decisions about complex technologies and whether or not to implement them very often give rise to controversy and even rancorous debate. This is true especially when the safety (or conversely the risk) of the technology is uncertain, and when there is little knowledge and less agreement about the potential social, economic, and ethical implications. Recent examples of this phenomenon include recombinant DNA research and, of course, nuclear power: uncertainty and disagreement characterize the "state of the art" of both of those technologies, and both have been hotly debated in the public arena.

In principle, the fact-value distinction traditional in the modern history of Western science permits a clear division of technical from political (or value) components of a public issue. In practice, their interaction prevents any clean separation. At the point where value concerns become salient, the specialized knowledge of experts does not confer on them any special privilege; in fact, it marks the limit of their competence qua scientists.[7] Potentially affected publics are the "experts" on their own perceptions and values. Since we encounter difficulty in achieving value consensus across the wide range of differences present in our society, there is a temptation to convert value questions into technical ones. Even if the science side of such questions were solved, the main task of public business would remain. Again to quote the conference prospectus:

> Much of the debate finally boils down to questions related to values. Disputes about the tradeoffs of risks and benefits often revolve around competing views about what the public will pay for what different groups say the public needs or wants. Other value-laden questions include the matter of the equitable distribution of risks (and benefits), particularly with respect to workers in reactors and residents of surrounding areas and the

compatibility of stockholder-accountable business and the safest possible operation of nuclear reactors.

Getting the right answer depends first of all on getting the question right. Not only is the public understanding of science at stake, but also the scientific understanding of the public.

PUBLIC INTEREST GROUPS

If intensional definition fails to yield a simple answer to the question of what public interest groups are, perhaps we will get farther by the extensional approach. Public interest groups are not part of the public sector, which is associated with government, but of the private sector. The latter can be divided between the corporate sector and that peculiar, but characteristic, American institution, the voluntary sector. Corresponding institutional organizations in the present context would be the Nuclear Regulatory Commission (NRC), the Atomic Industrial Forum (AIF), and, on the national level, the Union of Concerned Scientists (UCS). It is in the voluntary sector that we find public interest groups. Claims on the part of industry associations, such as AIF, to represent the public interest are somewhat vitiated by their pecuniary interest. Professional associations, such as the Human Factors Society and the Systems Safety Society, are more of a marginal case. While they possess a public service aspect, the basis of member participation is more vocational than volitional. Members of public interest groups are more likely to be paying than paid, except for small and typically underpaid (by industry standards) staffs.

For present purposes, the most clearly identifiable public interest groups are those of the antinuclear movement, such as the Nader affiliate Private Citizen, joined by numerous environmental groups—Environmental Defense Fund, National Audubon Society, Sierra Club, and so on. With one exception—the Union of Concerned Scientists—these national organizations played far less of a role in the aftermath of TMI than did spontaneously generated local groups.[5] Indeed, the antinuclear movement has long maintained a pronounced local character, mobilizing as threats appear and dispersing as they recede. In movement dynamics, legitimacy is derived from the concern of the potentially affected. Though episodic, associational vigor is strongest at local levels. A number of writers have discerned a consolidation of the movement from local to national levels, and this progression is believed to denote its success. This rather oversimplifies the picture;[8] in fact, the most conspicuous recent development has been the formation of regional alliances—loose federations of mostly local groups—patterned after the Clamshell Alliance and its confrontational tactics at Seabrook, New Hampshire.

As we enter the decade of the 1980s, it is apparent that energy development has become the major context of environmental concern. Compared with Earth Day, 1970, which many consider to have launched environmentalism as a national movement, the celebration 10 years later was a faint, almost inaudible, echo. It is an unfair comparison; the environmental movement had moved on. Where it had gone could be viewed in the light of Sun Day, 1979: alternative energy and appropriate technology. The TMI accident merely confirmed the apprehensions of these proliferating groups (of which the New Alchemy Institute is perhaps best known) and intensified their efforts to provide viable alternatives, particularly solar energy and conservation. It is also noteworthy that Earth Day, 1970, was preceded by NEPA; in one version or another, that bill had been in preparation for almost a decade. In our society, the structure of events and the structure of decision making are intricate and diverse.

The accident at TMI also influenced voluntary associations not focused on nuclear opposition and environmental protection, for example, the National Council of Churches (NCC). I am reliably informed that the accident was decisive in securing passage of NCC's 11 May 1979 resolution "Ethical Implications on Energy Production and Use," an extension of their earlier and influential statement "The Plutonium Economy."[9] A draft had been strenuously debated the preceding fall without reaching consensus. The effect of this resolution was to impugn further the moral reputation of nuclear power. A few smaller groups, such as Clergy and Laity Concerned, have been steadfast in their nuclear opposition.

The political representation of public concerns over nuclear safety is embodied—indeed, preempted—at the federal level in the NRC. The accident at TMI may well have accelerated a growing trend toward restitution of the federal system with respect to nuclear development (facility siting, transportation routing, and the like). A number of state governors had been claiming veto power over transshipment and disposal of radioactive wastes within their borders, for instance. Shortly after the accident, the governor of South Carolina declared Barnwell off limits to TMI wastes. Later, the City Council of Philadelphia was petitioned to resolve against the venting of krypton-85. Although NRC was criticized for regulatory laxity and has tightened its procedures, the political control of nuclear operations was widely perceived to have been overcentralized at the federal level. Counterpressures were exerted to nationalize the entire nuclear industry, however.

On the other hand, there has been an arrested "entry into politics" in the party system. Only Barry Commoner's Citizens Party made alternative energy sources and systems a central issue. Shortly before the New Hampshire primary, Ted Kennedy declared that the accident had swayed him to nuclear opposition. Independent John Anderson reversed his earlier support of nuclear development to emphasize solar and conservative alternatives; Jimmy Carter remained a "reluctant nuke"— nuclear energy as a source of "last resort"; while Ronald Reagan endorsed it as a source of choice. Unlike the intense party political activity in Sweden, or the formation of the Ecology Party in France, there is an "unpolitics" of nuclear energy. State referenda have been "orthogonal" to established parties and normal politics. If we regard politics as a chief means for societal problem solving, this omission is conspicuous.

EVENT ANALYSIS

In order to appraise the activity of public interest groups following the accident, it will be useful to review a short list of events. The first of note was the May 6 rally on the mall in Washington, D.C. Though hastily organized, upwards of 100,000 protesters turned out to hear the usual parade of media celebrities and antinuclear speakers. Critics of the movement characterized the crowd as "jazz lovers"; but as a "happening" to register outrage at the accident, at the NRC, and at the nuclear industry, it was fairly impressive. Van Liere, Ladd, and Hood wrote of this event:

> Another sign that opposition to nuclear energy is becoming less fragmented and locally oriented is the success the May 6 organizers had in attracting people through local groups and individuals, rather than through mass media. The results of this study suggest that people and groups in many parts of the country are forming an effective communication network useful in pooling resources and increasing political clout.[10]

Walsh also considers that "much available evidence suggests that citizen opposition to nuclear power may constitute the dominant social movement of the 80s."[5] This prospect was somewhat diminished by events of the fall, however: another mass rally, this time at Battery Park in New York City and with another good turnout, was followed however by two feeble confrontations at Seabrook, a split in the movement over nonviolent tactics,[11] and a harmless Wall Street "occupation" to dramatize investor responsibility.

Early in 1980, local concern around TMI focused on the proposed venting of krypton-85. The NRC held a series of heated public meetings, with local groups very much in evidence, but the anniversary of the accident passed without the massive demonstration some had expected, and which is still promised should TMI resume operations. The dumping of decontaminated water into the Susquehanna River, a water supply source for some downstream communities, will be the next event in the series. The restart of undamaged Unit 1 will likely provoke strong local reaction; recovery of Unit 2 is scheduled for sometime after 1983. It was during the venting controversy that one public interest group, the Union of Concerned Scientists, assumed a central role. The credibility of the operator, Metropolitan Edison (Met Ed), had been seriously compromised after the accident, due to conflicting information releases and the suspicion of covering up the accident's severity. After much initial confusion, the NRC emerged from the accident scene with moderate local support. As the prospect of venting loomed larger, its community relations deteriorated appreciably.

A survey of area residents (Middletown, Marietta, and Elizabeth-town) in early April 1979 found that 69% approved of how government officials handled the accident while 21% disapproved.[12] A follow-up survey in March 1980 showed approval to have declined to 32%, while disapproval climbed to 51%.[13] Nearly half (47%) of a sample of Lancaster residents surveyed in late November and early December 1979 responded that the NRC had not told the truth concerning the accident; the figure for Met Ed was nearly two-thirds (65%).[14] It would seem then that the major institutional actors in the situation had suffered a serious and continuing loss of credibility in the eyes of local residents. This did not translate directly into support for the antinuclear movement, however. Only a quarter (26%) of the Lancaster sample said they supported the antinuclear movement, while over two-thirds (68%) said they did not.[14]

On 28 March 1980, one year after the accident, Pennsylvania governor Richard Thornburgh asked the Union of Concerned Scientists (UCS) to conduct an independent assessment of various methods for removing krypton from the damaged Unit 2 and to evaluate the health risks associated with each. Area citizens had reacted angrily against Met Ed and NRC proposals for venting; UCS was a long-standing critic of the industry and agency, and a credible source for allaying residents' fears. While UCS confirmed that the health risks of the proposed venting method were minimal, it proposed four other alternatives designed to alleviate the widespread stress-related maladies diagnosed in the popu-lation. The USC report concluded that mental and physical stress should be considered "a medical problem of major public health importance."[15] On 16 May 1980, the New York Times editorialized that UCS, "usually one of the more responsible antinuclear groups," had issued an irre-sponsible report in finding that the Met Ed/NRC proposal was medically safe but publicly unacceptable. UCS replied:

> In our view, the Times' position amounts to saying that stress-related injury should not be taken seriously if its basis appears unfounded to radiobiolo-gists, and that no effort ... need be made to allay concern if no direct physical injury results. Stress-related injury has, in fact, a great reality and importance.[16]

Governor Thornburgh concurred that "by confirming now the accuracy of those figures and opinions relating to radiation safety, UCS not only appears to have helped reduce the stress over which we have all been deeply concerned but may have helped restore some of the public confidence U.S. regulators and industry spokesmen have lost in the past 14 months."[16]

The "reality and importance" of mental health impacts were not based on irrational fears of negligible health risks, however, nor was venting krypton the real issue. That was merely a first step toward decontamination and possible restart of Unit 2. What was stress-inducing was the prospect of resuming nuclear operations at TMI. Had decommissioning been the objective, the public reaction might have been quite different. The eventual fate of TMI is still officially unresolved, but UCS's sensitivity to the psychological stress of area residents is a valuable contribution to the accident assessment. In contrast, NRC has thus far failed to accommodate mental health impacts in its administrative system. The compensable status of such impacts in litigation is likewise a moot point.

That was not the end of UCS's involvement. The accident had raised serious concerns over prior NRC siting policy and over the continued operation of nuclear facilities in densely populated areas. On 17 September 1979, UCS filed a petition to decommission Indian Point Unit 1 and suspend operation of Units 2 and 3, located 30 miles upwind from New York City. UCS has been effective in expanding the boundaries of inquiry into the accident at TMI to problems generic to the industry and country, whereas local groups have confined themselves—logically enough—to local concerns. The only exception of which I am aware is a direct mail appeal from the Three Mile Island Legal Fund in Harrisburg.

I have dealt at some length with UCS because they were not present at the conference to speak for themselves. They were represented by counsel, however: Ellen Weiss.† The other scheduled public interest group speaker was Arthur Tamplin of the Natural Resources Defense Council. An added speaker, Judith Johnsrud, had been active at the state level before TMI through the Environmental Coalition on Nuclear Power, which formed the nucleus of one local organizing effort (among others) in reaction to the accident.[5]

THE CONFERENCE MICROCOSM

The composition and balance of conference speakers became the object of attack by Lorna Salzman, mid-Atlantic representative of Friends of the Earth. In a handout entitled "The New York Academy of Sciences Three Mile Island Conference: A Vehicle for Promoting Nuclear Energy and Suppressing Dissent?," she contended:

> Today's conference will, in all probability, successfully avoid talking about the real issues raised by the TMI "incident." Instead, it will, thanks to donations made by U.S. taxpayers and private utilities, serve as a platform for nuclear apologists . . . and the Atomic Industrial Forum, whose public statements and actions have in the main defended the right of the nuclear establishment to irradiate citizens without their knowledge or consent. . . .
> Today's one-sided conference will undoubtedly consist of the usual soothing pap purveyed by the nuclear establishment . . . it will not present any minority, adversary or dissenting views of what *really* happened at TMI,

† Ms. Weiss spoke at the conference but did not submit a paper for publication.

what the *real* effects will be, what the *real* radiation doses may have been, what the *gaps* in information are, and what *alternative findings and interpretations* are turning up. . . .

The political implications of nuclear energy are at least as grave as the technical ones. The nuclear establishment—utilities, scientists, regulatory agencies, research institutions—continues to utilize all tools at its disposal to falsify, suppress information critical of nuclear energy, harrass, intimidate and threaten dissenters, and exert control over all channels of public information. Today's conference is one of these channels. . . .

Yet the conference, in its makeup and participants, does not even pay lip service to public interest, adversary or environmental views; only two public interest participants are involved. The panel on public reaction has no independent local person from the Harrisburg area; the local public is represented by a Seattle, Washington, firm hired by the NRC to assess attitudes after TMI, and by Anne Trunk, member of the Kemeny Commission (and reportedly married to a member of the nuclear profession). . . .

We deplore the active participation by the New York Academy of Sciences in this successful attempt to exclude dissenting views. By such participation the academy has done a disservice to what should be one of its functions: furthering free open debate in the interests of a free society.

If this dissenting opinion strikes us as strident, perhaps that is because in the past its voice has not been heard. In the history of the nuclear debate, public involvement has been hampered rather than facilitated by existing institutions.[17] Public interest groups have waged a very unequal battle; the vast preponderance of resources has been on the side of nuclear advocates.[18] The result has been a growing mistrust of nuclear development and the alienation of its proponents from the general public.[19] Believing that their interests are unfairly excluded, it is understandable that public interest groups seize upon every possible occasion to urge their persuasion; they could hardly refrain from doing so in the case of TMI. At the same time, this may engender feelings of exploitation on the part of local publics, and perhaps it accounts for why rejection of NRC and Met Ed by local residents was not accompanied by greater acceptance of the antinuclear movement. Local groups may also resist the dilution of their immediate concern that would result from embracing the broader range of issues, e.g., nuclear weapons, that national groups typically contest.

The planners of this conference had been of two hemispheres in structuring its content and format. One read the conference charter as an attempt to compile a scientifically neutral record of the accident and its aftermath; the other was that it marked a point, if not a pivot, in the ongoing nuclear debate freshly rekindled by TMI. The former viewpoint was reflected in this statement:

> . . . we don't think of the conference as a debate either on nuclear safety or on nuclear power more generally. It is meant as a collective analysis by audience and speakers of the significance of a technical and social event. What we hope to do is to illuminate the issues, to clarify them, and to bring

important factual and descriptive information to the attention of the audience.[20]

The latter position, contained in a conference prospectus, contradicted this:

> ... we view the conference as part of the process of public debate about issues of science and technology, a process that the Academy regards as important and wishes to encourage and extend. This means not only that the conference should address matters that are of public concern, but also that the contents of the meeting should be appropriately conveyed so they can be assimilated into public discussions.
> ... we believe that the accident, like any sociotechnical event, has had diverse effects on a variety of different individuals and groups and that therefore, if the conference is to cover the spectrum of impacts, it must include representatives of the range of affected parties. By design, then, this meeting will bring together physical scientists, social scientists, public officials, leaders of the public, as well as including scholars from various fields who have studied the accident's effects.
> ... we know that conferences on controversial issues are often dominated by a single point of view or political perspective. Such a format, however, would not be appropriate for this meeting on Three Mile Island, which is designed to promote the exchange of diverse and perhaps competing positions. Accordingly, the participants have been selected to represent a wide range of views and perspectives on nuclear power.

This constitutional question of the conference's terms of reference in turn reflects the dilemmas of scientific institutions generally over their own involvement in transscientific issues. While much experience with handling these has accumulated in recent years, their proper disposition is by no means resolved. It does appear certain, however, that the transscientific nature of public policy questions requires some degree of public involvement. For the TMI conference, this implied a focus not only on what could be determined as to the causes, conditions, and consequences of the accident but also on what matters of public concern might be raised against this background and what avenues might be explored in working toward their resolution. Public interest groups help to formulate and articulate these larger concerns.

Thus the conference itself became a microcosm of the nuclear debate, and at times appeared to be heading for—in Cochairman Thomas Moss' apt phrase—a "social meltdown." The allegation of industry bias had been leveled against the Kemeny commission itself, in the mass resignation at their first meeting (24 July 1979) of the Citizens Advisory Committee. Formed as a counterbalance to the commission's Industry Advisory Committee, the group included representatives from the Interfaith Coalition on Energy, Critical Mass, Environmental Action of Colorado, and North Anna Environmental Coalition. The citizens committee felt itself discriminated against in being denied access to internal commission documents; only those in the commission's public documents room were to have been made available. The committee felt unable to carry out its review function under this restriction.

THE LARGER CONCERNS

It may be true of the accident at TMI, as Erikson has said of the Buffalo Creek disaster,[21] that the event itself is at least as important as the search for broader generalizations. Because of its specific charge and stringent time constraints for completing its task, the Kemeny commission confined itself to only the accident and its direct implications for regulatory and industry reform. It left to the "political process" such larger concerns as the level of acceptable risk and the evaluation of energy alternatives. From Kemeny's own testimony, however, that process is not up to the job; "the present system does not work," for TMI and many other problems of comparable or even greater complexity.[22] His diagnosis is that decision makers at the federal level—the president and Congress—can choose between alternatives but are incapable of formulating them. "What's principally lacking on the federal scene, it seems to me, is the existence of respected, nonpartisan, interdisciplinary teams that could at least tell us what is possible and something about the pluses and minuses of different solutions."[22] This is precisely the service that the congressional Office of Technology Assessment was designed to perform. But it remains a technocratic approach, begging the question of national consensus building, which Kemeny regards as part of any acceptable solution.

If there is to be an informed consensus, public issues must be debated before they can be decided. Public interest groups are instrumental in this process; while they cannot of themselves produce consensus, they may at least establish the *limits* of consensus. In reference to the nuclear debate, for example, it now appears that a rough parity exists between proponents and opponents, with approximately half the population between those extremes. Interpreting the results of 1976 state referenda, Weinberg has written: "Nuclear proponents point to the two-thirds who favor nuclear and consider this a mandate to go ahead. But I think this is unjustified: I do not believe that a primary energy system that is feared or rejected by 33 percent of the public can survive."[23]

Public interest groups have been instrumental in crystallizing the nuclear opposition and focusing nuclear development as a public issue. Groups such as the Council on Economic Priorities also have been influential in presenting alternative energy programs on which genuine national consensus can be forged, notably solar energy and conservation. (Starr contends, however, that once the social costs of conservation are reckoned, its public acceptance will decline.)[24] Whether these prove viable will require an industry "conversion" on the part of utilities. This is beginning to occur in isolated cases, e.g., California, and is more likely to set the national pattern than are top-down efforts at consensus building. In the process, locally and regionally based public interest groups are likely to play a constructive role.

In periods of stress, such as wartime, meeting national needs generally evokes efforts toward broadening the national consensus. In the case of the energy crisis, this means paying attention to equity impacts—both windfalls and wipeouts. In the TMI case, it means the incorporation of groups previously excluded from discussion and decision, and fully and fairly representing their central interests. Van Liere, Ladd, and Hood consider that the accident and its aftermath enhance "the importance of taking into account those people that oppose nuclear power in making energy decisions."[10] I believe this will prove to be a significant impact of the accident at TMI, and that public interest groups will contribute importantly to this end.

References

1. SANDMAN, P. M. & M. PADEN. 1979. At Three Mile Island. Columbia Journalism Rev. **18**(2): 43–58.
2. HARDING, S. 1977. Which interests are public interests? Paper presented at the Second University of Delaware Philosophy and Technology Conference, Newark, Del., June.
3. CAMPBELL, A., P. E. CONVERSE & W. P. RODGERS. 1976. The Quality of American Life: Perceptions, Evaluations, and Satisfactions. Russell Sage Foundation. New York, N.Y.
4. PITKIN, H. F. 1972. The Concept of Representation. University of California Press. Berkeley, Calif.
5. WALSH, E. J. 1980. Resource mobilization theory and the dynamics of local anti-nuclear coalition formation in the wake of the Three Mile Island accident. Paper presented at the 75th Annual Meeting of the American Sociological Association, New York City, August 29.
6. WEINBERG, A. M. 1972. Science and trans-science. Minerva **10**(2): 209–22.
7. HOLDREN, J. P. 1976. The nuclear controversy and the limitations of decision-making by experts. Bull. At. Sci. **32**(3): 20–22.
8. LADD, A. E., T. C. HOOD & K. D. VAN LIERE. 1980. Ideological strands in the antinuclear movement: consensus and diversity. Paper presented at the Annual Meeting of the Popular Culture Association and American Culture Association, Detroit, Mich., April.
9. BOFFEY, P. M. 1976. Plutonium: its morality questioned by National Council of Churches. Science **192**(4237): 356–59.
10. VAN LIERE, K. D., A. E. LADD & T. C. HOOD. 1979. Anti-nuclear demonstrators: a study of participants in the May 6 anti-nuclear demonstration. Paper presented at the Annual Meeting of the Mid-South Sociological Association, Memphis, Tenn., October 20–November 2.
11. MITCHELL, R. C. The anti-nuclear movement in the United States. (In preparation.)
12. Social Research Center. 1979. Three Mile Island: Local Residents Speak Out: A Public Opinion Poll. Elizabethtown College. Elizabethtown, Penn.
13. Social Research Center. 1980. Three Mile Island: Local Residents Speak Out Twice: A Public Opinion Poll, 1979–1980. Elizabethtown College. Elizabethtown, Penn.
14. KRAYBILL, D. B. & R. B. POWELL. 1979. Nuclear Energy: A Lancaster Sunday News Public Opinion Poll. Social Research Center. Elizabethtown College. Elizabethtown, Penn.
15. KENDALL, H. W. 1980. Decontamination of Krypton-85 from Three Mile Island Nuclear Plant. Union of Concerned Scientists. Cambridge, Mass.
16. Union of Concerned Scientists. 1980. UCS report analyzes TMI krypton problem. Nucleus **2**(4): 1–3.
17. EBBIN, S. & R. KASPER. 1974. Citizen Groups and the Nuclear Power Controversy: Uses of Scientific and Technological Information. M.I.T. Press. Cambridge. Mass.

18. NELKIN, D. & S. FALLOWS. 1978. The evolution of the nuclear debate: the role of public participation. Ann. Rev. Energy **3:** 275–312.
19. BUPP, I. C. & J.-C. DERIAN. 1978. Light Water: How the Nuclear Dream Dissolved. Basic Books. New York. N.Y.
20. SILLS, D. L. Personal communication.
21. ERIKSON, K. T. 1977. Everything in Its Path: Destruction of Community in the Buffalo Creek Flood. Simon and Schuster. New York, N.Y.
22. KEMENY, J. G. 1980. Saving American democracy: the lessons of Three Mile Island. Technol. Rev. **83**(7): 65–75.
23. WEINBERG, A. M. 1977. Outline for an Acceptable Nuclear Future. Institute for Energy Analysis. Oak Ridge Associated Universities. Oak Ridge, Tenn.
24. STARR, R. 1980. The Three Mile shadow. Commentary **70**(4): 48–55.

THE REACTION OF THE REGULATORY AGENCY AS VIEWED FROM THE PERSPECTIVE OF THE PUBLIC INTEREST ORGANIZATION

Judith Johnsrud

Environmental Coalition on Nuclear Power
State College, Pennsylvania 16801

I would like to begin by mentioning that I am—and have been since the inception of the licensing proceedings for Unit 2 as well as Unit 1—one of the two public interest intervenors in the operating license procedure for Three Mile Island (TMI). Since we operate without attorneys, I have conducted cross-examination concerning some of the issues before us today and perhaps have a familiarity with TMI specifically that is somewhat different from that of most of the speakers.

Now, the topic of interest today is institutional reactions to the Three Mile Island accident as a specific aspect of the societal reaction to the event. I would like to concentrate in particular upon the Nuclear Regulatory Commission (NRC) and its relationship to the public. As an intervenor in a number of ongoing licensing proceedings and as a participant in the generic rule-making proceeding concerning radioactive waste, I have observed some rather substantial changes in the NRC—changes that I term "lessons learned" from the Three Mile Island accident and that will differ substantially from those that the agency would tell you about.

First, I would say that over the 10 years that we have participated in the official legal proceedings relating to the licensing of nuclear power plants in Pennsylvania, we have found on the whole a remarkable degree of willingness on the part of NRC to cooperate—more so, I would add, than its predecessor the Atomic Energy Commission—until March 28, 1979. As intervenors in the still incomplete license for TMI-2 (and I say that because the license was granted with two outstanding major issues that were raised and litigated and are still unresolved), we found that as an immediate aftermath of the TMI accident, we were cut off from information concerning the accident and from all other data concerning TMI-2. Apparently, the NRC simply dropped the intervenors from their service list for information.

To an extent I can forgive them; I understand there was a substantial degree of chaos, and quite rightly so, in both Washington and Bethesda. However, we were particularly troubled by their continued lack of response. Within a month of the accident, Dr. John C. Kepford, the legal representative for our organization, filed requests with the NRC—with Harold Denton of the Office of Nuclear Reactor Regulation, specifically—for hearings on changes in technical specifications relating to TMI-2, because the reactor obviously had moved from its category of normal

0077-8923/81/0365-0257 $01.75/2 © 1981, NYAS

operation to that of a severely damaged facility in, if you will, an experimental condition.

It was not until July 6th that any response was received from the NRC. That response consisted of a copy of a letter from Harold Denton to Jack Herbein, station manager for TMI, conveying to Mr. Herbein these requests for hearing. There was no response to us—the formal intervenors requesting the emergency hearings under the code of Federal Regulation 2.206. To this date, we have not received a response from the NRC. We have received a recent notification of changes in the technical specifications for TMI-2. Our response to this was to request a hearing, and we incorporated our earlier requests for a hearing. We have since been told by the NRC staff that indeed this proceeding for changes in technical specifications is entirely different and that the earlier request is to be disregarded.

This cutting off of information from those who are legally involved in formal proceedings is one aspect of the change in NRC's approach. I would add that this is despite repeated requests, both written and verbal, to staff members and counsel of the agency to be restored to the service list and to obtain information.

Second, the process of discovery—which is of the essence in the preparation of any legal case—was honored previously by the regulatory agency. Apparently as a lesson learned since TMI-2, the NRC staff has now taken a very hard line against informing public interest intervenors that documents of the agency are available at various public documents rooms. Since I live roughly 100 miles from the nearest public documents room, this makes my particular situation rather difficult. But really quite troubling is the fact that there has been a cutoff of access to that information.

However, there has been an even broader shift in NRC policy in the aftermath of the accident. This was initiated with the August 6, 1979, Federal Register announcing a program of charging the public for NRC documents. Prior to this time, those citizens throughout the country who were interested in and concerned about nuclear power, about the functioning of the agency, and about the technical documents relating to nuclear energy and who were willing to do their homework had had access to the NRC's documents. There has been a very substantial tightening up; and unless one is a formal intervening party to an NRC proceeding, the lack of accessibility to those documents makes it increasingly difficult for citizens, especially those who live in the rather remote locations of a number of nuclear facilities, to have reliable information from their federal government.

I've discussed this with a number of NRC staff people and have been assured that if I ask for something, I will get it. However, we are finding that this actually doesn't work out in practice. On the other hand, we've received up to six copies of one particular report when we'd much rather have had a variety of the materials that we requested.

In addition, we have found that the NRC's licensing boards have cooperated rather excessively with the applicants for licenses in what I would call a form of procedural harrassment related to discovery and the answering of interrogatories prior to going to evidenciary hearing.

In the case of those reactors coming up for licensing in Pennsylvania farther up on the beleaguered Susquehanna River, the board has given *carte blanche* to the utility with respect to the request for response to some 2,700 interrogatories in relation to a mere 12 issues raised by this public interest organization. The sheer work load of responding within the context of the legal proceeding is operating in such a manner as to preclude our participation. The public interest groups involved were required to go to oral argument concerning a request by the utility to prohibit the public interest groups either from presenting direct evidence on their own contentions in the licensing case or from cross-examining witnesses presented by the utility and the NRC in opposition to those contentions.

If the Nuclear Regulatory Commission indeed desires the input of the public whose members participate in these legal proceedings within the framework provided by law, then their going out of their way to make it more difficult for us is defeating that purpose.

Do the public interest intervenors really have anything of significance to contribute? Well, I'll give some examples. In the case of the Three Mile Island Unit 2 license proceedings, the representative of the public interest, Dr. John C. Kepford (our representative), presented uncontroverted evidence that the NRC was failing in the code of federal regulations to account for what turns out to be the largest source of radioactivity over the full period of toxicity from the entire nuclear fuel cycle. This is the radon-222 that is released as a result of mining and milling of uranium ore. The procedure that had been used accounted for only the first year's release and traced it through some environmental pathways. In consequence of that testimony, more than a year before the TMI accident, the commissioners of the NRC admitted that error and, in April of 1978, removed the number for radon from the famous Table S 3—the summary of the environmental effects of the uranium fuel cycle—and opened reactor license cases for litigation concerning the amount of health effects of the radon gas.

Despite this fact—which means that the National Environmental Policy Act (NEPA) has not been satisfied with respect to these licenses— the license for TMI was issued and the reactor was allowed to go on line. We were back in hearings before the NRC a little over a month ago on this self-same, still unresolved issue.

A second and perhaps more significant issue in those proceedings had to do with the probability and consequences of the crash of a heavy aircraft into the containment at TMI-2. In that instance, the probability analysis of the NRC staff and the applicant, Metropolitan Edison Company, was found by the Appeal Board of the NRC to be defective.

This is a very important issue because the Harrisburg International Airport centerline for takeoff and landing lies within 1½ miles of the TMI reactors. Yet the probability analysis was based on the assumption that no planes ever fly directly over the reactor. We have found this not to be the case. I'm among those who have flown directly over the reactor, and within the last couple of months. This issue has not been resolved.

Now, TMI-1, the sister reactor on the same site, is up for hearings prior to either restart or revocation of license. One would think that the issue of a possible aircraft crash into TMI-1 would be of significance as well. Yet the agency refused to allow consideration of the aircraft crash issue in the restart proceedings for TMI-1.

Similarly, despite the fact that the NRC staff has declared the TMI-2 accident to be a class 9 accident, only those accidents that are very similar to that of TMI-2 may be considered in the restart proceedings for TMI-1 and in other reactor licensings. In the case of the Susquehanna reactor I mentioned before, only the precise events of the TMI-2 accident were allowed consideration as a class 9 accident. We were told by the NRC staff that because the NRC does not specifically define a class 9 accident, there was nothing to litigate: however, if we wished to devise a scenario, they would look at it; but of course we had to understand that such a scenario would be considered merely hypothetical and, as such, would bear no weight as evidence in the findings of the licensing board.

Above and beyond all of this, we have some serious concerns about NRC's posture with respect to the upgrading of evacuation planning. While we are looking with far greater concern—and, I would add, high time—at evacuation, we are now operating within the confines of an NRC document entitled *Draft Emergency Action Level Guidelines for Nuclear Power Plants* (NUREG 0610) issued in September of this year. We find it quite troubling that the announcement of a general emergency, that is, the first category that would involve evacuation of those off site, would be made when "events are in process or have occurred which involve actual or imminent substantial core degradation or melting with potential for loss of containment integrity." The document goes on to describe releases of more than a thousand curies of iodine-131 equivalent or more than a million curies of xenon-133 equivalent, with effluent monitors detecting levels corresponding to one rem per hour whole body or five rem per hour thyroid at the site boundary under actual meteorological conditions. We've checked back further on this and have discovered that the recommended action in such an event should be sheltering rather than evacuation. I'm really troubled that the NRC will allow the utility to act under those guidelines prior to the public's having knowledge of the actual degree of an accident.

In the same vein, we find that the utility, Metropolitan Edison Company, has requested of the NRC that throughout the cleanup procedures at TMI, they be allowed to use the existing criteria for releases to the public. These criteria involve a cost-benefit analysis relating to how much it will cost to reduce the person-rem exposure. As I read the Code of Federal Regulations, Appendix I to Part 50, we are still working within the context of the $1,000 per man-rem exposure level.

Now I would like to leave this topic and address some questions concerning the responses of those at Three Mile Island to the ongoing nature of the accident, enlarging on Professor Wolf's comments.

I think we need to draw a very careful distinction between those groups organized on a national or regional level that have continuities for their special concerns and those citizen groups—or, as they are sometimes called, grass roots organizations—that spring up in reaction either to an event such as Three Mile Island or to a proposal such as was made in the Commonwealth of Pennsylvania five years ago to undertake the national test case for nuclear energy centers. In the Commonwealth of Pennsylvania, we were proposed to receive energy plants, which would have been a combination of coal and nuclear reactors, of some 20,000 megawatts on a single large site. The response five years ago by middle-class citizens and local governmental officials, the pillars of communities, was a very strong and pointed political protest at a very early stage in which they were in fact capable of having a role in the decision-making process.

By contrast (and here I would certainly agree with previous speakers), the residents of the lower Susquehanna Valley, probably for cultural and historical reasons, had shown relatively little concern, except for those environmentalists who were willing to dedicate their efforts in the public's behalf. Thus, in the period prior to the accident, there had been very little support, concern, or interest. I'm not sure that there was a pronuclear attitude; I think they really had no idea what was involved at that site. Subsequent to the accident, we find a peculiar combination of situations—I believe the term used yesterday was "irrational fears" on the part of the public. I think we must put that term into the context of the public's uncertainty that the actual releases were ever properly assessed or the doses properly calculated.

Thus, at the present time, the population of central Pennsylvania faces an ongoing accident that they had not understood would continue after the beginning of April of last year. There have been citizen efforts with respect to the water releases last summer; the krypton venting was monitored by the local population in January and was only then admitted to by the NRC and the utility; the test borings on site, and this past week, were the consequence of input by the local citizen groups.

There is a very high level of continuing concern, in fact, a mounting concern, as these people recognize that the accident is still in progress, that it may well be getting worse and has a potential for recriticality (as cited by Herman D. Camp in his March 4th letter to the NRC commissioners). These are people who have received a radiation dose of unknown magnitude. They view the krypton venting—an immediate event—as potentially setting the precedent whereby further releases to the public will be permitted on this financial cost-benefit basis over the coming years during the cleanup of TMI-2. Finally, after a year of dismay, disorder, and something close to despair in their lives, they are coming to a strongly political and indeed militant position. Those who lie silently behind them, as I have found from my personal contacts and observations in the area, are persons who are suffering what I would call a kind of denial syndrome. People who live there are caught there, see no chance of leaving, and want to go on with normal lives. But they are acutely aware that the reactor at Three Mile Island has altered their lives quite permanently.

INSTITUTIONAL REACTIONS: PANEL DISCUSSION

Moderator: David L. Sills
Panel Members: Robert J. Budnitz, Judith Johnsrud,
Paul Leventhal, Allan Mazur, Robert A. Szalay,
Emilio E. Varanini III, and C. P. Wolf

R. J. BUDNITZ (*Nuclear Regulatory Commission, Washington, D.C.*):
I'd like to speak to you, Dr. Johnsrud. I've never met you before, and I
find your comments about my agency a little surprising. I really don't
know what to do with them. I'm one of the senior officers of the agency,
and what I mean by that is there are only half-a-dozen people like me
who are supposed to run the place—besides the commissioners. I was
pleased to hear you commend us for the openness of our assistance to
groups like yours before the accident, but your comments about the
hardening of that attitude since the accident are completely inconsistent
with what I know and what I have observed. I'll take it on its face, that
your observations must be true to you, and I'm wondering what you've
done to acquaint the commissioners or the senior officers of the agency
with these problems—whether you have done so and have had a
response from them—because it just strikes me as contrary to my
observations.

J. JOHNSRUD (*Environmental Coalition on Nuclear Power, State
College, Penn.*): Appalling isn't it? I really thought that you would think it
appalling if you knew about it.

R. J. BUDNITZ: I'm just surprised about it. It's completely inconsistent
with what I know.

J. JOHNSRUD: Precisely; and I personally feel that the commissioners
should have found it quite astonishing as well.

R. J. BUDNITZ: Do they know about it?

J. JOHNSRUD: Yes, they do. In the course of a request, dated March 14,
to the NRC [Nuclear Regulatory Commission] commissioners for expe-
dited consideration of actions of an atomic safety and licensing board
and other matters, a number of questions specifically relating to the items
that I have mentioned today are asked of the commissioners.

R. J. BUDNITZ: So that's just a month ago.

J. JOHNSRUD: That's a month ago. There has been no response
whatsoever from the commission. Some members of those citizens
groups that are less directly involved in litigation have met with the
commissioners a number of times. I have met with some of the commissioners and talked with others. But of course, under the *ex parte* rules
relating to such legal proceedings, we are hamstrung from discussing
matters that are in litigation in those proceedings. However, I do feel that
the commissioners themselves have altered very substantially in their
attitude and that if they were indeed aware, I would have expected
something different from them.

0077-8923/81/0365-0263 $01.75/2 © 1981, NYAS

J. S. MILLER (*Northwood, N.H.*): I have a question and a comment for Mr. Budnitz. But first let me say that my name isn't very well known in the nuclear field. I'm an electrical engineer, have been working for 25 years in studying reliability, performance and safety aspects of large and small systems, with the last 2 years devoted to nuclear energy. I've been a senior member of the Institute of Electrical and Electronics Engineers for the past 15 years or so. And I'm one of those engineers who have changed. I have gone from a nuclear advocate position to one recognizing the abuses of nuclear power. I feel very strongly about it—to the point that I've studied it and have come to the conclusion, very quickly and very obviously, that nuclear power as we know it today represents a very serious risk to the public and an unnecessary risk to the public.

Mr. Budnitz said that we have seen changes within the industry and the NRC. I'm afraid I have to disagree. The order of magnitude of the changes is miniscule compared to what it should be. The only change that I've seen has been in Dr. Rasmussen's reporting of statistics. Yesterday morning we heard him acknowledge the fact that his estimate for a Three Mile Island–type radiation release was 1 chance in 2,500; and he used his factor-of-10 range on it, taking it up to 1 chance in 250 per reactor per year. But he didn't translate this—as he had at the Oak Ridge conference last November—to mean that for a population of 70 or so reactors, we will see a Three Mile Island–type accident once every three years. Now that is a change in thinking, and it hasn't been acknowledged by NRC or the industry.

There are some constructive ways that we can acknowledge this change rather than pulling snow jobs on the public. From now on in the conference, I'd like to see meltdown defined as either a partial meltdown or a total meltdown. That would eliminate some of the problems that arose yesterday. Also, we're still snowing those people by using "dose rate" and "dose" equivalently, and they're not equivalent.

Now the question comes down to the last question that was asked by Mr. Budnitz, "How safe is safe enough?" You say that the NRC must answer that question. I agree with you fully. I hope I can help to answer that question. But I'd like to know whether that question will be answered on a risk-benefit analysis basis that has economics as the sole factor or whether they will bring in social and ethical questions as well. And shouldn't the question How safe is safe enough? apply not only to operating reactors but more importantly to waste disposal techniques?

R. J. BUDNITZ: Sir, that's quite fair. The NRC has committed itself as a commission to address that question. They haven't said when, but it will be in the next couple of years. To support that, the Office of Research—of which I am the director—has initiated a number of research programs and projects in the university communities and with some of our other contractors in order to gain insight into the way we should go about phrasing that question. Some of our work involves technical questions of risk, some is sociological or economic, and there are questions about the way it can be phrased in a politically neutral manner that nevertheless takes into account these other things.

I had mentioned before that it's a multiparameter question that requires a multiparameter answer. You just can't say that if you show by analysis that a core melt will only occur in 20,000 reactor years, that's fine, while 1,900 isn't fine. Something like that is completely inadequate. It requires a far more complex rationale than a mere one-dimensional figure of merit like that.

Our research program is intended to eliminate those issues so that the commission can begin its own deliberations. They've committed themselves to do that within the year, probably in the coming fall or winter. Concurrently with that, the Advisory Committee on Reactor Safeguards, which is our advisory committee, has been holding hearings and meetings on this subject for most of the last year, including a couple in the last month or two, in which people from European regulatory groups have come over and given their views on what they are doing or have done or are thinking about in this area.

J. W. JACOX (*Nuclear Consulting Services, Inc., Columbus, Ohio*): I'm speaking as a citizen and a member of the public, and I have a question for Professor Wolf. I've directed it to him since public interest groups were his subject.

Including the word "citizen" or "consumer" either in the title of these groups or in many of their pronouncements—such as "operating or intervening in the public interest"—implies that somehow there is legitimacy to them, that all of the public or all of the citizens or all of the consumers are included as if by some magical means.

Specifically my question is, How can one justify any legitimacy whatsoever to this approach when these are in fact special interest groups, when there's been no referendum and no voting? As a member of the public who dissents totally from the opinions of many of these groups, I don't understand how there can be legitimacy to them. I'd like a professional opinion as to how this can be legitimate.

C. P. WOLF (*Polytechnic Institute, Brooklyn, N.Y.*): Personally, I stick up for the usage of the term public interest. I thought it was fairly perverse of Mobil to attack public interest groups as special interests. The criterion here might be—as was applied in the case of environmental health—whether the advocate is a direct beneficiary. I don't believe that a numerical weighting scheme is adequate to an understanding of what the public interest is. There *is* a question about the legitimacy of a group or individual who speaks in the name of the collectivity; there is a question of representation. The same kinds of questions of legitimacy and representation can and have been addressed to elected officials and representative assemblies.

What is new in this situation, I think, is that the public as a category apart from the electorate has been acknowledged; they have been invited into deliberations. Their position in relation to decision making is unclear and needs to be clarified, but federal agencies like NRC are under obligation to engage the public. All right? And so they have this very same problem of, Who is the public? A generation ago, Walter Lippmann wrote a book called *The Phantom Public* in which he argued that the public wasn't real, it was incorporeal. Well, now the public has been mandated into some kind of social reality that still is trying to find out who it is. But to me, the significant change here is the intervention of the public and its involvement.

Now in the case of Three Mile Island [TMI], we could say that the public's involvement was involuntary. People woke up on the morning of the 28th and found that they were a member of an affected public without their choosing and without their consent. The criterion of informed consent in terms of what we're doing to people has become very important. Now this doesn't mean to say that we can answer definitively who the public is, because that's changing. That sort of interpretation has to be based on public opinion statistics, and they change as well. So when you interpret them, you have to interpret them for one time with the recognition that it's very likely to change.

D. A. KELLEY (*Merrill Lynch Pierce Fenner & Smith, New York, N.Y.*): I'm a stock analyst for the utility industry. This question is to the NRC representative, Mr. Budnitz. Since utility companies are investment vehicles, we have to look at this problem in order to see how viable they remain. We were told or led to believe that there was some sort of an action plan due out of the NRC in the earlier part of this year. It did not come when we thought it would. We are a little bit concerned because we feel that if the NRC does not put out the final word on what they intend to do since TMI, Congress perhaps will. We fear that that will make utility companies, at least those involved with nuclear power, less attractive as investments.

We would like to get an idea from you about whether an action plan is coming out. If you could, tell us a little bit about what's going to be in it. Also, in your original list of about seven items, you missed the item that dealt with utility management capability in terms of who should be building nuclear units and who should not.

R. J. BUDNITZ: On the last point, you're right, I left it out. It wasn't because I wasn't aware of it. There are half-a-dozen things like that that I didn't have time to touch on, though I did touch on the notion that there are stronger utilities and weaker ones and that the utility industry had not policed itself before the accident.

On the question of the action plan, the agency has been pulling together an action plan for almost six months now. This is a thick document that purports to be the total prioritized list of everything that we are going to do, or that the utilities are going to be asked to do, in response to Three Mile Island. If you want a copy of that, they're publicly available. The drafts are coming out about monthly, and all those drafts are available. If you ask me for it later, I can get one for you, or you can just write to me.

The most recent draft is probably not the final one. It will probably be one more month before this is done. There have been frequent—almost weekly—meetings between our senior management and the commissioners and the group pulling this action plan together to make sure that what's in it is important and to cull out the stuff that isn't important. This has turned out to be an excruciatingly difficult process because an awful lot of what we feel we have to do, we cannot do with our present resources. The agency doesn't have enough people and it doesn't have enough money to address all the things we want to do.

More important, the industry doesn't have enough people right now to do everything right away either. Just about everybody in the country that knows anything about this subject is working already on problems in response to Three Mile Island. To load another comparable amount of work on them right now and give them 90 days to do it is essentially impossible.

L. SALZMAN (*Friends of the Earth, New York, N.Y.*): My question is to Mr. Budnitz. You made the statement that the public pays too much attention to low-risk events. Well, I think the NRC probably pays too little attention to high-risk events, for example, the Union of Concerned Scientists and Indian Point. The NRC, Mr. Denton's office, admitted publicly at a meeting a month ago that Indian Point and the Zion plant in Illinois represent 40% of the total risk to the public for nuclear power in the country. Clearly that's an admission of high risk, and yet Mr. Denton and the NRC are refusing to come to grips with the problems at Indian Point.

My question is somewhat an elaboration on what Dr. Johnsrud said. If in fact the NRC is changing its spots as you say, why is it still doing the following? Why is the NRC refusing to apply the new siting policy— namely, to put reactors in low-population zones—to operating reactors like Indian Point? Why is NRC continuing to rely on the discredited Rasmussen report for accident probabilities? Why is the NRC still refusing to consider at all the possibility of catastrophic accidents in populated areas? And why is the NRC refusing to consider or require evacuation beyond the 10-mile zone even though the Rasmussen report says that catastrophic accidents would require evacuation beyond 25 miles?

R. J. BUDNITZ: I think that on each of those points, the agency is responding. The reason why we're not applying what you call our new siting policy to Indian Point is that we do not have a new siting policy yet. We're initiating just now a major rule-making public hearing on this, and it will take a couple of years before we're done. Whether or not that new siting policy will apply to sites already with operating reactors is an open question. Right now the commission is in the throes—and I mean the commissioners as well as the staff—of a complete new soul searching on that issue. It's not something that can be done overnight. And although it may not be rapid enough for some people—I guess it's not rapid enough for you—I think that there's very major progress in the siting area and that two years from now, we're going to have a rational policy that will take into account a lot of new wisdom.

As to the Rasmussen report, on one hand you discredit it and on the other hand you want to use it.

L. SALZMAN: No, I'm saying that's what the NRC does. They use it selectively. When it came to accident probabilities, they used Rasmussen; and when it came to consequences, they ignored Rasmussen. And I made those comments in my submitted comments on NUREG-396, on which emergency planning is based.

R. J. BUDNITZ: Actually, about the best thing you can do with the Rasmussen report is to use it selectively, because a lot of stuff in there is bad though a lot is good.

L. SALZMAN: Yes, NUREG-396 has lots of footnotes saying that there were many uncertainties in the Rasmussen report, yet they still go ahead and use it.

R. J. BUDNITZ: I suppose the answer is that the Rasmussen report's probabilities are being used as a guide for us to try to figure out which are the important accident sequences, but they are not being used numerically because the numbers on probabilities in the Rasmussen report are not thought to be reliable enough. I think that's fine.

I'm not really sure about evacuation beyond 25 miles. Along with the siting policy reevaluation, there is a thorough reevaluation going on within the agency of the whole question of evacuation.

S. SEELY (University of Rhode Island, Kingston, R.I.): I'd like to address my question to Mr. Budnitz. I just simply wanted to ask a few questions about the list you presented. Among the institutional flaws, there was a question of poor feedback. Then this morning, there was considerable discussion about communication gaps. I'm interested in knowing specifically what you might be prepared to do about some of the existing reports about which the public hears nothing. There are more than 3,000 reports per year, some of which are referred to as "abnormal occurrences" or "reportable incidents" or "unscheduled incidents;" that is, you have a jargon that in itself can be confusing to the public. The question is, Are these significant? and is it the intent of your organization ultimately to disclose these so that there might be community public inputs to that question?

Then there is the general question you raised, "How safe is safe enough?" So far as I know, there has been no discussion of the NRC judgment in authorizing the Indian Point plants to resume operation. I think this might be something that the public would be interested in. Another question that I think needs to be addressed is, What sort of on-site or off-site monitoring systems ought to be established by communities in order that they may feel secure that they will have instant alarm? At the moment, one has to depend largely on the utility, and Three Mile Island indicates that that is not the way to go. Lancaster County Civilian Defense, I know, is planning a separate, independent plan, because they do have TMI to their north and Peach Bottom to their south and they simply feel insecure.

These are just three of the things that seem reasonable for the public to hear about. I wonder if you actually are planning to address this kind of question in the future.

R. J. BUDNITZ: On the third point, about monitoring networks, I really don't know enough to answer. I know there are discussions, but I can't comment more on that.

On the first question, the agency has established a new office called the Office for Evaluation and Analysis of Operating Data. We've staffed it already with a senior, well-respected engineer that we brought in from the Tennessee Valley Authority and with a little more than a dozen of our own staff; we're still building up staff. The aim of that group is to analyze those abnormal occurrences and what we call licensee event reports of various kinds, to compile them, tabulate them, and publish the results. All that the group does will be public.

Secondly, we intend to establish liaison, in fact have already done so, between ourselves and the industry to do the same sort of thing. Although none of these groups are really fully running yet, that liaison has been established, and I'm sure it's going to be an important way of improving safety.

On your question having to do with how safe is safe enough, one of the problems that we're facing with Indian Point or Zion or any of the reactors is that the agency hasn't decided how safe *is* safe enough. For the time being, the operation of those plants is continuing. That's based on a judgment by NRC that whatever the risk is, the risk is acceptable in a continuing mode while these decisions are being made.

UNIDENTIFIED SPEAKER: Acceptable to whom?

R. J. BUDNITZ: Acceptable to the NRC of course. That's their statutory, you know; they have to make a finding that there's no undue risk. Whether that view will change when they ultimately decide somehow the criteria for how safe is safe enough is unknown.

E. MARMORSTEIN (*New York, N.Y.*): My question is for Mr. Budnitz. You made a very interesting point when you were talking; you said that some people even now claim that TMI was not an accident and, furthermore, that there have been no accidents, that it all depends on how you define accidents.

I have heard TMI and others referred to as "incidents," as "accidents," and as "events." Sometimes I've heard them referred to as accidents and incidents by the same person within the same paragraph. Now I must tell you that as a member of society, my responses to the words accident and incident and event are substantially different. I think it's important that you come to some consensus about how you are going to term these things, and I wonder if you could briefly define for me what the difference is between an event, an incident, and an accident.

R. J. BUDNITZ: Well, there's the old question about not knowing how to define a purple cow but recognizing it when you see one. I really don't know how they're defined, but I'll tell you that in my view, Three Mile Island was an accident; and I don't think there are very many people in our agency or the industry who are not going to call it an accident today.

On the other hand, there was a problem at Crystal River in Florida that was coped with successfully by the engineered safety features and by the operators. It came to a satisfactory conclusion, no one was hurt, the equipment has been restored, and they are going to start running again in a short while. I don't know what an accident is or what an incident is, really; but the Crystal River thing certainly was not an accident.

I think you're right. One of the great problems that we've experienced over the years—and it was alluded to earlier—is the question of jargon. You know, people talk about health effects. Well, why in the world don't they call them disease effects, which is what they are? There are health effects from radiation, and I mean positive health effects—very high doses can cure some cancers. We are not talking about health effects here but about disease effects; and people ought to start using that term. That's another example of this obfuscation that has resulted in a lot of confusion and, I must say, a lot of anger. And it is deserved.

A. P. HULL (*Brookhaven National Laboratory, Upton, N.Y.*): First of all, I want to make a technical correction of an impression created by someone this morning that somehow the dose at 10 miles or so, a considerable distance from Three Mile Island, was greater than it was close in. Since I had the responsibility for compiling and organizing the helicopter data and also had access to the ground-level dosimeters, I can tell you that our projection at Harrisburg was about one milliroentgen, give or take something, that there was a ground-level dosimeter right close to Harrisburg that read 1.3 millirem, and as you heard earlier, that the maximum dose to the individual close in was about 100 millirem. So I think that was just folklore that's crept in somewhere, and there's no factual basis to substantiate it.

My question is to Dr. Wolf, and it goes something like this. I would agree—somewhat more reluctantly than you, I think—that the public interest groups have served a useful function. They have raised questions that the nuclear establishment in some of its complacency—and I have to say even I was complacent—failed to raise for itself. I think this begs the question, Is this an efficient way to get at the facts?

Any group is entitled to have its say in the public, political arena but I think most of the public interest groups come to this debate from an antienergy, antitechnology position. But they can't argue that, or choose not to argue that, up front. Instead, they are arguing these technical issues, and the public is getting a lot of obfuscation or coloration that actually is given by groups who are so polarized in this matter.

C. P. WOLF: Yes, the alternative question is, Would the science court be a more efficient social mechanism for raising these kinds of issues? On the science court question, I would defer to my colleague Allan

Mazur or Dorothy Nelkin. Let me say that the way we do public business is not very efficient, but that's not so bad really. I think the public interest groups have a social agenda, and that's not bad either. I think the technical groups have a technical agenda. I would like to see the terms of reference improved so that social issues can be admitted into consideration along with technical issues.

To give you a local example, it's no good to argue about whether the Concorde should have landing rights at Kennedy Airport on the basis of decibel readings. But if those are the only available terms of reference, that's the way the game has to be played. Now we need better terms. For example, one of the big things that came out of TMI in my view was to get a hearing on mental health impacts. Before TMI, NRC would not admit that kind of question into licensing hearings. NRC changed its attitude on that in midsummer. We have substantiation from the Behavioral Effects Task Force of the Kemeny commission and others that those were the major health effects, or disease effects, of TMI. But they had not had a hearing in the licensing procedures up to now.

I think we have to improve on the way that we think about these questions. We have to understand that the technical arguments over an issue are not the real arguments, they just happen to be the admissible arguments, that we need not only institution analysis but institution building. There is the whole question of energy need: Why are we doing this? Who needs it? For what purpose? When NRC's interest is questioned, as it was here a minute ago, you see that we're still in confusion about the extent to which the agency interest reflects the public interest.

Now in a democratic society, we're not going to solve these questions neatly, we're going to solve them messily. They are messy problems, but they are necessary problems. These are the problems we ought to be having. And insofar as this forum can open them up and illuminate them, then I feel it's constructive.

UNIDENTIFIED SPEAKER: Excuse me, could I just add one point here with respect to the comment earlier in the proceeding about a higher dose at 10 miles than at close to the reactor. That is to be found in the evidence in the Population Dose Assessment Report. It is an inescapable conclusion if one actually plots the data by summing the NRC thermoluminescence dosimeter data in that report for the days March 31st through April 7th, wherein one will find that the dose recorded at the closest point—about $2\frac{1}{2}$ to 3 miles from the plant—is about $5\frac{1}{2}$ millirem, whereas at 10 miles from the plant, it is about 8 millirem.

UNIDENTIFIED SPEAKER: Dr. Hull raised an issue that is related to my question. Don't we need something parallel to the hearing process, where information can be put into the hearing to give perspective without necessarily challenging every issue. The reason I say this is that intervenors can delay proceedings for long periods of time, making it very difficult to distinguish between minor and major factors.

Related to this, I wonder whether the NRC has been looking at experience with different types of reactors in order to see which are the ones for the long-range future and how they would impact on safety and siting and such factors.

Both questions in a sense are directed to Mr. Budnitz, but Dr. Johnsrud might want to pick one up.

R. J. BUDNITZ: The second question is clear to me. No, the Nuclear Regulatory Commission is not investigating other kinds of reactors to decide whether they would be better to build. That's not our mission. Our mission is to respond to an application for a reactor and give it a license if it meets our criteria. Other reactors and their development are not our responsibility.

J. JOHNSRUD: In response to your first question related to delay, it's interesting to note that studies by both the Atomic Energy Commission and the NRC have shown that there is a greater delay factor associated with failure of delivered equipment, with labor unrest at sites, and, indeed, with the weather than can be attributed to the participation of the public interest intervenors in the license proceedings. For example, in several license proceedings, we're finding that it is the NRC staff that is overloaded, that is having trouble delivering the safety evaluation reports. It's not the intervenors; they want to get the thing over and done with, I assure you.

A. R. TIEMANN (General Electric Corp., Schenectady, N.Y.): My question is addressed to the entire panel, with special reference to the sociologists. In the past two days we have heard that a primary segment of the problem with high technology is social issues of one kind or another. Now from the people representing government and agencies, I should like to know whether post-TMI, some change in thinking has been adopted so as to bring social science professionals into the decision-making process; and to the sociologists represented on the panel, I should like to ask whether they think this might help.

R. J. BUDNITZ: I guess that the counterpoint must be a feeling that decisions had been made previously by technical people. Although I think that's probably accurate in part, it's certainly not accurate in total. Our commission, for example, has only one nuclear engineer on it now, and typically has never had more than one. There were long periods of time in the Atomic Energy Commission when it had none. Decisions are made at the policy level by nuclear engineers, decisions are made by a commission with a wide range of backgrounds, and decisions are made by Congress and by the public.

On the other hand, there haven't been very many of the sociological and social science community involved anywhere in this process, and I'm sure their participation would help.

E. E. VARANINI (California Energy Commission, Sacramento, Calif.): In California, other than with the Marin complex (that's an insider's joke: if you're not from out there you won't understand), we've been using multidisciplined approaches across the energy system for staff input, and we've been paying for some outside consultants—socioeconomic firms that are employing sociologists—particularly in rural decisions and the impacts on rural communities as well as the society's view of technology. We've used Ida Huss, and currently we'll use her again, in the confidence proceedings in front of the NRC. We find a certain humanistic debunking of the systems approach to have a salutary value in making some decisions.

D. P. SIDEBOTHAM (*New England Coalition on Nuclear Pollution, Putney, Vt.*): I've two very brief comments, really, and a question. My first comment is to thank you Dr. Sills and Dr. Moss and the New York Academy of Sciences for allowing the expansion of this panel and including an invitation to Dr. Johnsrud this afternoon to appear as someone who has actually participated in the Three Mile Island proceedings. That helps considerably in rounding out the view.

In regard to public interest groups, that's a designation of the New York Academy of Sciences. The New England Coalition on Nuclear Pollution is a private, independent citizens group for public education. We are chartered as a public education organization. We also participate in licensing proceedings. The criterion for participation in the nuclear regulatory licensing proceedings is to identify concern: how one would be affected. Of course, the reason for the licensing proceedings of the NRC is to raise all questions relative to the licensing of that facility and to have them resolved.

There's been a good bit of discussion this afternoon about population and siting criteria, and I want to ask Dr. Johnsrud about a change in thinking—the possible factoring in of social factors—in regard to siting criteria and population, particularly as referred to by the Advisory Committee on Reactor Safeguards [ACRS].

J. JOHNSRUD: Yes, I think this was referred to earlier by one of the other speakers. With respect to the report of the Siting Policy Task Force of NRC, which is looking at potential alterations, there appears to be an increase in the thinking on the part of the agency and some public interest groups that remote siting of reactors in areas of low population density would represent an improvement over the policies of most of the last 20 years.

I would, on the basis of our experience in the Commonwealth of Pennsylvania five years ago, point out that the residents of those rural population areas found themselves quite incensed as the recipients of large energy complexes meant to supply the urban areas. They viewed the problem there as relating to distributive justice, that is, those who wish the electricity should also be those who bear the risk, which in turn would take us to a load-center siting.

I have been troubled by what appears to be the acquiescence of ACRS in its report on the NUREG-0625 report of the Siting Policy Task Force, acquiescence with the much earlier conclusions of the ACRS back in 1960, when the Part 100 siting criteria were first adopted. I would quote here one thing that I find quite troubling—the fourth point made by the 1960 ACRS group, and apparently agreed with by the current group: "Even if the most serious accident possible (not normally considered credible) should occur, the numbers of people killed should not be catastrophic."

Now that raises in my mind a very deeply disturbing question. We're not talking about no deaths from a nuclear reactor accident, we're not even talking about latency effects. Now we're going to have to ask ourselves what constitutes catastrophic numbers and at the same time, as this same report goes on to describe, what constitutes acceptable risk, which in turn takes us back to a previous question, Acceptance to whom and determined by whom?

D. L. SILLS (*Social Science Research Council, New York, N.Y.*): Thank you very much. The last word this afternoon will be from Mr. Robert Szalay.

R. A. SZALAY (*Atomic Industrial Forum, Inc., Washington, D.C.*): Thank you. I've been waiting my chance. There are several impressions that were left from the questions, and they relate to siting, to the problem of degraded core cooling, and to the safety of plants. We have to remember some of the lessons that we learned from all the studies that have been done in the last year, such as the Kemeny and Rogovin reports. Basically, they move toward more simplicity in information given to the operator, so that he will have a better understanding in order to cope with and respond to situations and prevent them from getting worse. That's one of the major things that we've learned, that we need to concentrate on. It seems that the design of the plant is fundamentally sound in terms of its general resilience to fairly large insults but that we need better information and display of what's going on in the plant.

Now, the question of what we've done so far and the question of what is acceptable risk kind of come together. We've concentrated our efforts on training the operator to deal better with those things that weren't considered previously in the design. This includes looking at degraded core cooling conditions, taking steps to decrease the chance of the core going to a worse situation beyond the design basis, and, even if it does go beyond there, preventing the containment from failing.

So the efforts that we've made have been aimed at making the probability of containment failure smaller and smaller, not in a quantitative sense as yet, but at least we are approaching a way to prevent these postulated major releases from ever happening. This is relevant to the siting study, it's relevant to the Con Edison situation—their specific design situation, it's relevant to this remote siting versus population density, and it's relevant to emergency planning.

I think you have to include the sociologists in the decision making; but while you're including them in, I think they also have to understand that the engineers have looked into the best information that we've developed and have come up with general conclusions.

HIGHLIGHTS OF THE CONFERENCE THUS FAR, FROM THE THREE SESSION CHAIRMEN

THOMAS H. MOSS* (*Subcommittee on Science, Research, and Technology, U.S. House of Representatives, Washington, D.C. 20515*): I have been nearly overwhelmed by the rate with which insights, ideas, contradictions, arguments, and issues have come in from the speakers and also from the audience. I especially want to express my appreciation to the audience. I value very much the role that you have played. I think your questions and comments from the floor have added much to the conference and have helped keep it from being just an academic or intellectual exercise, and instead have helped make it a true learning analysis of the Three Mile Island experience.

I said at the beginning of the conference that I thought it was most important for us to take this accident, which was so traumatic, and try to analyze its implications for our technological society—particularly those implications that go far beyond nuclear issues and even beyond energy issues to the general question of how to organize a complex, technologically based society. I felt that the accident was a test of those institutions and communications systems and that we shouldn't lose the results.

I have tried to test that hypothesis by writing down a list of issues raised here and then trying to see which of those seemed to apply to nuclear questions particularly and which were far more general.

Of the 18 issues I happened to list on the first day of the conference, 17 seem to be applicable to society as a whole. The chief impact of viewing them in the nuclear context is that it puts those issues into a very sharp perspective, one that makes us see their importance more so than we would in almost any other technological system at this stage of our society. Nonetheless, I believe they are more general and will perhaps be as vexing in other areas in the future.

I'm going to give you a couple of examples from my list and point out why I feel that they are such general issues and not particular to the nuclear industry.

If there was any one theme that came out of the presentations on the first morning, it was the need for a systematic and reliable use of operating history in nuclear technology, to learn from and improve on in the future. A corollary of that was the ability to communicate freely about failures of the system and to make sure that the public and the industry itself perceived the lessons of those failures. The major issue here was the need to establish that kind of a system in a situation where either economics, historical commitment, or some illusory strategy on the part of either an individual, group, or company had led to a defensive posture, which then blocked free communication about the lessons that were taking place. This really is a general problem—how to learn from

*Dr. Moss is science advisor to this committee, not a member of Congress.

275

our failures and not let our defensive reactions to those failures inhibit communication about them.

The second major theme that came out of the sessions on the first day was the difference between theory and practice. There's a vast difference between the diagrammed, systematic plan of a complex system and its actual performance, especially under stress and with people involved. There were many examples of that, but I was struck by a very curious juxtaposition. One speaker mentioned that part of a failure of the kind that we saw at Three Mile Island was the tendency of operators to deviate from prescribed procedures under certain conditions, perhaps causing the accident to become worse. On the other hand, a different speaker pointed out that in many cases, an accident may be prevented by the intuition of a very intelligent operator who breaks through the standard procedures, sees what is really going on, and corrects the situation using his skill.

Another main point was the scientific problem of low-level health effects, which I think is far broader than just radiation effects. Here, we are trying to define a level of public safety, trying to convey an understanding to the press and the public of how that public safety level has been derived, trying to develop regulatory action, and trying to develop legal action in a realm of science that is based on incomplete statistics and probabilities and in which the basic mechanistic theory is not fully developed. The difficulties of trying to fit chronic, low-level health problems into our regulatory mechanisms is a problem that is going to be more and more pervasive in the next decade in many areas other than nuclear industry.

Now, I should leave the summary of the second day to my cochairmen, but I can't resist a couple of observations. Obviously on the press day, the key question was information—the accuracy or reliability of information conveyed to the press. However, I thought there were many other aspects of communication, which were all related. There were questions of information to citizens and intervenors, of information injected into the legal process, and of information overload in the regulatory agencies—the work load of the Nuclear Regulatory Commission, which Mr. Budnitz described.

I put all of these into a general class of problems that will force us to make a fundamental choice: Do we try to change our institutions, society, legal system, and people to deal with the issues of this very complex technology, to make them work under all the regulatory pressures? Or do we decide instead to take as a criterion for acceptable technology that which fits into our normal social and institutional system?

I was very struck by Dr. Wolf's comment in discussion that at least we are beginning to move social issues into technological decision making and that, in fact, we ought to be sure that the framework for technological decisions isn't confined by the legal system or by historical precedent to purely scientific questions. Perhaps people should look at the broader issues of a technology like nuclear energy: Does it tend to polarize a community? Does it put unreasonable burdens on a population? Or does it, in fact, serve to release tension all over the world by relieving pressure on fossil fuel resources?

My last impression was that our current proceedings in the social sphere are analogous to the unfolding fault tree Dr. Rasmussen described for the sequence of events in a reactor accident. Certainly the accident still is not over in the social sense. There are choices to be made; whether the krypton should be vented, the unit restarted, and others. For each choice, we have a branch; but because we don't have the fault tree laid out, we don't know exactly where each of those branches will lead. The end might be a "social meltdown," or perhaps it will be the quenching of the accident. I don't know which one it will be, and I guess we'll have to leave it to the sociological side of the house to make that judgment.

DAVID L. SILLS (*Social Science Research Council, New York, N.Y. 10016*): I was fascinated by the province of measurement, inadequacies of measurement, and the failure of communications having to do with measurements, not only between scientists and nonscientists, but among the scientists themselves. One example is the whole question of probability and of what a complicated notion this is in the fault tree that Dr. Moss just mentioned. Professor Rasmussen says that at every juncture of the fault tree, both the mechanical failure, such as metal fatigue, and the probability of human error are taken into account. Well, I'm very curious to know just how these two problems are in fact correlated, since one of the things that comes out so clearly in the reports on the Three Mile Island accident is that human error itself is a very, very complex thing. It's not easy to say that the reason someone does one thing rather than another has to do with training, has to do with instrumentation, and so forth. Assigning the probability of future accidents on the basis of combining so-called mechanical failures with human failures is a challenge that I'm sure has not been met.

The second example is radiation. There was a comment the first day on whether or not the public understands, or can ever understand, radiation. It was quite clear that most of us really don't understand either radiation or how it is measured. Perhaps, as Dr. Eisenbud has mentioned, it's easy for all of us to be very confused about any report of radiation simply because we can't cope intellectually with the problems of scale involved.

The third has to do with two things that are said to have occurred at Middletown after the accident: one is that the traffic accidents that weekend were lower than normal; and the other is that the crime rate was not only lower than normal, but was zero the week after the accident. This is a very nice thing to think about because even the people who are convinced that Three Mile Island was the best possible thing for the nuclear industry are certainly not advocating nuclear accidents as a way of lowering crime rates or accident rates. It's an intellectual problem that fascinates me, and you have to look at what the meaning might be of these two facts.

The positive side, the upbeat side, is that there were fewer traffic accidents because people were concerned with safety. After all, they were evacuating what they perceived to be a dangerous situation, so they drove more carefully. The other interpretation is that with half of the people out of town, the traffic accident rate had to be lower simply because of the cars that weren't there. You'd have to look at how many miles were driven over the weekend in the area.

The crime rate is again fascinating. It's nice to think that it's an example of community cohesion, that people were all together in the face of this near tragedy. The other explanation is that with half of the people gone, half of the criminals were gone too, so of course you were going to have a lower crime rate.

These problems of measurement have to be looked at. I find that one of the consequences of this conference should be a fresh new look at some of these problems.

MERRIL EISENBUD (*Institute of Environmental Medicine, New York University Medical Center, New York, N.Y. 10016*): I believe that the unpleasant atmosphere of a meeting like this keeps many people away. Many people who are very involved either on the technical side or the government side do not want to get involved in the advocacy that dominates the present scene, not only in the field of atomic energy, but others as well.

I regret that there has been unpleasantness here. Anne Trunk's husband was alleged in print to be employed by the nuclear industry. A reference to my own inability to measure I-131 and correct for the 8½-day half-life was erroneously made in something that was handed out. These things are unpleasant; we have other things to do, and so many of us don't come to these meetings. This, I think, may explain why many of the invited speakers just came and left; they did what they were asked to do; few of them remained.

I think we must learn how to maintain a more constructive environment when we get together—people from both ends of the spectrum and those in the middle—to discuss these issues.

I also think it's too bad that the press has such a special relationship to the rest of society nowadays. Who wants to tangle with the press? At a meeting like this, if you give good reasons why the press did something wrong, they're not going to publish it in newspapers. David Burnham, one of the better atomic energy reporters of one of the major newspapers in the world, just came and left. He wasn't available for general discussion for any length of time, and in his comments, he raised matters that were totally irrelevant to the Three Mile Island accident.

I could make make allowance for the fact that the press may have been carried away during the days when the Three Mile Island episode was at its maximum. However, I must mention a front page photograph that came out of a newspaper just last week, with a big headline talking about a radioactive release of 53,000 millicuries of krypton-85. Incidentally, it was 15 mCi that they were talking about, not 53,000. And to put the 53,000 into perspective, it turns out that a 1,000-megawatt coal-fired plant puts out the equivalent of 400,000 mCi of krypton-85 a year. That the press would devote a full page of a tabloid-type newspaper read by several hundred thousand people in a large, well-populated county in New York State to 15 mCi of krypton-85 seems to me a great deception.

Finally, I should like to mention my own disappointment, which is chronic. I dwelt on it at great length in a book I published about a year ago. My disappointment has to do with the fact that there are such enormously important environmental matters that need to be dealt with—dealt with by people like the Friends of the Earth, the Natural Resources Defense Council, and other similar organizations. They should focus on teaching young people not to smoke, on getting the legislators to increase highway control so that we can cut down on our accident deaths or, more important in my opinion, to put some money into health education. If a fraction of the money that is being wasted on unnecessary protective features simply to satisfy the people at Three Mile Island were put into health education, particularly maternal health, the appalling infant mortality rate in the city of Harrisburg, which is due to the inverse relationship between socioeconomic status and infant mortality, could be eliminated. The whole Health Department budget here in New York City is only $60 million a year. Give them a few hundred thousand dollars, and they could save hundreds of lives a year. Yet in Middletown, they are going to spend millions of dollars to conduct an epidemiological study to see whether after 20 years the number of cancers is 0.7 or 1.2; and they are never going to know. So there are these inconsistencies.

In summary, I think we have to learn how to debate these issues in a more pleasant environment. I think we have to be able to attract people who are the statesmen of the movement. I'm sorry that Arthur Tamplin, for example, was only here for a short while and that we did not have representatives of other important public interest groups that should have had a voice in this symposium.

THE ACCIDENT AT THREE MILE ISLAND AND THE PROBLEM OF UNCERTAINTY

Cora Bagley Marrett

President's Commission on the Accident
at Three Mile Island
and
Department of Sociology
University of Wisconsin
Madison, Wisconsin 53706

THE PRESIDENT'S COMMISSION: LESSONS LEARNED*

It would be presumptuous of me to offer an analysis of the lessons others have learned about the accident at Three Mile Island Unit 2 (TMI-2) from the report of the president's commission, for the views on that report are as diverse as is the readership of the document.[1] One newspaper highlighted the findings on radiological releases and concluded that there had been no disaster, either for the local populace or for the future of nuclear power.[2] Similarly, the *New York News* printed in bold type in its edition of November 1, 1979, that the commission uncovered "no inherent fatal flaws" that would justify giving up on nuclear energy.[3] In contrast, other editors saw the report as dooming the future of nuclear power. According to one, the commission really recommended the end to nuclear power "but didn't have the nerve to state it in those terms. It should have."[4] Still another reached the conclusion that, as evidenced by the commission investigation, nuclear power is neither safe nor cheap. Yet, this is the same report that Ralph Nader described as "milquetoast" and that a Pennsylvania paper termed unoriginal. The latter source held that

> the final report of the Presidential Commission on the accident at Three Mile Island has little new to tell the people who lived through the nightmarish days from March 28 on—and little more about which they might cheer. As a post-mortem, it is doubtless an excellent piece of work, but the report falters when it attempts to apply what was learned from TMI to what this nation's policy should be in regard to nuclear power in the aftermath of the nation's worst nuclear accident.[5]

As all of this should indicate, it would be hazardous and impracticable

---

*I was assigned the topic "lessons from the perspective of the commission" for the New York Academy of Sciences conference on Three Mile Island. The other members of the President's Commission on the Accident at Three Mile Island were John Kemeny, chairman, Bruce Babbitt, Patrick Haggerty, Carolyn Lewis, Paul Marks, Lloyd McBride, Harry McPherson, Russell Peterson, Thomas Pigford, Theodore Taylor, and Anne Trunk. The commission report was issued in October 1979.[1]

0077-8923/81/0365-0280 $01.75/2 © 1981, NYAS

for me to speak in generalities about the conclusions the public has drawn from the report of the president's commission.

There is yet another reason for my avoiding a discussion on the lessons the citizenry has learned: such a discussion is likely to be premature. As other analyses appear and as events at TMI-2 continue to unfold, the report undoubtedly will be reexamined and reinterpreted. Just as it is still too early to assess the long-term effects of the accident, so is it impossible to determine which of the commission findings and recommendations will endure.

What I propose to discuss are not the deductions the public has or should have developed from the report. Rather, I give my attention to lessons about TMI that the commissioners themselves learned. Of particular interest to me are the problems we found to be associated with the search for clarity and certainty in the midst of incomplete information. The efforts to create certainty came from all sides—from the public and their representatives who wanted to know not simply the status of the plant but also the actions they should follow; from utility, governmental, and other analysts who sometimes offered statements on the basis of insufficient or controversial data. Our commission chronicled communication difficulties and drew some conclusions about ways in which information could be transmitted better than it was at Three Mile Island. What we did not explore is an area I wish to address here: the likelihood that a polarized political environment made the handling of uncertainty even more problematic than would have been the case had the hazard been of another type. The accident took place at a time of considerable controversy over nuclear power, and the presence of that controversy affected both the questions raised and the assessments made of events at Three Mile Island.

The polarization had its effects on those who represented the public, the utility, and the industry at large. These representatives were pressured to give definitive answers in order to demonstrate their veracity and competence. Yet often the questions could not be answered, because the data were not all in, the conditions were unstable, or the consequences of known conditions were in dispute. To illustrate these themes, I will discuss the problems surrounding the relationship between the office of the governor and the utility, and comment briefly on the evacuation advisories from the nuclear regulatory commissioners.

THE STATE AND THE UTILITY: THE DEVELOPMENT OF A CHASM

On the morning of March 30, Governor Richard Thornburgh conveyed to a number of state and federal officials the urgent need he felt for reliable information about plant conditions. According to Thornburgh, what he required most was an authority who could diagnose the situation, advise his office on appropriate actions, and above all disentangle the web of conflicting reports. His request for "one good man" brought a telephone call shortly after 11:00 A.M. in which President Carter informed him that Harold Denton, director of the Office of Nuclear Reactor Regulation at the Nuclear Regulatory Commission (NRC) would come as the White House representative.

As Thornburgh has described events subsequently, his principal problems from Wednesday, March 28—the first day of the accident—on were informational ones: communication between his office and the site was inadequate. He indicated to the president's commission, "I was very much concerned about the information we were getting and [felt] that we were going to have to keep pushing to get to the bottom of this."[6] I would propose, however, that the difficulties between the governor's office and the utility involved more than the existence of blocked communication channels. The state officials as well as those from the utility had to seem believable to a large and alert public, many members of which were skeptical of or hostile to the nuclear option. Consequently, both sets of officials took special pains to provide unambiguous assessments to the public; their strategies for doing so differed, however. The governor and his staff relied quite heavily on their legal training to elicit unanimity from the scientific and technical experts on whom they relied. In contrast, the utility representatives did not attempt to reconcile conflicting facts; they tended instead to deemphasize in their public statements any inconsistencies in the facts. Neither approach made it obvious that, as in most accidents, complete and indisputable details were not immediately available.

Creating Certainty: A Legal Strategy

Before Harold Denton arrived, Governor Thornburgh, Lieutenant Governor William Scranton, and Executive Assistant Jay Waldman turned to technical experts from the utility and from various state offices for appraisals of events. Often the appraisals varied widely; in these cases, the officials cross-examined the appraisers to secure greater accord among them. Waldman portrayed their efforts in this way:

> . . . it was the lawyers who were able, because of their training and experience, to force the facts out, to separate what was relevant from irrelevant. [The scientists] would come with recommendations based, I would assume, on their scientific expertise, but under questioning would actually change their own minds.[6]

The governor confirmed this depiction, adding that Waldman had been relentless in his search for the pertinent information from experts who often seemed rather abstract.

Had the informational problems that the governor faced resulted solely from utility intransigence, then a cross-examination strategy might have been appropriate and effective. But in some instances, the essential knowledge simply did not exist; in other cases, the facts were available but analysts disagreed about their policy implications. The governor's concern about core damage serves to highlight the problems associated with incomplete technical knowledge.

Governor Thornburgh reported to the president's commission that as early as Wednesday evening, he had become concerned about the integrity of the reactor core. He had recalled from his reading of *We Almost Lost Detroit* that a core meltdown was probably the most serious aspect of a plant accident.[19] What had disturbed him that evening was the realization that none of the briefings given that day had described the condition of the core at Three Mile Island. The governor had made this one of the first topics he raised on Thursday morning.

Why was there no analysis of the core at the meetings on March 28? Did this indeed indicate that utility briefers would not report unfavorable information unless pressed to do so? That conclusion seems unwarranted if we trace the development and diffusion of knowledge about core damage. Accounts of an uncovered core actually did surface on Wednesday, but those accounts did not reach some key individuals on that day. Reportedly, a Metropolitan Edison (Met Ed) employee informed a control engineer quite early that the measurements pointed to core damage, but his assessment was not relayed to Gary Miller, station manager at TMI-2. Miller informed our commission, "I don't think we analyzed in our minds whether [there was] coverage or uncoverage or the amount of damage, at least I don't think I did."[7]† Conversely, Victor Stello of the NRC decided on the first day that the core had been uncovered; yet his conclusion did not reach some other parts of his own agency. As a result, an NRC technical operations specialist told an assembled group on Wednesday evening that there had been no core uncoverage. In sum, at the time that Governor Thornburgh was wondering why he had no information on the core, the analysts themselves were unaware of or unclear about core damage.

The governor's interest in the core was not strictly technical. Waldman described the concern in this way:

> ... it was less important whether there was damage to the cladding or damage to the core ... [than] it was to get direct answers in simple layman's language: what could happen, what might happen, and how much lead time do we have if it does happen?[6]

The state officials wanted direct and simple answers, but the "what if?" questions they posed were extremely complex. Even after the analysts acknowledged a damaged core, they disagreed on the health and safety consequences of the damage. This was the case for at least two reasons. First, estimates varied widely on the extent of damage, on the kinds of gases that could be released, and on the likelihood that in fact there would be releases. Second, the biological scientists were engaged in a

---

†Miller and several others have given inconsistent statements as to when they first learned about core damage. For a fuller review on this subject, see Reference 8.

long-standing debate on the health effects of low-level radiation. Given these circumstances, no matter how intensely the public officials cross-examined the experts, they could not have obtained identical advice, for during this tense period there were endless uncertainties.

Relationships between the office of the governor and the utility began to deteriorate as of Wednesday. The deterioration could not be attributed totally to Met Ed ineptness in transmitting information to the governor and the staff. Contributing as well to the tension was the fact that Met Ed was viewed quite skeptically by a wider public, including an aggressive press. As better information evolved over the course of the accident, many of the media representatives became convinced that Met Ed was inclined to give reports that were unduly optimistic. The governor, as a public official, could not ignore the growing disenchantment with Met Ed. He himself contended:

> Our principal concern was to remain credible with the public who looked to us for advice and leadership. And all of this process of laborious cross-examination and cross-checking . . . was designed to make this office an island of credibility to which people could look for reliable advice. That is political in a sense, because you don't last very long in government or business if you are not credible.[6]

To insure their own credibility, the state officials worked to establish their independence from Metropolitan Edison. Following a meeting with the lieutenant governor on Wednesday, Met Ed vice president John Herbein proposed that they hold a joint press conference. According to the lieutenant governor's press secretary, he personally rejected the idea of a Scranton-Herbein press conference because he wanted to preserve the credence of the state officials. In essence, it would appear that in the given climate, the elected representatives had little to gain from a close affiliation with the utility. They were especially inclined to scrutinize closely any actions of and statements from the company, the result of which was the creation of a chasm between the site and the statehouse.

Creating Certainty: A Public Relations Strategy

In his testimony before the president's commission, Walter Creitz, then Met Ed president, reflected on the image problems his company had confronted during the accident. He commented on the mistrust expressed by state officials and others as follows:

> . . . I assure you and everyone, in every instance, we tried to tell the public as well as the state and NRC about significant events as they occurred. At first, we attempted to tell and explain the events to the media. But the fluidity of this type of developing accident changed facts a number of times, resulting in what was interpreted to be conflicting comments. We told the facts as we saw them at the time. The final record should show the citizens of this Commonwealth and the nation that the Met-Ed people were straight-forward.[9]

Subsequent evaluations have indeed suggested that utility officials were expected to produce clear-cut statements on conditions, even when the factual basis was inadequate for such statements. But those evaluations also show that Met Ed took actions that deemphasized the uncertainty that often prevailed. The company seemingly did not distort known facts, but it sometimes presented material with such a sense of conviction that observers felt misled when newer and different facts were presented just as persuasively. I have termed the approach of the utility a public relations strategy to indicate the emphasis given to creating an impression of certainty.

On Wednesday afternoon, several plant personnel, including the emergency director, went to Harrisburg to meet with the lieutenant governor, who at the time was in charge of the state's response. Gary Miller, the emergency director, has since indicated that he left the plant rather reluctantly, for he knew that conditions still were unstable. The NRC Special Inquiry Group, in its review of events at Three Mile Island, adjudged Miller's departure to have been ill timed, for it reduced the technical support at the plant and erroneously implied to state officials that the situation there was under control.[10]

The Special Inquiry Group was not critical of the lieutenant governor for requesting the briefing; he had no reason, given the extant reports from the utility, to think his request inconvenient. Moreover, continued the group, the utility could have refused the request or dispatched personnel who were less central to the ongoing diagnoses at the site. I would maintain that the utility responded as it did because of its interest in conveying an image of competence and control.

That overarching interest led officials at times to offer their own assumptions as if they were facts. An incident cited by the Media Institute exemplifies this point. The following exchange took place between Herbein and an NBC reporter on March 28:

HERBEIN: We do have our crews out and we're monitoring for airborne contamination. The amount that we found is minimal.
REPORTER: We understand that some of the workers did get some radiation.
HERBEIN: I'm sure that some of them got exposure but I'm positive that no one was over exposed.[11]

The Media Institute called attention to the fact that the reporter did not probe further, that he did not push to determine how Herbein could have been positive. In many of the reports that the institute covered, newscasters who supposedly were investigating nuclear energy placed undue stress on peripheral matters. But there were cases, such as the one just cited, when figures central to the investigation provided incomplete or unsubstantiated material.

The public relations strategy of Met Ed extended beyond the use of reassuring tones and statements; the company also recruited a public relations firm to manage its interface with the press and the larger citizenry. According to the *Wall Street Journal*, Met Ed took this action because it faced a public relations nightmare: a hostile press, an edgy public, and an industry that was looking the other way.[12] With the assistance of the Hill and Knowlton firm, Met Ed set up what was to be a coordinated news center within two days of the start of the accident. The plans for a coordinated system were not fully realized, however, primarily because those most knowledgeable had too little time to keep the news staff apprised of the rapidly changing conditions at the plant.‡

Metropolitan Edison officials were criticized at the time of the accident and have been since for being excessively confident. It seems quite likely that the officials, when asked about information that was still under analysis or being disputed, would provide more reassuring than alarming scenarios. This I would interpret as an effect of operating in a public and highly charged arena. Neither the utility representatives nor the state officials could behave as if the environment in which they operated was a benign one. Neither side had the luxury of acknowledging uncertainty.

The NRC and Evacuation

The state and plant officials were not the only actors searching for clarity. Now legend is the statement on Friday from Joseph Hendrie, then chairman of the NRC, that "we are operating almost totally in the blind."[13] He continued, ". . . it's like a couple of blind men staggering around making decisions."[13] The NRC, as was true of the governor, was expected to make recommendations on which the public could act. Indeed, it was on Friday that Governor Thornburgh asked Chairman Hendrie and his associates for their evaluation of existing conditions and of the actions he could take. Here, as occurred again and again during the first week of the accident, decision makers had limited and inconsistent information to work with; they did not have the benefit of the highly detailed analyses that now fill the annals on TMI. Given the absence of high-quality data and the presence of ever-shifting parameters, Hendrie and his colleagues could not base their recommendations on technical information. Complicating the matter even more were the errors they made in calculating the technical details. In particular, their assumptions about the possibility of a hydrogen explosion in the reactor vessel conflicted with other assessments, but entered into their own deliberations about ways for protecting the public.§

Under the circumstances, the commissioners had two options on Friday morning, as they saw the situation. On the one hand, they could

‡See Reference 10 for more details on the operation of this center.

§For information on the discussions concerning the hydrogen bubble, see various staff reports for the president's commission, especially References 14–16. Thomas Pigford has concluded that the NRC miscalculations regarding a reactor explosion had serious consequences for the management of and public response to the accident. See Reference 17.

take a highly conservative approach—one based on a "worst-case" analysis—and recommend the evacuation of the area; Commissioner Peter Bradford favored that approach. Alternatively, they could identify less sweeping measures and await fuller details. Commissioner Hendrie chose that course when he suggested to Governor Thornburgh that he advise residents to remain indoors. Obviously, all concerned hoped that if they were to err, it would be on the side of safety. Quite unclear, however, was exactly where that side was located. In uncertain situations—such as the commissioners, the governor, and other officials confronted—the wisdom of a decision made can only be determined retrospectively. It is no wonder, then, that those who must propose actions probe continuously and deeply for consistent interpretations of events. Again, the controversy that swirled around nuclear power no doubt intensified the probes.

The initial Thornburgh-Hendrie conversation was followed soon after by another call from Chairman Hendrie apologizing for the evacuation recommendation that the governor had received earlier from some members of the NRC staff. As the conversation continued, the matter of advisories to local residents again surfaced. The governor noted that his secretary of health had recommended the evacuation of infants and pregnant women from the area, to which Hendrie supposedly replied, "If my wife were pregnant and I had small children in the area, I would get them out because we don't know what is going to happen." Following that discussion, officials in the governor's office discussed ways for implementing the recommendation and announcing it to the public. At 12:30 P.M., the governor issued his advisory for preschoolers and pregnant women within a five-mile radius of the facility to leave the area.

Perhaps the term "collegial" would best describe the approach the commissioners used for handling the mass of incomplete and inconsistent data they compiled. The approach was collegial in that the commissioners often relied on one another for confirmation of analyses or impressions; they carried on some of their work rather independently of efforts being made by other NRC personnel. The way in which the matter of an evacuation was dealt with supports this conclusion. The five officials wrestled with the subject of evacuation long before they made any recommendation to the governor. They were unprepared, however, for the news on Friday morning that NRC had recommended an evacuation, for at that time, the commissioners had not decided that the conditions warranted such a response. Only later did they learn that the advisory had been transmitted to state officials directly from members of the NRC senior staff.

The approach was also collegial in the sense that the five made their decisions jointly. On Sunday, the four commissioners who were at their headquarters reached the consensus that given the uncertainties about the hydrogen bubble, a precautionary evacuation was appropriate. The four reported this view to Chairman Hendrie, who was then at the site, but noted that they would defer to his decision as he was more likely to have firsthand knowledge of the situation. Commissioner Richard Kennedy later told the Special Inquiry Group:

> We were not telling him as a collegial body, you have just been given an instruction by a majority of your peers. We were telling him, the majority of your peers, from its own perception, sees it this way, but recognizes that there may well be factors which it doesn't know or comprehend in the same way as you do on the ground there.[10]

Hendrie did indeed see the circumstances differently, for on that day, he received information that alleviated the concern about a hydrogen explosion in the reactor vessel. With the changed situation, there seemed to be no need for an evacuation.

The president's commission criticized the NRC for the internal communication problems it experienced and for its use of a collegial approach to emergency management. Was there any indication that this collegiality led the commissioners to take the wrong actions regarding the evacuation of the public? To answer this question, one must decide first if the NRC commissioners made the right decision. The president's commission made no such assessment and in fact was challenged by NRC commissioner Peter Bradford for failing to do so. Bradford wrote of the report:

> Much is said, and validly, about the inadequacy of the process for appraising and conveying the need for evacuation. Nevertheless, the TMI Commission said nothing about the validity of the actual recommendation that was made. This seems to me to be an oversight of some magnitude, for such decisions are often likely to involve the allocation of unquantifiable uncertainties.[18]

Even though we had the advantage of hindsight, we on the commission did not reach a consensus about the evacuation advisories. We, as had been true earlier of the NRC commissioners, were faced with a situation in which one's assessment depended on whether one stressed actual damage to physical health that had taken place or the potential health problems that might have emanated. We struggled, too, with this question: Should efforts be made to develop clearer criteria on which public officials could act to protect public safety? We could offer no such criteria and concluded that there will continue to be situations in which our elected representatives will have to act without the benefit of unambiguous guidelines. That has always been the case; quite possibly the political context of TMI made the representatives particularly determined to find consistent support for their decisions and concerned about developing ways for determining the courses they should follow.

Whatever the quality of the decision, what about the process through which the NRC commissioners handled the matter of evacuation? How appropriate was collegial management? Clearly, the approach was inefficient, for it demanded considerable involvement from all of the commissioners. Nevertheless, in light of the uncertainties that prevailed and the clamorous climate that surrounded the accident, a hierarchical decision structure might have been precluded; no single individual would have wished to take responsibility for decisions that later events could disparage. Even if the commission had not been structured collegially, those in authority would have turned to others for advice and confirmation. Such a strategy would have spread the risks that almost inevitably are associated with actions taken when the data are incomplete. Probably, the management of the uncertainties involving a controversial issue calls forth decision-making methods that would be avoided under other conditions.

The NRC commissioners realized that they were operating in the public limelight. One of their responses was to spend time drafting and redrafting public releases; another was to vacillate in their directives to the plant. The evidence shows that NRC halted the dumping of wastewater into the Susquehanna River in part because it feared that the press would exaggerate the matter. Such dumping was routine at the time of the accident, and the material transported into the river was not in fact produced by the accident. Following negotiations among utility, federal, and state personnel, the dumping was resumed.¶

The commissioners were concerned about their image, but in their treatment of the evacuation issue, they accepted the reality of uncertainty. When finally they formulated a recommendation, they did so not because they had technical data that showed high levels of radioactivity, but perhaps because they recognized that they were in one of those situations in which the only known truth is that conditions are ambiguous.

ACCIDENTS AND UNCERTAINTY: FURTHER REFLECTIONS

I have proceeded as if two easily identifiable and irreconcilable camps exist at present. That admittedly oversimplifies the views held, as the public opinion polls demonstrate. But the most active elements do tend to end up either criticizing or defending nuclear technology. In that kind of highly emotional setting, there may be an exaggerated tendency by the defenders to take a stance that, as our Public's Right to Information Task Force argued, interprets the uncertainties far too positively. Likewise, there may be an inclination for the other side to treat inconsistencies as clear signs of bungling or of gross incompetence. The result is the reinforcement in both camps of the view to which allegiance already has been paid.

Significantly, considerable planning for radiological emergencies has

¶The governor described the involvement of his staff in the dispute over the wastewater discharge. He made this assessment of their interaction with NRC: ". . . there was a feeling we were getting nudged a little bit, that they were trying to hang this one on us, in effect."[6] Ironically, the press gave little coverage to the dumping. See Reference 16.

taken place since the accident. Much of the planning seeks to overcome the communication difficulties that surfaced during the period. The president's commission concluded:

> Federal and state agencies, as well as the utility, should make adequate preparation for a systematic public information program so that in time of a radiation-related emergency, they can provide timely and accurate information to the news media and the public in a form that is understandable.[1]

Yet, no plans for communication will eliminate the problems associated with uncertainty. No matter how well developed the channels of information or how strong the commitment to their use, questions about the character and consequences of an accident will remain. Decisions will have to be made, and when the accident involves a controversial technology, such as nuclear engineering, uncertainty will be equated with surreptitiousness and incompetence. Consequently, no one— advisor or decision maker—will want to be accused of being uncertain either of the facts or of their subsequent effects.

The recommendations post-TMI stress, too, the need for improved diagnostic and response capabilities at nuclear power plants. More intense operator training and better control room design can make a difference, but they cannot guarantee clarity and certainty. First, in some respects, the nuclear technology as applied to power plants is still evolving. Consequently, there will continue to be certain unknowns and certain conditions for which the data will be hypothetical only. In the case of Three Mile Island, a highly trained cadre quite likely would have picked up signals about core uncovery that the TMI engineering personnel missed. But that cadre perhaps would have expected greater releases of radioactive iodine into the atmosphere than in fact took place, for the technical data that existed at the time certainly supported that expectation. Second, some of the decisions to be made will not depend strictly on technical criteria. Some uncertainties will involve judgments, and we would be misled to assume that such judgments can be resolved solely by scientific facts. Third, by definition, one cannot anticipate and thus plan for all uncertainties. Had there been no uncertainties, no encounters with events never seen before, there would have been no accident at Three Mile Island. The same uncertainties may not reappear elsewhere, but other ambiguities will still make difficult the task of explaining conditions to those who wish the information to be unequivocal and thorough. The planning now under way may be appropriate for accidents that happen within geographically limited areas that conform to the proposed emergency planning zones or for which there is knowledge that can be transmitted in a highly systematic manner. Such planning may not fit the needs of a public faced with a lingering threat whose dimensions are not fully known and whose nature and course need not remain constant. Finally, we cannot lose sight of the political context in which interpretations and analyses even of known data on nuclear power must now take place.

REFERENCES

1. KEMENY, J. G., et al. 1979. Report of the President's Commission on the Accident at Three Mile Island. U.S. Government Printing Office. Washington, D.C.
2. State-Journal-Register. 1979. Commission report merits top priority. November 5: 10.
3. New York News. 1979. November 1: 59.
4. Oregon Statesman. 1979. Disguised death sentence for nukes. October 31: A7.
5. Patriot. 1979. TMI report: little new and little to cheer. November 1: 5.
6. THORNBURGH, R. L. 1979. Deposition to the President's Commission on the Accident at Three Mile Island, on August 17: 42, 49, 63, 71.
7. MILLER, G. 1979. Deposition to the President's Commission on the Accident at Three Mile Island, on June 9: 287-288.
8. Staff. 1979. The Role of the Managing Utility and Its Suppliers: 174-185. Staff report to the President's Commission on the Accident at Three Mile Island.
9. CREITZ, W. 1979. Testimony before the President's Commission on the Accident at Three Mile Island, prepared May 18: 5-6.
10. NRC Special Inquiry Group. 1980. Three Mile Island: A Report to the Commissioners and to the Public 2(part 3): 836-837, 982. U.S. Government Printing Office. Washington, D.C.
11. The Media Institute. 1979. Television Evening News Covers Nuclear Energy: A Ten Year Perspective: 35. Washington, D.C.
12. PETZINGER, T., JR. 1979. When disaster comes, public relations men won't be far behind. Wall Street Journal (August 23): 1, 27.
13. HENDRIE, J. 1979. Statement made at press conference on March 30. Harrisburg, Penn.
14. President's Commission. 1979. Report of the Technical Task Force 2. Chemistry. U.S. Government Printing Office. Washington, D.C.
15. President's Commission. 1979. Report of the Office of the Chief Counsel: The Nuclear Regulatory Commission. U.S. Government Printing Office. Washington, D.C.
16. President's Commission. 1979. Report of the Public's Right to Information Task Force. U.S. Government Printing Office. Washington, D.C.
17. PIGFORD, T. 1980. Kemeny commission conclusions. In An Acceptable Future Nuclear Energy System. M. W. Firebaugh & M. J. Ohanian, Eds.: 31-41. Institute for Energy Analysis. Oak Ridge, Tenn.
18. BRADFORD, P. 1979. Letter to Frank Press, Office of Science and Technology Policy, Executive Office of the President, on November 9.
19. FULLER, J. G. 1975. We Almost Lost Detroit. Readers' Digest Press. New York, N.Y.

THE THREE MILE ISLAND NUCLEAR ACCIDENT:
THE OTHER LESSON

Chauncey Starr

Electric Power Research Institute
Palo Alto, California 94303

I was asked to speak on the lessons learned from the Three Mile Island (TMI) accident from the standpoint of the utility industry. As you all know, TMI highlighted organizational and engineering weaknesses in our nuclear power operations—weaknesses that clearly called for remedies.

This subject, from a technical and management point of view, is a familiar task for me and for hundreds of my peers in industry, because it concerns an internal reorganization of resources, engineering changes, and matters of science—areas in which we've spent long careers and are professionally qualified to identify problems and develop remedies and fixes. This is the way technology improvement has always worked.

The utility industry clearly has been responsive to these needs. It has:

1. Formed an Ad Hoc Nuclear Oversight Committee to marshal the resources of all the nuclear utilities;

2. Formed and organized the Nuclear Safety Analysis Center (NSAC) to develop technical modifications where needed;

3. Conceived and organized an Institute of Nuclear Power Operations (INPO) to improve the quality of management and operation;

4. Formed the Nuclear Electric Insurance, Ltd. (NEIL) to share the burden of accident costs not otherwise covered;

5. Formed a Committee on Energy Awareness (CEA) to improve public understanding of the nation's energy dilemma.

The first four steps are straightforward technical and business actions that any prudent manager would take to solve problems of management and operations. In this case, the novelty arises from the number, organizational variability, and regional dispersion of the nuclear utilities. The nuclear utilities' cooperative response in these actions arises from their recognition of their individual responsibility for an issue of national significance.

These four steps taken by the industry are problem specific and positive, and will result in more reliable and safer operation of nuclear power plants. I have been involved personally with the industry's responses, and I am convinced they will be constructive and effective.

I had prepared a paper on the four technical and operating subjects, entitled "Post-Three Mile Island Nuclear Utilities' Initiatives."* I've

*See Appendix for the text of this original paper.

0077–8923/81/0365–0292 $01.75/2 © 1981, NYAS

decided not to read it today. You've heard the essence of it already, and copies are available in the press room.

But last week it occurred to me during my daily jog that there are other less obvious lessons that we have learned from TMI but that we in the industry cannot solve with all our accumulated technical, scientific, and industrial abilities and experience. They are problems that, if left unsolved, will negatively affect our society and its people more than anything I can imagine. They stem generally from our mix of public attitudes on economic growth, on technological development, on our social structure, and on our nation's international status.

Specifically, TMI showed that we do not know how to communicate to the public the need for nuclear power in this country and its role in keeping our society alive and viable—even when so many comprehensive professional group studies have reaffirmed this need, as, for example, the extensive CONAES study by the National Academy of Sciences. This is the focus of the fifth industry step, the Committee on Energy Awareness.

TMI has made it clear that, like everyone else, the industry is dependent on the media for communication to the public of its views, facts, and functions. Information, which by its very nature is often technical and complex, is media summarized, filtered, and oversimplified in leads and headlines that often result in half-truths or inaccurate or misleading public information. Typically, a question answered by a utility executive on television and taking many minutes to complete is reduced to a few seconds of message. A balanced answer is molded into a "pro" or "con" by the editing. A full disclosure is molded into an evasive partial response. My adrenalin frequently has been stimulated by a comparison of the complete and edited tapes of such interviews.

The growing acceptability of editorializing and advocacy in journalism has also put powerful public influence on energy questions too often in the hands of articulate and fervent writers with minimal technical qualification but with some control of the media output. It is increasingly difficult to separate the reporting of facts from journalistic opinion, or from even the more cynical result of creation of the news by journalese. As with every profession, journalism has standards and ethics, but these are not always met.

This laxity in media journalism is profitably used by publicity-seeking scientists and engineers, often with pernicious effects on public understanding. Instead of their work first being filtered by professional review and criticism, they use eager and often gullible journalists to opportunistically jump into headlines. With regard to our energy options, the public is thus confused by premature or exaggerated developments or philosophic cure-alls, all offering enticing escapes from the inexorable realities. The professions of science and engineering have standards and ethics, and clearly they also are not always met.

These comments are not intended as a tirade on an imperfect world, but rather as a description of some of the obstacles to developing a balanced public understanding of issues and options in such a complex technical field as energy.

We have also learned that competition for the "news break" within the media themselves results in unpredictable and fearful consequences when an operating accident such as TMI occurs, and when factual information is not instantly available. This was amply demonstrated at TMI by the proliferation of speculative scare headlines about the "hydrogen bubble," "evacuation plans," "the death of spring robins," "clouds of radiation," and "brink of catastrophe"—all highly exaggerated at the time and since proven time and again to have been without factual basis.

Although this is a general communications problem in our society, specifically in this context the utility industry doesn't know how to handle this kind of fear stimulation about nuclear power in an effective manner. Such fears are long-lasting once instilled, and effectively block public acceptance. Innovative ways of developing a full public understanding of nuclear power must now be tried. Without public acceptance, nuclear power will be unable to proceed. It has gotten this far (12% of all electricity) because of the enthusiastic public and political support during the 1950s and 1960s, when nuclear power was generally considered a godsend and the utilities were pushed to deploy it.

We are living in a generation that has been beset with unrest and instability. We've had two distasteful wars in 20 years, with powerful inflationary results. We're in the middle of a technical revolution that may be more influential than was the industrial revolution of 100 years ago. We're in the middle of a racial revolution, a sexual revolution, a population revolution, and some kind of revolution in personal morals and concepts of social and national responsibilities. And the most recent of these foundation-shaking changes is our current energy vulnerability. If solutions are not found soon, this may result in the greatest revolution of all. Energy malnutrition will clearly force many major social changes. The consequences are not all foreseeable, but they will be large. All this instability has stirred and alienated a great many people in our country. They are confused and irritated—and they strike out, haphazardly at times, searching for answers and ways to get their lives under control, to bring back a sense of security, to once again feel at ease and safe.

We in the utility industry believe we understand the state of mind that makes the public receptive to anxieties. We recognize its impact on the public acceptance of nuclear power. We try to cope with it by making speeches, giving interviews, and writing papers and developing scientific and technical information on the subject we know best. But we don't control what happens to our information after it leaves our hands or lips.

Without public understanding of the facts and real options in resolving our energy problems, we can expect unwise legislation and bad public decisions. The consequences will be a continued threat to our economy and to our social fabric, extending for decades.

Another lesson we have learned is that everyone's innate fear of the invisible and the unknown is a major impediment to the appropriate development and use of nuclear power. It's comparable, perhaps, to the fear of witchcraft, which was so intense in early American history. Witches were summarily burned at the stake as a form of American justice. We've progressed emotionally only a little from our witchcraft era. Today the invisible and intangible hazards exist in UFOs; in carcinogens in our air, our food, and our water; in cholesterol in our traditional foods; and, of course, in radiation. And these fears have been fanned and stimulated by a surprisingly wide variety of special interest groups who, although diverse in many ways, have found in nuclear power a common focus for their various antagonisms.

What can we do? For example, would it help to convince the public if nuclear scientists and their families set up camp on TMI grounds during the venting of TMI's krypton gases? I volunteer. I do so because I understand the nature of low-level nuclear radiation, and because the existing controls provide a degree of safety during the venting that far exceeds the daily natural exposure we all receive in our own homes, wherever they are. But will that convince the anxious local residents and the institutionalized alarmists?

It would be a great mistake to think that public acceptance is the concern of the energy industry alone. Herman Kahn of the Hudson Institute, for example, points out the difficulty that the public would have in getting the benefits of aspirin if that substance had been discovered last week. It is all too easy to visualize the field day that the antibusiness and antitechnology critics would have in proclaiming the thousands of deaths that would result from the sale of those deadly white pills of acetylsalicylic acid. Such a fantasy might be humorous if it were not for the fact that just this kind of "protection" also withholds from the public the benefits of new drugs. Just this kind of "protection" from nuclear power may also cause the national disaster of a major energy shortfall.

These lessons are not new, and they were only emphasized for us by the travesty of public communication that surrounded—and which still surrounds—the accident at Three Mile Island.

Who has the responsibility for the development and presentation of balanced information on critical energy problems? It is obvious that the utilities alone cannot assume this responsibility. In the first place, it is a national, not solely a utility industry, problem; and secondly, we're not very good at it, as everyone knows by now.

As a matter of fact, under present circumstances, the myriad problems of utility management would be greatly eased if they did nothing more to explain, demonstrate, and remonstrate with the public on these matters. Public opinion has already set the scene for a minimum or no expansion of electrical capacity for the next 10 years. If this status quo attitude continues, the utilities would be spared the onerous and demanding work of financing, siting, securing environmental approvals and licenses, and finally performing construction in a period of inflation and 20% interest. The financial pressures on most utilities would certainly be eased by a long period of inaction, with increased returns to stockholders. But with 10-year lead times for new plants, what of the future?

The obligation of the utilities to supply electricity on demand has been a professional credo from the beginning of the industry. Most franchises require service to the public, if electricity is available. But it is not clear that utilities are obligated to prepare for a speculative future demand and to go through all the agonies of building facilities for the future under current circumstances. Traditionally, utilities have built for the future as a matter of social responsibility and business prudence, and will undoubtedly continue to do so. However, in the past it has been with public acceptance taken for granted. Obviously, this is not the case today.

In today's environment, it would be easier for the utilities to sit back and let the public pressures run their course. It wouldn't be pleasant for knowledgeable and responsible people to watch, but it would make the near-term job of managing utilities a great deal easier.

I don't believe the utilities intend to abrogate their perceived responsibility to prepare for the future, but the diminishing public acceptance of so many expansion plans is affecting their present positions.

Perhaps this planning for the future should now become a shared concern with the political overseers of the industry: the regulatory commissions, the governors, the Congress, and the executive branches of government. And perhaps, most of all, it should be the concern of the business community and labor unions, whose existence depends on energy supply.

My summary conclusion to all this is that the most important lessons we've learned from Three Mile Island are the ones we can't resolve. The technical, scientific, and management lessons have been recognized and are being handled expeditiously and effectively. But the creation of an informed public opinion is beyond the ability of the utility industry alone.

I'm sure you've all heard the story of the businessman who paid $1,000 for a parrot that could speak five different languages. He gave instructions to the pet shop to deliver the bird to his home that very afternoon. Arriving home, he asked his wife if the bird had been delivered. Yes, it had. "Where is it?" asked the businessman.

"In the oven," said his wife.

"My God, in the oven? Don't you know that bird spoke five languages?"

"Well," replied his wife, "if he was so damned smart, why didn't he speak up?"

It's for this reason that I've taken the unusual step today of discarding my prepared and assigned subject and am talking to you about the overriding problem of public perceptions.

Certainly, the responsibility for developing public understanding cannot rest solely with the utility industry. This would not be sufficient or credible. Only a concerted effort by all involved, regardless of position, is called for.

Some are speaking up. The former secretary of energy, James Schlesinger, told the Washington Press Club, as he left his position and could speak freely: "The energy future is bleak and is likely to grow bleaker in the decade ahead. Quite bluntly, unless we achieve a greater use of coal and nuclear power in this decade, this society just may not make it."

I could make another speech—and I have done so many times—that would present evidence to prove that:

●Nuclear power development is in the best interests of the consumer and of the future of this nation;

●It is the lowest cost power that can be produced from available fuels;

●It is a benign source of power environmentally;

●It provides the greatest independence of fuel supply;

●Its public risk has been, and continues to be, the lowest of all available sources of energy.

I can prove these statements to any qualified professional group. Furthermore, with singular modesty, I'm qualified to make these statements as a tested lifetime scientist, engineer, and manager who professionally continues to be active in all alternative sources. These views are also held by many electric utilities that are free to choose alternatives. One major utility with about half coal and half nuclear capacity already has seven nuclear plants operating, six under construction, and two more in their plans. Such choices are not made lightly.

What more can I say?

I don't have any better answers on how to publicly transmit my convictions through the media than have other members of this industry. And, like them, I'd be happier and more productive back at my desk in Palo Alto dealing with technical improvements and with the new ideas we'll need for all the forms of electricity production in the future.

The time has come for thoughtful businessmen and people outside the utility industry to assume a partner's share in the responsibility of changing public opinion to full support of the development of nuclear power. For all business interests, nuclear development is in their *special* interest. It is time to remember that public interest *lets* things happen, but only special interest groups *make* things happen.

Finally—and perhaps most important—we must have some credible national spokesmen on energy policy generally, and on nuclear policy in particular. I believe this is the only way to bridge the public information gap between decisions based on professional analysis of complex details and the need for journalistic simplicity and clarity in public communication. The NASA moon program of the 1960s, for example, successfully utilized such credible spokesmen. Perhaps we can find a way to establish a similar relationship for nuclear power.

POST-THREE MILE ISLAND NUCLEAR
UTILITIES' INITIATIVES

INTRODUCTION

Two weeks ago we marked the first anniversary of the accident at the Three Mile Island nuclear plant. TMI, without doubt, has been a watershed event for the electric utility industry. The industry will never be quite the same. Before I discuss the lessons learned by the electric utilities from the accident and the specific actions initiated by this industry, I would like to discuss the utilities' present view of nuclear power. I believe that the recent utility responses—which I will describe—represent a radical step for improving the nuclear option.

Recently, a transient, its origin not unlike the one at TMI, occurred at the Crystal River nuclear plant. This transient was terminated quickly, efficiently, almost by reflex, so much so that it deserves only passing mention. I would like to think that we as an industry are today much better prepared to cope with or prevent another large-scale accident than we were a year ago.

New analysis of the TMI accident now being completed indicates that the potential catastrophic consequences to the public were highly exaggerated at the time. It is now clear that sufficient time and a large number of countermeasures existed to control the damage to the reactor core in such an accident. Furthermore, there exists a substantial safety margin between a core melt and the containment building failure that could pose danger to the public.

The events at Three Mile Island have taught us that many things can be done to improve the safety and operational reliability of nuclear power plants. The utility actions described below demonstrate the utility industry commitment to achieving this goal. The driving motivation is the basic perception of this industry that nuclear power is an essential part of our future energy mix. It is relevant, therefore, to understand the basis for this position.

WHY ARE UTILITIES PRONUCLEAR?

The electric utilities know better than any other group the problems of nuclear power because of their firsthand exposure. Financial, regulatory, public opinion, and operational problems are today strong disincentives to the use of nuclear power. With these disincentives, it takes substantial balancing incentives for a utility to build nuclear plants. Positive incentives are few, but they can be decisive. First, utilities desire to diversify their source of supply to reduce their vulnerability to the many disruptions that always seem to arise unpredictably from a variety of causes. In this respect, the coal mines strike of 1978 and the Iranian oil supply interruption of 1979 are cases in point. Second, comprehensive cost analyses usually reveal that in most regions, nuclear is the cheapest source of baseload power, despite its many uncertainties. Additionally, from an air quality viewpoint, nuclear power is far more attractive than is coal, the only alternative baseload source. But the final, and clinching, factor is that in general, the utilities cannot foresee how to meet their future baseload requirements without nuclear.

THE UTILITY CONSENSUS ON NUCLEAR

Given the list of incentives and disincentives just described, many utilities for the past several years have found that the short-term disincentives outweigh the incentives, and as a result, many plants have been postponed or cancelled. Yet, as estimates of long-range supply and demand repeatedly indicate, the utilities have good reason to believe that a return to nuclear power construction is generally inevitable in the long term, and in some regions, needed now.

It must be obvious that the utilities have a tremendous stake in maintaining reliable and safe nuclear operations, and certainly they believe that they can do so. This is the central issue of the nuclear debate. What is really at issue is the validity of the basic belief upon which the utilities' position on nuclear power depends, namely, that nuclear power is a controllable technology. Despite the a priori assumption by many nuclear critics that the risks are uncontrollable, my own view and that of most nuclear professionals and of the industry are precisely the opposite: we are confident that the risks to the public from nuclear power are controllable to acceptable levels. I further believe that the international nuclear experience, even including accidents such as TMI, bears out this viewpoint.

The TMI accident focused the attention of the utility industry on the following issues:

- The controllability of nuclear risks.
- The relative complexity of nuclear generation.
- The economic consequences of nuclear accidents.
- The required training of nuclear plant operators.
- The need for effective public communication regarding the dangers and benefits of nuclear power.

These issues are discussed next, and the industry actions derived from the "lessons learned" are then described.

THE TECHNICAL CONTROLLABILITY OF NUCLEAR RISKS

It always has been known by professional experts in the nuclear industry that should low-probability accidents occur, their consequences could be controlled and contained, so that the danger to the public at large would be avoided. This is based on the multiple barrier approach to containment design, on the relative ease of detecting nuclear radiation, and on the redundancy in built-in safety systems in an operating reactor, which allow corrective actions during the course of an accident that will minimize its consequences.

The containment technology is a part of the defense-in-depth philosophy that underlies reactor design, and has been successfully demonstrated in all aspects of nuclear power generation. This ability to isolate the hazardous materials from the environment, coupled with amazingly sensitive measures for detecting radioactivity, has enabled the industry to demonstrate almost arbitrarily low levels of routine releases and very low probabilities of accidental release to the public, approaching the probabilities of cosmic catastrophies to the earth. Most important, the technology of containment is flexible and adaptable to any real need.

The detectability of radiation is a necessary condition for controlling its risk. In this regard, the nuclear technologist is fortunate, because radiation is easily detectable with simple instruments at levels that are thousands of times lower than the levels at which harmful effects occur. The effect of this detectability is to permit sizable safety margins between the level of routine operation and the level at which the public health risk becomes significant.

A generic analysis of a small-break loss-of-coolant accident (LOCA) in a pressurized water reactor (PWR) that uses the TMI-2 system as a base PWR design has recently been performed by the Nuclear Safety Analysis Center (NSAC), which is operated for the electric utility industry by the Electric Power Research Institute.[1] This analysis has demonstrated a dramatically lower probability of a reactor core melt occurring, as compared with the *Reactor Safety Study*, WASH-1400 (the Rasmussen report).[2] This results directly from stepping beyond the conservatively simplistic WASH-1400 assumptions and considering the many actual options that are available in a PWR system for providing effective countermeasures in stopping a small-break LOCA progression.

The significant features of a PWR small-break LOCA that provide the ability to apply effective system responses include:

•Sufficient time is available to effectively select and apply countermeasures that can control and even reverse the approach to, or progress of, core damage at virtually any stage in the accident sequence. For example, at TMI-2, core damage did not begin until two hours into the accident.

•The magnitude of required responses necessary to control a small-break LOCA are well within the capacities of each of several coolant injection and heat sink systems available in a PWR, a condition that provides inherent redundancy and high probability of availability of more than a minimum required response.

•The developing deterioration of heat sink capabilities in a small-break LOCA provides many observations that indicate both current accident state and trends, on a time scale that permits rational selection of effective countermeasures and bases for conservative projections of potential public danger for emergency planning decisions.

Assuming a small-break LOCA has been initiated, the WASH-1400 study predicts a 2 out of 100 chance of core melt occurring. For the NSAC study, and assuming only the availability of the existing cooling systems properly deployed by the operators, the probability of a core melt and subsequent containment building failure could be reduced by a factor of 10 to 1,000. The probability of the initiating event occurring is 1 in 1,000 to 1 in 10,000 per year.

The most conservative assumption in the WASH-1400 studies appears to be that core melt would directly cause containment building failure by one of several modes, without any possibility that the operator could prevent reactor vessel failure by positive actions and, if necessary, even improvise the cooling of the core in the reactor cavity to prevent either containment building failure by overpressure or basemat penetration.

The quantitative evaluation of the degree to which WASH-1400 overestimates the probability of progressing from a badly damaged core to core melting and containment failure is still under study by NSAC, but it is clear that, in general, there is a substantial margin of safety between core melt and containment building failure and that, specifically, the perception of an imminent massive release of radioactivity at TMI-2 was incorrect.

THE RELATIVE COMPLEXITY OF NUCLEAR AND COAL-FIRED GENERATION

The accident at TMI has reinforced the original utility perception that nuclear power is a high technology compared with fossil fuel burning for the generation of electricity. This is not because nuclear equipment is not sturdy enough, but because the system response to an accident-initiating event is not linear but highly complex, with complicated interactions that have to be studied as a whole. We would define a linear response as one that stops the entire mechanism if one component breaks down. An example of such technology is the power train of an automobile.

Nuclear power reactors are nonlinear systems in that their response to an initiating failure is the triggering of successive failure of subsystems in very complicated chains. A simple example of such technology with complex feedback response mechanisms is the television set. In both cases, the dynamics of the internal feedback interactions are complex and depend on the state of each subsystem at the time of the failure.

This awareness of the inherent complexity of nuclear power generation and the need for attention to detail was quite well known to the pioneers of nuclear power development in this country and abroad. The TMI accident has brought home again to the commercial manufacturers and the utilities the perception of the complexity of the nuclear endeavor as compared with fossil fuel generation. Not only is the technology itself more complicated, as discussed above, but the process of mastering it is more involved. There is a basic need to test nuclear components individually, then as subsystems, and then the entire system several times over, in order to insure reliable operation of the entire reactor plant.

As a result of this heightened awareness, many utilities are establishing separate nuclear power divisions within the company structure. Such reorganizations have been announced recently by Pennsylvania Power and Light, General Public Utilities, Consumers Power, and Pacific Gas and Electric Company. Such reorganizations would enhance management control of nuclear power generation and would improve the internal coordination between all the different utility components associated with the operation of nuclear power plants, such as quality control, radiological health and safety, and fuel cycle engineering. Together with the centralized utility industry organizations established to deal with nuclear generation problems as described below, these utility company reorganizations represent a major internal restructuring aimed at improving the overall reliability of nuclear power generation.

THE INSTITUTIONAL AND ECONOMIC CONSEQUENCES OF NUCLEAR ACCIDENTS

The accident at TMI has demonstrated again the enormous economic incentive utilities have in maintaining safe operation of nuclear power plants. As most nuclear plants have a relatively large capacity of 1,000 MWe and operate at the lowest range of incremental generation costs compared with any other baseload power technologies, the loss of such plants from the generating mix would bring severe economic penalties to the affected utilities. Not only would a large block of capacity have to be replaced, but the differential in incremental costs between nuclear and outside replacement power is the largest, as compared with replacing other fossil-fired plants by emergency power. Due to the loss of electrical production from TMI Unit 2 and its sister plant TMI Unit 1—which, although undamaged, has not been permitted to operate—Metropolitan Edison Company has had to purchase replacement electricity from other utilities for distribution to its customers. Purchase of this replacement power is costing General Public Utilities Company (GPU) an additional $80,000 each day. This replacement power is generated by coal-fired and oil-fired power plants, and the added cost is in itself a clear demonstration of the economics of nuclear fuel.

Beyond the immediate problems of GPU, which may be symptomatic of the consequences of large-scale accidents to the operating utility, two general trends affect all other nuclear plants. These trends are related to increases in financing costs of nuclear power projects and increased regulatory and institutional requirements that will translate into delays in plant commissionings and additional safety features at the reactors, both of which will increase nuclear generation costs. While the temporary increase in investor nervousness with this technology may have diminished, an overall effect of 1.0 to 1.5% increase in financial costs may persist. The advantage in stock prices for nuclear utilities (higher prices-to-earnings ratio) has also decreased.

The costs of regulatory delays and additional mandated safety measures related to both plant equipment and operating procedures are harder to quantify now but are, nonetheless, quite real. Evidence of this is the recent cancellation of six east central nuclear power projects due to the perception of increased regulatory and economic uncertainties affecting nuclear power at this time.

An example of such counterproductive institutional requirements is the preparation of detailed evacuation plans for the nearby population as a precondition to the operation of nuclear plants. The popular wisdom that if an accident occurs, one should immediately "run for the hills" is not necessarily the best course of action. It deprives us of time to evaluate the situation more carefully, and does not recognize that even a quite serious accident is likely to result only in a small, radioactive release. In most cases, evacuation may, in fact, expose the public to greater danger than would less drastic measures, such as just staying indoors. The shielding ability of structures in itself would offer substantial protection. Even a simple frame wood house reduces the dose rate from a passing radioactive cloud by a factor of 2. A masonry structure may give dose rate reductions up to a factor of 10 on the first floor and 50 or more for a person staying in the basement. These values are for isolated structures. A town where 35% of the area is covered with buildings may provide another factor of 3 protection. In fact, the greater the concentration of people, the more protection is afforded by buildings, and the more difficult is evacuation. Evacuation, on the other hand, is likely to increase exposure due either to changes in meteorological conditions or to the fact that evacuation may be in the direction of the plume travel. Such complex considerations would indicate that hasty regulatory changes in response to public clamor do not always serve the public interest.

OPERATOR RESPONSIBILITY AND TRAINING

The fourth major lesson that the utility industry learned as a result of the TMI accident is the greater importance of rigorous training of nuclear plant operators. This is due to the fact that the same people who handle the routine operation of the nuclear plants will be the ones who will have to respond immediately in case of an emergency. All other external review and control groups that may be brought in to augment the operating personnel will be of limited help because of their lack of familiarity with the actual plant. At TMI, the technical capabilities of some of the review and monitoring groups that descended on the local utility in the emergency were found to be doubtful, and communication between all these groups, as extensively documented in the TMI case, proved faulty. This point increased the utilities' awareness that ultimately they are responsible for handling the total effort of accident prevention, control, and mitigation whether or not outside groups will perform properly.

All sectors of the industry are now focusing their efforts towards a reduction of the effects of both technical and human fallibility on nuclear risks. The continuous audit of nuclear plant performance—either through the collection of operating data on off-design events or through analytic and experimental studies of nuclear systems—is now a major ongoing effort of the entire industry. This review is guiding the approach to modifications of plant design and to the reduction of operator errors.

As operating events occur in "real time," the approach taking shape is to seek to upgrade reactor operator training and qualifications to reduce the likelihood of judgment error and to improve the human response to abnormal events. In addition, a significant effort will be made to reduce both the likelihood and the impact of operator errors by making malfunctions easier to detect and correct, by designing reactors to be less sensitive to operator error, and by seeking to design systems in which the consequences of such error are as low as possible.

The full scope of the actual measures taken by the utility industry to cope with nuclear emergencies is described below. As demonstrated in the recent event in the Crystal River plant, the current industry response works properly, and this is only a beginning.

NEED FOR EFFECTIVE COMMUNICATIONS REGARDING NUCLEAR RISKS AND BENEFITS

As mentioned above, two major lessons emerge from the analysis of the TMI accident and its aftermath: (1) many improvements could be made in reactor operation in order to assure reliable nuclear power generation; and (2) despite the public hysteria at the time, there existed only a very small likelihood that the TMI accident would have progressed to complete core melt or that the ensuing radiation would have escaped from the containment building. The public thus was not really at risk during the course of this accident, although it was the most serious in the history of civilian nuclear power. These very important observations should be stressed, because they directly relate to current public perceptions regarding the risks inherent in large-scale nuclear power utilization.

Risks that may be accepted for a few experimental units, or for a small number of production units, are not acceptable for a hundred such, or for a thousand. We should not defend the status quo as the pattern for the future. The status quo is what was adequate for the past.

I believe that the Kemeny report does encapsulate the public concerns in these matters, and the industry has to be responsive as best it can to achieve the objectives of reliability and safety that are implied in the recommendations of the Kemeny report.[3]

In order to assure the public that nuclear power is a safe and controllable technology, the utilities have organized a major "energy awareness program" and are trying to communicate these ideas to the public by "truth squads," media channels, talks, and debates. They are motivated by a deep sense of commitment to their professional responsibilities. It is still the utilities' perception that a safe and reliable nuclear option is required to protect the nation's electricity future, which is half of our year-2000 energy problem. Communicating this need and the real conclusions from the TMI accident to the public is the major challenge with which the utilities are now faced.

SPECIFIC UTILITY ACTIONS

One of the major results of the TMI-2 accident has been recognized by the utilities as the fact that they must assume a national and leading responsibility to ensure the reliable and safe operation of nuclear power stations. We believe that these steps are completely consistent with the experience from other fields, which has demonstrated that reliability and safety cannot be established solely by the regulatory process but require the dedication, motivation, and participation of the operating group. It is also evident to the utilities that corrective measures will have to be carried out on an industry-wide, coordinated basis and not solely at the individual company level.

The utility industry has established an Ad Hoc Nuclear Oversight Committee at a chief executive officer's level to coordinate utility activities related to the TMI accident. This committee is chaired by Floyd Lewis, the president of Middle South Utilities, Inc. A schematic organizational chart is shown in FIGURE 1. The Nuclear Oversight Committee coordinates the activities of the Atomic Industrial Forum and the utility associations Edison Electric Institute, American Public Power Association, and the National Rural Electric Cooperative Association. The oversight committee also coordinates the contacts of its organizations with the U.S. Department of Energy and the Nuclear Regulatory Commission.

Three major TMI-related activities have evolved under the guidance of the Nuclear Oversight Committee:

●The Nuclear Safety Analysis Center (NSAC), which investigates and generalizes the technical lessons learned at TMI, in order to suggest improved safety measures to be incorporated in all reactor designs and operating procedures.

●The Institute of Nuclear Power Operations (INPO), which will provide enhanced training for reactor operators, benchmarks for excellence industry wide, and the regular review of utility operations.

●Nuclear Electric Insurance, Ltd. (NEIL), a mutual insurance organization that will cover member utilities against costly power replacement in the event of an extended outage following a major nuclear accident.

As the discussion will indicate, the electric utility industry is addressing the consequences of the TMI accident on the technical design level, at the utility operations level, and on a financial risk-sharing basis. The combined effects of these measures should result in the creation of a self-help organization within the electric utility industry, which will go a long way to ensure more reliable and safer nuclear power generation.

Nuclear Safety Analysis Center

The Nuclear Safety Analysis Center was established by the electric power industry at the Electric Power Research Institute (EPRI). Though a separate administrative entity from EPRI, it reports to EPRI president Floyd Culler and receives guidance from the NSAC subcommittee of the EPRI Research Advisory Committee.

NSAC was established in May 1979 in order to provide a detailed technical study of the TMI-2 reactor accident so as to:

●Determine and report precisely what happened during the course of the accident.

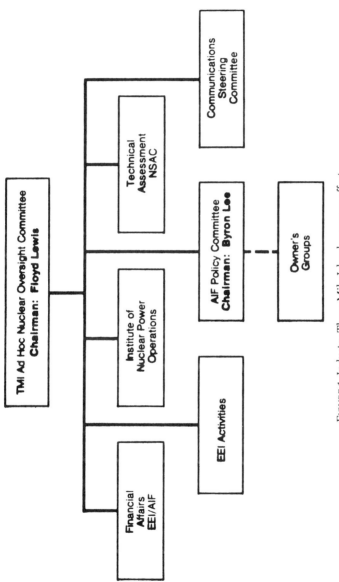

FIGURE 1. Industry Three Mile Island response effort.

●Analyze the various failure modes encountered and evaluate their relative contributions to the overall event.

●Evaluate proposed technical fixes and modifications and recommend priorities of suggested responses.

●Survey and clarify the low-level radiation effects upon the local population.

●Collect and exchange technical information related to the accident.

In addition, NSAC coordinates its evaluations with 60 different individual utilities, with the six utility owners' groups, and with TMI-related committees established by the Atomic Industrial Forum. NSAC interacts with the Nuclear Regulatory Commission and provided technical support to the Kemeny commission appointed by the president. Technical support is also provided to GPU in evaluating the TMI-2 reactor reentry and recovery alternatives.

In its first major activity, NSAC completed a detailed analysis of the TMI-2 accident.[4] The analysis includes a second-by-second and then a minute-by-minute reconstruction of the events that took place at TMI over the six-hour period until forced circulation cooling of the reactor core was reestablished. This sequential listing is based on verified hard data obtained from the reactor control instruments, recorders, and process computers and from interviews with plant operators. The report also includes 17 appendices, which discuss in detail the technical performance of various nuclear plant subsystems.

NSAC is now concentrating on three major topical areas:

●*Significant Events.* Evaluating those events from nuclear reactor operating experience that appear to have special safety implications for others and alerting any affected utilities.

●*"What If?" Studies.* Determining what additional margins of safety were available at TMI before more severe plant damage or radiological releases occurred.

●*Communication and Coordination.* Keeping everyone informed as to current problems, incidents, and plant/operations modifications on a timely (immediate) basis. This function includes working with committees to define interfaces and set priorities.

As an example of such coordination activity, NSAC has been developing and using a methodology based on decision theory for evaluating the relative priority and value of various response actions to nuclear accidents proposed by the Nuclear Regulatory Commission "Action Plan" reports. The results of this NSAC analysis have been incorporated in part into the new *Draft Action Plan* issued in March 1980.[5]

Two documents now being completed by NSAC are the *Defense-in-Depth Study* and the *Probabilistic Risk Assessment.*[6,7]

The essential elements of the *Defense-in-Depth Study* include (1) a description of the observable characteristics that are present in a variety of hypothetical accidents involving core damage; and (2) the successive options and backup responses that the operator can deploy to terminate the progression of damage prior to the point of challenge to the integrity of the containment system.[6]

The second report analyzes the behavior after postulated core melting has occurred, and for alternate sequences with various lengths of time following postulated core melting without operator intervention.[7] This includes analysis of cases in which core melt breaches the reactor vessel.

Given all its functions, NSAC is projected to continue operation through 1980, at an estimated budget of $8 million.

Institute of Nuclear Power Operations

In parallel with the technical evaluation of lessons learned from the TMI-2 accident conducted by NSAC, the electric utility industry has established the Institute of Nuclear Power Operations, which is dedicated to ensuring the high quality of operations in nuclear power plants. INPO will thus become a self-help instrument of the utility industry in regard to plant operations. It will not seek to preempt utility management responsibilities. Rather, as an industry-sponsored institute, it will provide utilities with the means to improve their own operations.

The institute will establish industry-wide benchmarks for excellence in the management and operation of nuclear power plants. These criteria will be evolved through active participation of experienced plant operating staffs, and by professional consultants. Independent and periodic evaluations of operating practices at nuclear plants will be performed by INPO in order to assist the utilities to meet the established benchmarks.

INPO (in close cooperation with NSAC, as mentioned previously) will review nuclear power operating experience for analysis and feedback to the utility industry. Lessons learned will be incorporated into training programs available to electric utilities.

The institute will perform studies and analyses in order to support the development of operating criteria, for personnel training, and for human factors design consideration.

INPO will establish educational and training objectives for reactor operations and maintenance personnel and will develop screening and performance measurement systems. In developing training programs and practices, the best available teaching techniques will be used. As part of its operators' training practices, the institute will accredit existing programs and will certify their instructors. The educational training provided by INPO will also include seminars and generic training programs conducted for various utilities employees, including executives, upper management, and reactor operators and supervisors. Finally, INPO will provide emergency preparedness coordination for nuclear power utilities and will act as the clearinghouse for information exchange with foreign nuclear power programs.

An organizational chart for INPO is shown in FIGURE 2. The organization was activated in January 1980 with the selection of Rear Admiral Eugene Wilkinson as its first president. It is estimated that the functions of the institute will require a staff of 200 professionals. Annual operating costs of the fully established organization are conservatively estimated at $11 million. An advisory council and an electric utility industry review structure have been planned, modeled on the EPRI structure.

Nuclear Electric Insurance, Ltd.

Nuclear reactor operators are insured for public liability through the Nuclear Energy Liability Insurance Association (NELIA), the Mutual Atomic Energy Liability Underwriters (MAELU), and the Price-Anderson Act; and for property damage through the Nuclear Energy Property Insurance Association (NEPIA) and the Mutual Atomic Energy Reinsurance Pool (MAERP).

An additional kind of insurance covering electricity replacement costs in an event of extended nuclear reactor shutdown following an accident is now being planned as a result of the TMI-2 reactor accident. By establishing outage cost

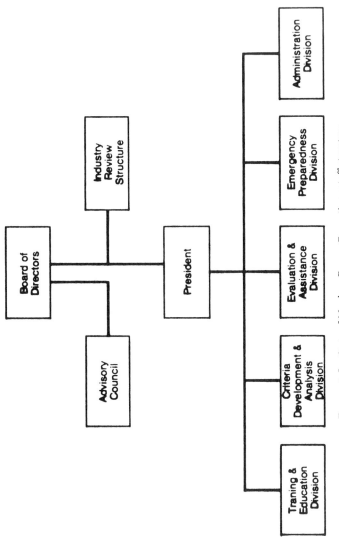

FIGURE 2. Institute of Nuclear Power Operations staff structure.

insurance, the utility industry will be provided with a full range of coverage against all the potential consequences of a large-scale nuclear accident. Membership in the insurance organization would assume cooperation with INPO and its recommendations. In this way, a strong economic incentive to insurance excellence in reactor operations will be built into the coordinated utility response plan.

In order to provide financial incentive to utilities to upgrade safety procedures, and in order to limit the range of maximum assessments to prospective pool members, it is now established that the policy would not cover power replacement costs during the first 26 weeks of the outage. A limit of maximum recovery will be imposed, such as $2 million per week per unit for a 52-week period, followed by another similar period at half the weekly indemnity level. The final details of the proposed policy of NEIL were worked out during December 1979, and invitations for membership were issued. Current applications amount to $68 million of anticipated premiums subject to the approvals of the related public service commissions, and cover 50 operating reactors in the U.S. It is estimated that the insurance pool organization will be operational in the next few months.

The combined effect of the protection against very high outage cost of the high qualifying threshold and of the INPO cooperation requirement would make the idea of power outages mutual insurance attractive from both the public and the utility operators' points of view.

SUMMARY

It is evident that the Three Mile Island accident stimulated the electric utilities to reexamine the institutional performance of the nuclear industry and the respective roles of government agencies, manufacturers, and utility operators. The utilities have recognized their lead responsibility to ensure the reliable and safe operation of nuclear power stations. Further, the TMI accident strongly illustrated the broad-ranging impact of the performance malfunctions of any one utility. A very strong motivation for the concerted utility response to the TMI accident is the need to protect their industrial investments. The major lesson learned from that event is that the industry should do everything in its power to protect its very large capital investment in nuclear power as a major source of future baseload generation. These perceptions have caused the utilities to undertake a nationwide, mutual self-help approach to improving their performance and to maintaining the quality of all nuclear utility operations.

To implement these objectives, the utilities have initiated three major programs. The first, NSAC, addresses the technical aspects of nuclear plant design and its relation to reliability and safety. The second, INPO, addresses the quality of operation of nuclear stations and the coordination of industry-wide emergency preparedness. The third, NEIL, is a mutual insurance arrangement for sharing the large financial risks of major accidents. These three programs, although organizationally separate, are intimately related to each other, provide mutual support and motivation, and are complementary in objectives and functions.

The utilities have reaffirmed their confidence that nuclear power technology allows the achievement of the highest levels of public safety if this objective is made a major institutional and cooperative focus of the utilities, manufacturers, and government agencies. While much is left to be done, it is clear that the risks to the public at TMI have been highly exaggerated. A high redundancy in engineered safety features built into the plants, and proper design of containment buildings, will reduce the likely radiation risk to the public from the future operating nuclear power plants. Ultimately, the utilities are on the front line of performance and, therefore, have the strongest motivations to ensure reliability and safety. We believe these recent utility initiatives represent major steps toward effectively achieving this goal.

REFERENCES

1. NSAC. 1980. Mitigation of Small Break LOCA's in PWR Systems. Report No. NSAC-2. Nuclear Safety Analysis Center. Palo Alto, Calif.
2. NRC. 1975. Reactor Safety Study: An Assessment of Accident Risk in U.S. Commercial Nuclear Power Plants. Report No. WASH-1400 (NUREG-75/014). U.S. Nuclear Regulatory Commission. Washington, D.C.
3. KEMENY, J. G., et al. 1979. Report of the President's Commission on the Accident at Three Mile Island. U.S. Government Printing Office. Washington, D.C.
4. NSAC. 1979. Analysis of the TMI Unit 2 Accident. Report No. NSAC-1. Nuclear Safety Analysis Center. Palo Alto, Calif.
5. NRC. 1980. Draft Action Plan. Report No. NUREG-0660. U.S. Nuclear Regulatory Commission. Washington, D.C.
6. NSAC. 1980. Defense-in-Depth Study. Nuclear Safety Analysis Center. Palo Alto, Calif. (Internal study.)
7. NSAC. 1980. Probabilistic Risk Assessment. Report No. NSAC-7. Nuclear Safety Analysis Center. Palo Alto, Calif.

LESSONS OF THREE MILE ISLAND FOR THE INSTITUTIONAL MANAGEMENT AND REGULATION OF NUCLEAR POWER

Harold P. Green

Fried, Frank, Harris, Shriver & Kampelman
Washington, D.C. 20037

Before I talk about the institutional lessons learned from Three Mile Island (TMI), I would like to discuss some fundamental institutional facts that should have been learned before TMI, but only now seem to be coming into focus in the light of TMI.

Since 1954, when nuclear technology was opened to private enterprise, the industry has been the beneficiary of substantial government support in the nature of subsidy and benign regulation. The results have been dramatic. As James R. Schlesinger noted shortly after he became chairman of the Atomic Energy Commission in 1971, the remarkable growth of nuclear power from 1954 to 1971 was "similar to the entire history of commercial aviation from Kitty Hawk to the Boeing 747 being compressed into less than a score of years." Although Dr. Schlesinger did not say it, Boeing 747s would not be nearly as reliable and safe as they are today if they had come into service only 18 years after Kitty Hawk. Technological advance in the civilian economy normally proceeds in small, incremental steps, as feedback from experience with each step shapes the next successive step. When, however, as in the case of nuclear power, technological advance is compressed in time as a consequence of government intervention to force its advance, that technology leapfrogs experience. In nuclear power, for example, we have moved to more, larger, and more sophisticated successive generations of reactors without the benefit of substantial feedback from previous experience.

Second, the nuclear power regulatory structure was created and has evolved in a most abnormal manner. Laws intended to protect the health and safety of the public usually are enacted only after years of legislative effort, and only upon a clear showing of palpable dangers to the public interest. In the case of nuclear power, however, the basic regulatory statute came into effect in the Atomic Energy Act of 1954 at a time when there was no nuclear power technology and therefore no record of any injury to the public. Indeed, the basic purpose of the Atomic Energy Act of 1954 was not to protect the health and safety of the public against imminent hazard, but rather to provide a vehicle for the birth, development, and growth of a nuclear power industry. The regulatory provisions of the Atomic Energy Act were designed to ensure that the desired rapid development and growth of the industry would be compatible with the health and safety of the public. One important consequence of this is that, whereas most regulatory agencies undertake to enforce new health

311

0077-8923/81/0365-0311 $01.75/2 © 1981, NYAS

and safety legislation with the zeal of a "cop," in the case of the Atomic Energy Commission, the implementation of the regulatory program was always balanced against the need not to discourage the development of the technology.

A third important factor is the fact that for the first 15 years or so under the Atomic Energy Act of 1954, there was no particular public constituency significantly concerned about nuclear safety. As a consequence, there was little congressional interest in nuclear safety except on the part of the powerful Joint Congressional Committee on Atomic Energy, which had a very strong pronuclear promotional bias. It is a remarkable fact that, notwithstanding the obvious national commitment to nuclear power as a primary energy source for the future, there has never been a great national debate or even significant public discussion of the benefits and risks, advantages and disadvantages, pros and cons of nuclear power out of which this commitment arose. Indeed, one cannot identify a single piece of legislation in which this commitment is made. The fact of the matter is that we owe the present commitment to nuclear power, the present status of nuclear power, and the present plight of nuclear power to the abnormal political environment out of which nuclear power emerged—a political environment in which Congress permitted the Joint Committee on Atomic Energy to make most important decisions with very little control and virtually no accountability.

Let me turn now to the lessons learned from TMI.

As I observed a few moments ago, nuclear power has progressed quite far in a relatively short period of time. It is worth noting that nuclear power plants are licensed for a full 40-year period despite the fact that no nuclear power plant in the United States has actually operated for very much longer than half that period. We have very little in the way of experience to tell us how the equipment and components used in a nuclear power plant will behave over three or four decades of operation. If there is any validity in the maxims "there is no substitute for experience" and "experience is the best teacher," what have we attempted to substitute? We have substituted the best scientific, the most conservative, computer-aided analysis that money can buy. We have assumed that what the scientists and engineers say has been designed to work reliably and safely will function reliably and safely because there is no competent scientific evidence to the contrary. And just to add a margin for unlikely error, we have built in redundant fail-safe backup systems that we are assured by the scientists will also work.

To place complete reliance on predictive judgments of scientists and engineers is to assume the omniscience and infallibility of the human beings who design the systems, who build the parts, and who operate them. We all know, however, that human beings are neither omniscient nor infallible, and this is one of the major lessons—or reminders—of TMI, as we all watched the nuclear power establishment grapple with what appears to have been a problem that previously had not been regarded as possible. I suspect that now, in retrospect, we all knew that someday there would be a Three Mile Island episode. Another lesson of TMI is that it is fatuous to speak of "no more TMIs." So long as human beings remain imperfect, there will be nuclear power mishaps, and some will be worse than TMI.

A second major lesson of TMI is that no regulated technology can prosper unless the general public has confidence in the government regulatory institutions. The fact that nuclear technology involves hazard does not mean that nuclear power is not acceptably safe, any more than the fact that from time to time large commercial aircraft will crash into population centers means that air transportation is unacceptable. Indeed, there is an interesting juxtaposition of the Three Mile Island episode and the saga of the DC-10. It is self-evident that a large commercial aircraft may crash into densely populated metropolitan areas—or indeed into Yankee Stadium in the midst of a World Series game—with catastrophic consequences. Discussion of such a phenomenon does not require esoteric scientific and engineering inputs, and if the public were told that such an accident was incredible, such assurances would not be believed. Still, even after last summer's Chicago DC-10 crash, there has been no public clamor to restrict flights over such population centers, and the flying public, according to press accounts, seems to have only slight and diminishing reluctance to fly DC-10s. On the other hand, the Three Mile Island occurrence seems to have dealt nuclear power a mortal or near mortal blow in terms of public acceptability. I would suggest that the difference in public attitude between nuclear power plants and the DC-10 is based primarily on the fact that the public has confidence in the folks who regulate aviation safety but lacks confidence in the Nuclear Regulatory Commission (NRC).

A third lesson of TMI is that a civilian technology must make its own way in the world and that efforts by government to shield it from public criticism, suspicion, and concern are counterproductive. We all know about loving parents who are so overprotective of their children, shielding them from contact with the risks of the real world, that the children are not able to deal with these risks when they are grown up and must face them. This is the fate of nuclear power, overprotected by its loving parents—the nuclear industry, the now defunct Joint Committee on Atomic Energy, and the now defunct Atomic Energy Commission, which, however, lives on in its progeny, the Nuclear Regulatory Commission and the Department of Energy. For 25 years since 1954, these entities have sought to protect the tender infant and the adolescent technology from the consequences of disclosure and discussion of its risks, because they believed the public might react with an irrational and emotional response that would retard the growth of the technology.

Within the licensing process, the risk of a Three Mile Island–type incident has been consistently characterized as almost impossible, as incredible, or as so extremely remote that it was not worth discussion. The public has been encouraged to believe that a little radiation exposure was not undesirable, that systems designed to override human error *would* override human error, and that because a particular natural phenomenon had never before been known to occur, it never would occur. To put it in a nutshell, the public has been promised a degree of safety and a freedom from risk that are unattainable and could not be delivered. It is that promise of near-zero risk that causes the public to subject nuclear power to much more critical scrutiny and much more stringent standards than any other technology.

There have been two consequences. First, this approach has turned off and antagonized intelligent citizens who interpret this overprotective attitude as a cover-up. Second, and of greater current importance, it has left no latitude for public acceptance of a Three Mile Island episode, even though it appears that at Three Mile Island there was only minimal damage to the environment and to the public health and safety. The forceful demonstration that the scientific priesthood is not omniscient and infallible has shaken public confidence in the basic premises underlying public acceptance of nuclear power to date.

Before TMI, nuclear power's prospects were grim; today they are desperate. The fact of the matter is that we need nuclear power as a matter of national policy, and I am not prepared to write off something we cannot do without. But we cannot have nuclear power without radical reforms of government institutions for its licensing and regulation.

The present plight of nuclear power is a heritage of a grotesquely complicated licensing process, a promise of complete safety that cannot be delivered so long as humans are human, and a pattern of dissembling rather than truth telling. To salvage nuclear power, we must start telling it like it is—acknowledging openly that nuclear power involves real risks, that there will be accidents, and that the public may be injured. I have little doubt that the American people, if given the true story in a spirit of open candor, will gain confidence in the institutions of government that deal with atomic energy. With such confidence, the American public will respond by recognizing that the risks of nuclear power are in fact no greater than those they willingly accept in other contexts in their daily lives and, on balance, than those incident to the realistic alternatives to nuclear power.

Moreover, the nuclear power licensing process urgently requires a massive simplification. The licensing process now is grotesquely over-complicated in terms of the number of procedural steps required by statute, the manner in which each of these steps is burdened by so-called public participation, and the measures adopted by the Nuclear Regulatory Commission to encourage such public participation while at the same time preventing it from strangling the industry. This excessive complexity not only is a principal cause of lack of public confidence in the licensing process, but is also enormously and unnecessarily costly to everyone involved in terms of time, money, effort, and emotion. The licensing process must be simplified and rationalized.

In my view, Three Mile Island provides an opportunity to revive and resuscitate a moribund nuclear power industry. Let me use, not invidiously, a bankruptcy analogy. It provides an opportunity to wring the water out of the ill-conceived, paternalistic public relations gimmickry of the past quarter-century and to make a fresh start. This approach must be based on a realistic and candid assessment of risk. We must simplify the licensing process and abandon the practice of the past 25 years of dealing with licensing problems through an approach akin to slapping on Band-Aids®, often one on top of another, to treat a patient suffering from a malignant disease. We must reverse the perception that the NRC is more a big brother than a tough cop to the industry it regulates. In such a new regime, we could and should also consider economic incentives to encourage the revival of an industry that may well have lost the capacity to make it on its own.

In other times and under other circumstances, we might well forego the benefits of nuclear power because of the peculiar risks inherent in the radiation generated in its production. These risks are not insubstantial, and some do not appear to be amenable to entirely satisfactory resolution. Indeed, it might even be asserted with good cause that our present reliance on nuclear power is the heritage of past mistakes that should not now be perpetuated. But given the present, very real energy crises, availability of more nuclear power—at least for the next 25 to 50 years—seems to be imperative for the preservation of the kind of society most Americans seem to want. What is now needed is a candid explanation by our government and a recognition by our public that in the context of the energy crisis and available technologies—both physical and social—for coping with it, nuclear power is the least worst alternative.

COMMONWEALTH OF PENNSYLVANIA EMERGENCY PREPAREDNESS AND RESPONSE: THE THREE MILE ISLAND INCIDENT

Oran K. Henderson

Pennsylvania Emergency Management Agency
Harrisburg, Pennsylvania 17120

INTRODUCTION

This paper discusses the Commonwealth of Pennsylvania's emergency response mechanism and legal basis in being at the time of the Three Mile Island incident. It recounts the sequence of events as they directly affected the agency and the methods whereby the agency discharged its responsibilities. Finally, some of the problems and experiences accruing from Three Mile Island are listed as lessons learned.

I am not taking sides in the debate about our need for nuclear energy or how easily we can do without it. It is sufficient to say that many knowledgeable people argue that this country's use of nuclear power to meet a small portion of its energy needs cannot be eliminated without very serious and damaging economic consequences. During the past year, small percentage reductions of imported oil have reminded us of this fact. It is, therefore, reasonable to assume that we will continue to utilize nuclear power to generate nuclear energy and may be more dependent upon it in the future. Logic dictates that if we must have it then we should try (1) to minimize its risks; and (2) to be prepared to meet the consequences should all else fail.

PENNSYLVANIA EMERGENCY MANAGEMENT AGENCY

The Pennsylvania Emergency Management Agency functions as a semi-independent agency under the office of the governor.

It is the organization through which the governor exercises his emergency responsibility for the protection of the health, safety, and well-being of commonwealth citizens faced with disaster from either man-made, natural, or enemy attack causes. The Pennsylvania act for emergency preparedness and response is patterned after the model act developed by the Council of State Governments. Many of the states have used the model act as a guide, modifying it as necessary to meet special needs and to accommodate respective organizational arrangements.

The act designates a council arrangement and prescribes its membership. The council, chaired by the lieutenant governor, is responsible for developing overall policy and providing guidance to the agency. The council membership includes the governor, lieutenant governor, four

0077-8923/81/0365-0316 $01.75/2 © 1981, NYAS

members of the State Senate and House and ten secretaries of departments having major emergency responsibilities.

The law prohibits the Pennsylvania Emergency Management Agency from duplicating the functions of any other state agency. Consequently, the Pennsylvania Emergency Management Agency task is that of maintaining close liaison with all state agencies, determining gaps in missions and roles, developing integrated plans to assure continuing identification of state resources, and the prompt assignment of these resources towards reducing the vulnerability of people involved in emergencies.

In the conduct of emergency operations, the Pennsylvania Emergency Management Agency Emergency Operations Center and its three area centers are augmented, at times of disaster, with representatives from those state agencies having an emergency role. These representatives are given the authority to perform in behalf of their respective departments and agencies and commit resources in accordance with predeveloped procedures.

The Pennsylvania Emergency Management Agency's day-to-day role is one of analyzing problem areas and vulnerabilities, planning for all eventualities, and the training of staff and county coordinators. Quarterly exercises are conducted with the Pennsylvania Emergency Management Agency and agency response team members to assure the maintenance of understanding and knowledge of the agency's standing operating procedures.

The Pennsylvania Emergency Management Agency Emergency Operations Center is located in an underground, protected facility in the Capitol complex in Harrisburg. Three area underground and protected facilities are strategically located to extend the Pennsylvania Emergency Management Agency coordination and management role.

The Harrisburg Emergency Operations Center is in communication with its three Area Emergency Operations Centers by telephone, radio, and a dedicated teletypewriter system.

Each political subdivision is required by law to have an emergency management organization. The commonwealth has a total of 2,636 political subdivisions (67 counties, 52 cities, 966 boroughs/towns, and 1,551 townships). Each emergency management coordinator is appointed to his or her post by the governor, based upon recommendations of the elected heads of government of the political subdivision. Information flows from boroughs, towns, and cities to county, then to area, to state, and vice versa.

Within the commonwealth, there are four fixed nuclear sites. Two additional sites are under construction and scheduled for activation in 1982 (Berwick) and 1983 (Limerick).

Planning for commonwealth response to an emergency nuclear incident at a fixed site has, in the past, been the responsibility of the Bureau of Radiation Protection, Department of Environmental Resources. In 1975, the Bureau of Radiation Protection forwarded its plan to the

Nuclear Regulatory Commission for concurrence. The Nuclear Regulatory Commission did not concur in this plan. The plan was revised in September 1977; however, it was not formally submitted to the Nuclear Regulatory Commission. The 1977 plan provided the basis for the Pennsylvania Emergency Management Agency's issuing an annex, "Nuclear Incidents, Fixed Facility," to the Pennsylvania disaster operations plan, which provided guidance to state agencies and political subdivisions. All planning was based upon a five-mile protective-action distance. County plans in support of the state disaster operations plan were "in place" and had been reviewed and updated during the summer and fall of 1978.

THE EVENT: SEQUENCE

At 7:02 A.M. on the 28th of March, the Pennsylvania Emergency Management Agency watch officer received a telephone notification from the Three Mile Island plant supervisor that Three Mile Island had declared a site emergency. In accordance with the Pennsylvania Emergency Management Agency standing operating procedure (SOP), the watch officer notified, in order of priority, the following:

1. Duty officer, Bureau of Radiation Protection, Department of Environmental Resources.
2. Dauphin County (host county).
3. York County (within 5 miles).
4. Lancaster County (within 5 miles).
5. Pennsylvania Emergency Management Agency operations officer.

At 7:20 A.M., the duty officer of the Bureau of Radiation Protection advised the Pennsylvania Emergency Management Agency watch officer that, following his discussions with Three Mile Island, he had concluded that there were no off-site consequences. The watch officer, in turn, informed the affected counties. In the meantime, the operations officer had arrived at the Pennsylvania Emergency Management Agency Emergency Operations Center and relayed the report to the Pennsylvania Emergency Management Agency director. The operations officer, with the early staff arrivals, relieved the watch officer, and subsequent actions were handled through the Pennsylvania Emergency Management Agency Emergency Operations Center.

At 7:35 A.M., the Three Mile Island plant supervisor notified the Pennsylvania Emergency Management Agency operations officer that conditions were worsening and that a general emergency had been declared. The plant supervisor recommended that the Pennsylvania Emergency Management Agency be prepared to evacuate Brunner Island and the Borough of Goldsboro. The report stated that a radiation release to the atmosphere had and was occurring and that winds were in the direction of 30°. The Pennsylvania Emergency Management Agency operations officer passed this information to the Bureau of Radiation Protection, alerted York County to be prepared for evacuation of the above-cited locations, and advised Dauphin and Lancaster Counties of the situation. The director of the Pennsylvania Emergency Management Agency notified the governor and lieutenant governor of the reported conditions. Shortly thereafter, the Bureau of Radiation Protection advised that the release had been halted and that no evacuation was required.

It was subsequently learned that the conditions leading to the Pennsylvania Emergency Management Agency notification had actually occurred around 4:00 A.M. With this exception, all reporting and notification procedures were in accordance with established procedures and all telephone circuits were functioning.

During the 28th and 29th of March, the Pennsylvania Emergency Management Agency and the three affected counties spent considerable effort in fleshing out existing plans. As a precaution, the Pennsylvania Emergency Management Agency maintained a scaled-down 24-hour operational capability in its Emergency Operations Center. Reports received during the 28th and 29th of March, although frequently conflicting, generally reflected a movement towards a cold shutdown of the reactor. Nuclear Regulatory Commission representatives repeatedly acknowledged the superb job being done by Metropolitan Edison personnel and the dedication and professionalism being demonstrated. National news media were seeking stories away from the scene, and several articles treated speculative events. The film *China Syndrome* was showing in several area theaters.

At 8:40 A.M. on the 30th of March ("Black Friday"), two simultaneous telephone calls were received at the Pennsylvania Emergency Management Agency from Three Mile Island. Three Mile Island reported a general emergency condition due to a radiation reading of some 1,200 mR/hour. The report contained the information that the facility was preparing to evacuate nonessential personnel and recommended that the Pennsylvania Emergency Management Agency be prepared to evacuate personnel from downwind. This information triggered an alert to the affected counties and the initiation of the full activation of the State Emergency Operations Center.

At 9:15 A.M., the Bethesda, Maryland, Office of the Nuclear Regulatory Commission advised the Pennsylvania Emergency Management Agency that it, too, was recommending that Pennsylvania evacuate downwind of the Three Mile Island facility and out to a range of 10 miles. It was reported that this recommendation had the support of the senior personnel at the Nuclear Regulatory Commission.

Subsequent discussions between the Bureau of Radiation Protection, Department of Environmental Resources, and Three Mile Island personnel and between Governor Thornburgh and the commissioner of the Nuclear Regulatory Commission determined that existing conditions did not warrant such severe action. Consequently, the governor issued an advisory at 10:00 A.M., March 30th, for all personnel within 10 miles of the Three Mile Island facility to remain indoors until further notice. This was followed by a further advisory at approximately 12:00 noon for all pregnant women and preschool-aged children to evacuate the 5-mile area.

Selected units of the Pennsylvania National Guard were placed on a "white alert" and directed to be prepared to support each risk county with one battalion and to provide a backup battalion in support of each committed battalion. The Pennsylvania State Police were prepared to bring maximum force into the area. Both the National Guard and the Pennsylvania State Police were asked to be prepared to assist with any evacuation and provide traffic control and security for any area evacuated. The local police, fire, and emergency medical forces in the area at risk were, for the most part, fully activated and in an advanced readiness posture.

PROBLEMS AND EXPERIENCE

Following the governor's advisory for partial evacuation, a number of people commenced an orderly and voluntary movement out of the area. Commercial banks and savings institutions were immediately besieged with people withdrawing funds. Hospitals commenced reducing their patient load by discharging some patients and rescheduling elective surgery cases. During this period, the Pennsylvania State Police aerial traffic observer reported no abnormal traffic patterns. Counties in the affected area subsequently determined reportable traffic accidents to have been approximately 25% fewer than for a similar period prior to and a corresponding period after the Three Mile Island incident.

Immediately following the governor's 10:00 A.M. take-cover advisory, a telephone overload occurred, and until approximately 1:00 P.M., March 30th, a 30- to 40-minute time delay was experienced throughout the Harrisburg exchange. Both the news media and public inquiries totally saturated the 100 + lines servicing the Pennsylvania Emergency Management Agency Emergency Operations Center.

As a result of the Nuclear Regulatory Commission's recommendation to evacuate out to a radius of 10 miles, the Pennsylvania Emergency Management Agency, the Department of Transportation, and the Pennsylvania State Police initiated a hasty traffic analysis and reassigned major route priorities to the risk counties. Personnel from the Pennsylvania Emergency Management Agency and the Defense Civil Preparedness Agency were assigned to the risk counties to assist in the planning effort. Direct telephone lines between the Pennsylvania Emergency Management Agency and the affected risk counties were installed, and the Defense Civil Preparedness Agency volunteered and placed a radio system in operation between the Pennsylvania Emergency Management Agency and the counties at risk.

At approximately 8:30 P.M. on Friday, March 30th, the Nuclear Regulatory Commission advised that it would be prudent to have plans for an evacuation out to 20 miles from the Three Mile Island facility. This increased the counties at risk from 3 to 6 and enlarged the population at risk from 30,000 to 750,000. Neither the existing county 5-mile plans nor the preliminary efforts towards a 10-mile evacuation scheme were compatible with the problems associated with a 20-mile evacuation scenario. For all practical purposes, all earlier plans and evacuation guidance were of necessity cancelled, and a new course of action was set in motion. The pace of planning took on an increased degree of urgency. In addition to the 6 counties at risk, 21 additional counties were alerted for an evacuation-hosting role.

A considerable number of new problems surfaced, i.e., movement of seriously ill patients, those on life-support systems, and babies in neonatal status; legal issues were raised; pets and livestock care and disposition; a degradation in the availability of medical personnel and volunteer forces; lead-time requirements of business and industry; security of prisoners and other institutionalized persons; responsibility for associated costs; stockpiling and issuance of radioprotective drugs; movement of the seat of government; sounding of sirens both inadvertently and necessarily; and, throughout the handling of the news media, treatment of rumors and misquotes were extremely time consuming. Considerable credibility as a source of information was retained by Governor Thornburgh and Mr. Denton. Other Nuclear Regulatory Commission federal agency spokesmen received little acknowledgment, as did Metropolitan Edison. A large number of people reported that they didn't believe anyone.

The Pennsylvania Emergency Management Agency experiences during the Three Mile Island incident were in many respects similar to those encountered in previous natural disaster events and differed primarily in degree or magnitude. The single exception is perhaps that people's perception of a danger (aura of mystery) that cannot be seen, felt, smelled, or tasted resulted in an attitude not generally shared in other emergency responses. Antinuclear groups and concerned individuals and groups were extremely outspoken in this criticism. The Pennsylvania Emergency Management Agency was required to convey an attitude of restraint and caution during a possible catastrophe to the general public.

These are some of the lessons learned:

• We can no longer take for granted that nuclear power plants are safe. Our planning should not treat nuclear accidents as merely an extremely remote possibility.

• We need to obtain and plan for the use and dissemination of potassium iodide.

• We need to increase our attention to the handling of people under special care in homes, hospitals, and institutions.

- We need to improve the educational level of all segments of the public as it pertains to nuclear radiation.
- We need a fully integrated and adequately redundant communications system in place.
- We must do a better job of bringing county and local government officials into the decision-making process.
- We must recognize emergency management's dependency upon volunteers and volunteer forces and understand that they may not always be available.
- We must improve our capacity for handling the news media and assuring their continuing availability.
- We must develop a more formalized system for tests and exercises. Our planning effort must assure total integration at all levels of government and between these levels of government.
- We need, from the national level, a uniform emergency nuclear incident classification system.
- We need to standardize procedures for the systematic study of the social, economic, and health aspects of an accident.
- We need to improve our record keeping.

Finally, we need to seek and pursue solutions to the problems associated with fixed nuclear power facilities as a matter of urgency, not on the emotionally charged rhetoric of the moment, but upon dispassionate, reasoned analysis.

MAJOR LESSONS AND ISSUES, I: GENERAL DISCUSSION

Moderator: Thomas H. Moss

Subcommittee on Science, Research, and Technology*
U.S. House of Representatives
Washington, D.C. 20515

E. WEISS (*Sheldon, Harmon, Rosman, and Weiss, Washington, D.C.*): Colonel Henderson, there are a couple of questions I'd like to get your opinion on. I've been curious about these for a while. The first is, What is your opinion of what would have occurred had you gotten the word at 6 A.M. on Wednesday, the 28th, to evacuate?

O. K. HENDERSON (*Pennsylvania Emergency Management Agency, Harrisburg, Penn.*): To evacuate?

E. WEISS: Let's say 20 miles, which is the area that was decided on.

O. K. HENDERSON: I'll admit that we were in no position to execute a 20-mile evacuation around Three Mile Island, but I would also say that we are not in a position on a day-to-day basis to evacuate as a result of floods. A plan in itself is not going to make an evacuation work. It's been my experience that a plan is a fine tool and one that people should be knowledgeable in, but when it comes to the actual execution, very few people run to a plan and start thumbing through it and say, What am I supposed to do now?

I don't know what would have happened; I haven't tried to extrapolate back on that. We would have had a difficult time getting the warning out, there's no doubt about it. Again, how many hours notice or how much time we would have had, I can't answer.

E. WEISS: Let me ask you the second one then. You've related how the area of interest changed from 5 to 10 and finally to 20 miles during the few days. I'm sure you know that the Nuclear Regulatory Commission [NRC] seems to have settled on a 10-mile zone for emergency planning. What's your view about whether that's adequate?, and could you give me your reasons please.

O. K. HENDERSON: Well, I don't agree with the 10-mile circle, because nobody knows whether they live inside of that circle. It's fine for industry and maybe for my headquarters to have as a point of reference, but it's not a good reference for the public.

We are not satisfied that 10 miles is the single solution. In our taking 10 miles as a guide, we've included all municipalities that are in that 10-mile area, and in some cases it goes out to 12 or 13 miles.

What I'm concerned about is that during the Three Mile Island [TMI] accident, we requested that NRC conduct a site-specific survey for us around each of our nuclear facilities, taking into consideration the

*Dr. Moss is science advisor to this committee, not a member of Congress.

0077-8923/81/0365-0323 $01.75/2 © 1981, NYAS

source, the demography, the geography, and so forth of the area. We've been unable to get that action from them as of this time, and until we get it, I'm not settling on any 10-mile zone.

J. S. MILLER (*Northwood, N.H.*): My question is to Colonel Henderson. Could you tell us a little bit about the booklet that's been prepared by your agency to warn the citizens about what to do to prepare for a nuclear accident, and how it's going to be distributed. Would you recommend that all states develop such a booklet?

O. K. HENDERSON: We hope to do several things in this area. Three or four years ago, we prepared a document for dissemination to the public as a whole. *What You Should Know about Radiation* was the title of the document. We sent it around through the technical community, the University of Pittsburgh, Pennsylvania State University—to everybody that we felt could contribute to it. Everyone concurred in the technical parts of the booklet, although there was some concern expressed by one commission—I don't recall which—that we were picking on the nuclear industry and that if we wanted to publish such a booklet, we should include floods and treat all hazards in this document.

In the introduction of this booklet, we had stated that a nuclear incident was extremely remote. We removed the words "extremely remote," and three months after TMI, we required all of the nuclear industries in Pennsylvania to disseminate this document to everyone within a 10-mile area around each plant. That has been done, and the nuclear industries were happy to comply.

We have a program in our schools called Your Chance to Live, and it's an old program; it's been in the public school system for a great number of years. We've included another lesson in that program, one on nuclear radiation. This is going to be mandated by the Commonwealth of Pennsylvania to be taught in all schools in the 8- to 10-year age bracket.

T. H. MOSS: I might observe that we have a relatively new federal agency, the Federal Emergency Management Agency, that is trying to pull together the various civil defense, earthquake preparedness, and various other emergency planning and responsibility categories.

However, like all new agencies, it has an enormous responsibility suddenly thrust upon it in a time of budgetary and other kinds of restrictions, when it may be almost impossible for it to fulfill the hopes that are defined for it. Would you want to comment on that, Colonel Henderson?

O. K. HENDERSON: I would comment only to the extent that the Federal Emergency Management Agency is in the process of organizing; it's been in the process of organizing for the last 10 months and is having difficulties. It's trying to determine what its role is. As far as the subject that we're talking about today, all we're getting from that agency is conversation, not any special assistance.

T. H. MOSS: Perhaps this is one of the many instances where the institutions are not really prepared for the kind of responsibilities that are thrust on them.

F. BRESCIA (*City College, New York, N.Y.*): My question is directed to Colonel Henderson. What were your plans for the evacuation of persons without automobiles and for persons who were mobile but who do not drive?

O. K. HENDERSON: Certainly our planning at that time was not that detailed. Today, we are requiring municipalities and counties to define all unmet needs in the way of ambulance transportation requirements and so forth; these will be consolidated by our agency. We, in turn, will press other counties surrounding the areas at risk to provide services to the degree that they can, and will also utilize all state resources.

Any still unmet needs we have will be referred to the Federal Emergency Management Agency, which has promised that it will take a look at this and determine what it can provide. Many of our municipalities have gone to great lengths to determine individuals who are at home on life-saving kinds of apparatus and who need special kinds of vehicles. Some of the plans include taking seats out of buses so that people who are on litters or people in wheelchairs can be moved in that fashion.

Our plan includes the use of all means of transportation—railroad cars coming into the area, commercial and school buses, etc.

W. MAIBEN (*Columbia University, New York, N.Y.*): Four brief questions for Colonel Henderson. First, has the potassium iodide dosage problem been solved since the incident, in case you have to deal with it again? Second, in view of that problem, did you tell people to eat iodized salt, or did you consider that inadvisable? Third, do you think Middletown and the communities around there are more ready to evacuate again if they should have to? The fourth, I'm afraid, is a little rhetorical: Wouldn't it be better to evacuate before the major release of radiation, rather than two days after as happened last time?

O. K. HENDERSON: To your last question, yes, it would have been advisable to have evacuated before. But I will say that we did not have the information that there had been a release of any significance, and still have no such knowledge.

As far as the KI is concerned, except for the study by our Department of Health on the cost associated with the KI and the resulting recommendation for how we should store the KI, the decision has not been made. The decision is with the legislators at the present time as to whether we're going to require the storage of it around each nuclear facility or whether we're going to concentrate on it at the state level and send it wherever it is needed at the time.

A. P. HULL (*Brookhaven National Laboratory, Upton, N.Y.*): This one is to Chauncey Starr, and anyone else who wants to comment on it is free to do so. If you look at the reports of the Three Mile Island incident, the radiation doses were very small. Unless you have some almost incredible circumstances, you can't predict an immediately lethal dose from the released gases, even to the nearby surrounding populations.

What I'm getting at is I think all this talk about evacuation has a certain amount of hype in it. Aren't we making a special case here in protecting people against a sort of hypothetical, long-range risk? We don't adopt this stance in most other emergency situations: we move to evacuate people only when we think life and limb are at risk, and not for some long-range, hypothetical risk.

C. STARR (*Electric Power Research Institute, Palo Alto, Calif.*): Basically, I can summarize the answer this way: The 10-mile area that's been mentioned is a kind of upper boundary, which is based really on the lack of analysis of whether a lower boundary would be adequate. That lack of analysis is in what happens to the fission products when they're released from a core in the building and their likelihood of getting out. That study, the technical study, is under way now. The empirical information— which is spasmodic, mostly from radiochemistry work that has been going on in the last years, but never from an accident—is that the heavy things like plutonium oxide actually would get out of the building very rarely. They'll deposit very quickly because they are very heavy. For example, probably very little of the radioiodine in TMI would get out, because of the water vapor. So you end up with just the volatile gases getting out, in which case the activity levels are low. The point you're making—that the radiation levels to the outside area are extremely low and the question is whether the saving to the public is worth the cost of evacuation—is a different kind of question than gave rise to the 10-mile number.

My guess is that as time goes on and more work is done on the rate at which fission products can possibly get released from the building, there'll be some readjustment of that information to the emergency preparedness people. But what's being offered now, the 10 miles, is a sort of upper-boundary protective position that the regulatory commission is taking.

T. H. MOSS: Did you want to comment on that, Colonel Henderson?

O. K. HENDERSON: The only comment I would make, and it's a negative one in a way, is toward the present guidance we have under 0654, which requires the public to be notified within 15 minutes. It seems to me that the NRC may be turning toward wanting the operator at the plant to be in a position to push the button to notify the public, and to take these actions regardless of the elected officials and any emergency management apparatus.

A. P. HULL: Just a quick comment. I would agree that should be done. It's better to have a false alarm, because people who are running facilities never seem to believe that they're in trouble until they are really, really in trouble. It's like failing to notify the fire department when the fire is still small: a lot of people don't believe it's going to get out of control. I think that's a problem we have to deal with in all high-risk ventures.

The second part of my question to Dr. Starr is, If iodine is the real problem, what would you say about a technical approach to better provide against the possibility of containment rupture? For example, providing a larger emergency standby treatment facility for iodine?

C. STARR: This is a highly complicated question; the answer really depends on the nature of the accident. In the case of TMI, there was so much water in the building, the building was under a continuous internal rain, that there is probably almost no free iodine gas circulating in the building, it's probably all in the water. Even if the building were to go or open up in any way, even in the early days, very little would have gotten out to the public.

It could vary from accident to accident: when the core gets dry and the building is dry, or where there's a lot of moisture. There has to be a more technical analysis of the variety of things that go wrong. In the case of the TMI situation, the likelihood is that even if the building had opened up, all the activity would have been right around the plant and probably would never be around the plant boundary, except for the noble gases.

But that's speculative, and work on that is still on the way. I think the NRC just took the most protective position it could. I have to remind you that the Rasmussen report, maligned as it is, was a very revealing document in terms of much of the technology. They assumed arbitrarily that if anything went wrong with the core, if there was a core meltdown, the building would open up and half the fission products would get out. They assumed this in order to set an upper boundary; but the experimental information indicates that this is really an extreme upper limit and it's unlikely to be the case.

UNIDENTIFIED SPEAKER: My remarks are addressed to Harold Green. He said a good deal of what I would have said, and I'd just like to fill in some information on my personal experience. At the time this whole question of nuclear power expansion was coming in, there was something in this country known as the McCarthy hearings, and I was unable to get clearance. It took seven years, but I did get cleared without any further complication.

But in that time, I had to go through an intermediary and sort of express views and not have any interchange—and basically the responses were this: If you have ideas for improved nuclear reactors and criticize the poor thermal efficiency, the poor neutron efficiency, of the present ones, too bad; the submarine reactor is highly successful, and it's going to be the basis for our design. We have no time to consider other systems. And furthermore (if you talked about heavy water), just think of the economic loss if we should have a leak. That was the context in which it was done, and I think we have to keep this in mind as we now consider the overall view of where we go for the future of nuclear power. We have time to consider it, and I think this is a start. I thank you for your presentation, Professor Green.

UNIDENTIFIED SPEAKER: My question is to Colonel Henderson. I think it's a very difficult thing for us to think about evacuation because most of us admit that it probably wouldn't be possible given the two- or four-hour time period. Was the scenario considered that if there was a large release or meltdown or hydrogen explosion, the people in the immediate vicinity could have been exposed to very high levels of radiation, in which case evacuation for those people may not have been desirable because they would have then spread the radiation with them?

Now, I'm not just making this up. This was brought up in the intervention hearings for the restart in Harrisburg of the other units. I'd like to know what position your office took on this scenario and also on the idea of limiting traffic beyond the evacuation radius. In other words, if people within a 30-mile ring started evacuating, the roads would have been blocked so that the people could not have gotten out. Was this considered by your office? and what plans did you have to deal with it?

O. K. HENDERSON: We had no provisions for blocking roads to anybody that wanted to leave. Very fortunately, in the Harrisburg area, we have one of the finest road networks in the country. That is the capital, and some of our representatives in past years who wanted to get down to their offices very rapidly had the foresight to prepare for such eventualities.

We developed our plan very conservatively based on the number of people living in any given area and on an average family size of 3.2. We know we've got enough automobiles for that. For a very conservative figure of 35 miles per hour, the road network is sufficient for anybody coming into any one of these exit routes. Also, each county had its own specific routes going out; there was no crisscrossing. We also were conservative in that, in the four-lane roads, we used only the two lanes heading out; we left the other lanes open for emergencies or any diversions that might be needed.

Now, on Sunday, we received the first real scenario from Mr. Denton. It suggested that if anything more serious were to occur, we had a minimum of 30 hours in which to conduct any precautionary evacuation or any other protective actions that may have been necessary. We were fully confident that we could have evacuated that area in that time frame.

C. STARR: I just wanted to comment quickly on this matter of spreading radioactivity. Most of the heavy particles that settle out and would cover an area, do so very locally, and that has been measured. The question is whether it's within 100 feet or 200 feet or 1,000 feet, but most of them settle out. The worry about covering a wider region applies to the windswept plume, which is at a fairly high altitude and which doesn't settle out. It's the radiation from that that people worry about in terms of being exposed.

So the issue of carrying radioactivity, of its settling out on automobiles, or spreading it to other areas is an issue not of great magnitude. It's easy to measure (presumably in any kind of emergency preparation situation, there are going to be radiation monitors) and it washes off with just a hose and water, so that the issue is a minor one in the total evacuation plan. But I wanted to ask Colonel Henderson another question. In Toronto, there was an evacuation of, I believe, 250,000 people as a result of a gas effluent release, which has many of the characteristics of a slow release from a nuclear station. Do you know how that was handled? and did your people look at how Toronto took care of that number of people successfully?

O. K. HENDERSON: This is one of the things that our agency is going to do. They have sent a team up there, and we're waiting for that report at the present time. But we're not inexperienced in our agency with evacuations. In Wilkes-Barre back in 1972, 80,000 people were evacuated. The decision was made at 4 o'clock in the morning. However, we could see the rivers coming up, and we decided that rather than awaken the people at 4, we had plenty of time and would wait until 6 o'clock to notify the people to move up the hill.

The difference there was that you had time on your side: you could see the waters rising and you had a relatively short distance to go. The same was true up in Toronto: we were only talking about moving two or three thousand feet away from this debris in order to be safe, so you didn't have 5 or 10 miles to run. Two weeks before this incident, we had to evacuate 1,200 people up in Jefferson County because of a chemical spill. The total community, 1,200 people, was moved without incident. Just three days before this incident, we had another chemical incident in downtown Gettysburg, and a couple of blocks of people were moved. So we're not totally inexperienced with the moving of people. The problems of a Three Mile Island–type incident, though, are a little bit different. We don't know how long the people are going to be gone. We've got to move the people into facilities and maintain them. We're talking about great numbers of people, 650,000 people. You have to get cots and blankets and all of the other good stuff, including diapers and the moving of food from the normal transportation channels up to another. It's not as simple an exercise as some others where you don't have to change the whole system around.

UNIDENTIFIED SPEAKER: It is obvious that the professional risks of the nuclear industry are not directly related to the risk of actually exposing the population, but still there is interesting and important information about this. Could I ask some of the speakers to address the question of the safety records of the nuclear industry as compared to other energy industries.

C. STARR: There is a classic answer to this that is technically correct, namely, that the nuclear industry per se has had one of the most remarkable safety records as an industry in terms of occupational hazards and occupational risk and so on. There are many good reasons for this; but of course, from the point of view of the public, that's not the issue. From the point of view of society, it is an issue.

The difficulty in the public's accepting this answer is that the occupational hazards are under constant monitoring and the people who go into those occupations are really doing so voluntarily. The real issue, of course, between the public and the industrial exposure is this issue of voluntary versus involuntary exposure. Now recognize this, but I have to tell you also that the people within the industry are deeply concerned for obvious reasons with the public exposure. The public is their customer in effect, and this whole issue is one of How do you handle the unknown? As Colonel Henderson just mentioned, when you have a flood situation, you can see the water rising and falling. When you have the kind of thing you had at TMI and you don't know what's going to happen down the road in the next several hours, you end up with precautionary moves. And those precautionary moves tend to be frightening. That's the issue we face, I think; but as far as the industry internally is concerned, it's had a remarkably good record of safety.

I might make one other comment. There's a tendency to confuse the military programs with the civilian programs. The military programs have a completely different attitude. You read about the fallout in Utah as a result of the soldiers being exposed to testing and so on. The whole attitude in the early days of the military program was that national defense had a higher priority than did anybody's life, and so it was a risk that, on a national basis, was acceptable to some fraction of the population of the military forces for military purposes and for military information. Those tests in the military scene are at a different level of decision making than in the civilian scene.

R. C. MORRISON (*University of New Haven, West Haven, Conn.*): I'd like to address a comment and a question to Chauncey Starr on the communications issue. I agree that there is a difficulty in communicating with the public who are not technically or scientifically trained. However, I have background training experience as a nuclear physicist, and the industry can communicate with me, and I have avidly devoured all of the material that has been made available to me over the years and have identified myself as a long-time, cautious advocate of nuclear power, or critical advocate of nuclear power, and have used that material as presented.

The other day, Professor Lamarsh talked about the layered safeguards in preventing release of radiation, and I have discussed those in teaching my classes. So to say the least, I was really annoyed to discover that at Three Mile Island, a steam valve was used to routinely bypass one of these layered safeguard systems. Not only that, but using that in turn allowed two of the other layers of safety to disappear, the fuel and the clay. My question is, What is the industry going to do to close the gap between what they are saying and the way that the reactors in plants are actually being operated?

C. STARR: Your analysis is absolutely correct. The intent of the containment building and the intent of the various levels of safeguards were violated in the technical design of the machine. The various ways in which these machines were reviewed by the NRC staff, the manufacturer, and so on missed the fact that the original intent of these various layers of safeguards have been violated by this process of design. Of course, it should have been apparent from other precursor events, as you may have read, but because of a lack of internal communication, it was not disclosed. TMI has shown this up, and the industry is making technical changes in the design to restore that level of protection.

And you're quite right in your perception of what happened. It was, in my point of view, a major technical flaw in the review processes. You have to understand that the people who are concerned with these issues are only a handful—I can almost name them—in terms of the basic philosophy of designs. The designs actually are constructed by a huge multiplicity of manufacturers and engineers and designers; and the wholistic review that is supposed to go on inside the NRC missed this one. That's one of the things we learned technically out of TMI; and I said in my speech that of all the things to fix, those are the ones that we know how to fix. But you're right, and it is being fixed.

H. ETZKOWITZ (*In These Times, Chicago, Ill.*): A question for Colonel Henderson. In interviews conducted with residents of Hershey, Pennsylvania, about a month after TMI, we found that there were significant differentials among the population between those who left and those who stayed; and there was, of course, a substantial evacuation from the area. For instance, elderly people were unwilling to leave based on information they had that if there was a serious release of radiation, the area could become uninhabitable. They were simply not willing to leave the area and face starting life over in another area. How would your agency, in the event of an ordered evacuation, deal with people who simply refused to leave the area?

O. K. HENDERSON: Only our governor has the emergency powers to order and to compel evacuation. I don't believe that the governor would have used that authority, though I can't speak for the governor. We had many farmers in the area whose entire livelihood was in their dairy herds and who insisted that they would not evacuate the area totally. They would leave the area provided we could get them back in to take care of their stock. This is true, the surveys that I have seen indicate that there were elderly people who were unwilling to leave. We don't have that many Old Order Amish in the area, but there are those a little bit farther out at Peach Bottom who would be more than likely unwilling to leave.

C. STARR: I want to ask this of Colonel Henderson. Considering the actual radiation level exposure, if you were 50 years old or more, the likelihood is that you would die of many, many other things before cancer would ever show up. Do you think you have an obligation to evacuate those people, regardless of whether this is or is not a significant factor in their lifespan?

O. K. HENDERSON: I would say that we have the obligation only if it endangers other life; for example, if we would have to keep forces in the area in order to care for these people in some fashion. Otherwise, I do not believe that it is our obligation or that the governor would necessarily have made that kind of decision.

T. H. MOSS: I wanted to ask one question of you, Colonel Henderson.

You mentioned that there was going to be some meeting of the emergency preparedness community in the next couple of months. Is it under discussion whether there should be practice evacuations on any scale that involves the public as part of the general trend towards more emergency preparedness?

O. K. HENDERSON: I think it's generally accepted, at least in the emergency management community, that the actual practicing of the evacuation itself—the physical movement of the people—doesn't solve anything. The few times that we've tried it in the past, we have struck out. I think that what we need to do is perhaps representative-type evacuations to test out our mechanism. But the physical movement of people from point A to point B in my estimation is not worth the effort.

J. R. JACOBSON (*New Jersey Department of Energy, Newark, N.J.*): I'd like to ask Mr. Starr a question. You make a persuasive case for the safe operation of nuclear generating stations. In view of the extraordinary safety record, why does the industry insist upon Price-Anderson government subsidy of its insurance? Why don't they insure themselves like every other industry in the free enterprise system does?

C. STARR: Let me answer that question on Price-Anderson; it keeps coming up frequently. The industry is not monolithic on Price-Anderson. Price-Anderson was passed—Congressmen Price and Anderson, both of whom are national figures, were involved—by Congress as an encouragement in the early days to get the utilities to go into a new technology with which they had no experience.

Once that was in, the utility industry was getting what amounted to low-cost insurance by a responsible insurer, the U.S. Government, and an insurance that is essentially no-fault insurance too, which is an easy one for any industry to accept. The industry was very loathe to give it up. When it came up for renewal, there was a debate on this, and I'm in the *Congressional Record* as one of the industry persons who said: We don't need it. The industry can self-insure for the rare occasion that it's needed.

And in fact, the Price-Anderson renewal as now in effect disappears over a period of years. The government's fraction of it disappears, and the private sector part builds up. So eventually, Price-Anderson will not be a government-provided protection. The Nuclear Electric Insurance Limited, which is being set up to pick up the other costs, is evidence that the industry could have self-insured. So, in fact, my answer to you is that it is a remnant of some goody that the government had given the industry and the industry was loathe to give it up. But I don't believe the industry needs it, and I think as time goes on, there are going to be more people in the industry who feel that its psychological relevance to the public makes it a cost burden rather than a benefit.

J. S. MILLER: I have just a technical comment to clear up something. I believe Mr. Starr mentioned that some of the heavy radium nuclides would be limited in their fallout to just a short distance from the plant. But the three reports that I've read on reactor safety study indicate that the distribution of these particles could be much farther. One report on the nuclear disaster in the Urals comes up with concentrations of heavy particulates on the level of millicuries per square meter over a distance of about 1,000 square miles.

C. STARR: I haven't read this report. All I could tell you is that the experimental information on plutonium oxide and what happens to it would show a fairly rapid fallout. Maybe Dr. Hull would want to answer that question for me.

A. P. HULL: I can't claim to be completely informed, but I have heard of the articles and looked into it to some extent, because I started out as very much of a skeptic. There's no question in my mind that some radioactive material got abroad in Russia. There have been a number of suppositions, including an article in *Science* not too long ago that suggested it was a nuclear explosion that came down in Russia and deposited its material. I also find that unpersuasive. I've talked about this to Dave Coates, who's been in this business longer than I have and is somewhat more knowledgeable about what the possibilities are. His hypothesis is unpublished at the moment, but I don't think he'll mind my saying this: it seems that the Russians were in a crash program to produce as much material for weapons as they could, and they simply didn't store the waste in any fashion. They put it out somewhere, and it got abroad and ran down into a river and got spread out that way. The numbers in the report you mentioned, all of them, had to do with what's deposited and what's in lakes and rivers. But the numbers become rather preposterous in short order.

LESSONS FOR THE RESOLUTION OF TECHNICAL DISPUTES*

Donald B. Straus

Research Institute
American Arbitration Association
New York, New York 10019

In his opening remarks to the conference, Cochairman Thomas H. Moss asked, "Are we organized to deal with the problems and stresses of a complex and technological society?" This question was posed again as the starting point for discussion of the topic "Lessons for the Resolution of Technological Disputes," and a tentative answer of no was given.

The discussion leader then suggested that the current conference provided a convenient case history for an analysis of this problem. The session concluded with a brief demonstration of one alternative model for the discussion of technical, complex, and controversial issues.

To open the discussion, the following premises were advanced:

• Environmental energy issues require the invention of new solutions. One obstacle to such invention is the difficulty of understanding the huge overload of relevant factors, the interrelationship of these factors in predicting outcomes, and the resulting complexity of the total problem. New tools are needed for managing complexity.

• When facts and scientific opinions about a complex public issue are in dispute, it is easy for strongly committed advocates to promote extreme views as "proven truths." Uncertainty reinforces polarity.

• Under such circumstances, public debate is rooted in prejudices and emotional preferences, not in analysis.

• Ordinary citizens today are insisting that they can understand complex and technical issues sufficiently to have a valid opinion regarding their resolution and therefore that they should participate in policy decisions.

• Present procedures for public participation in such complex decisions are inadequate. Better procedures for involving citizens and better tools for helping to understand complex issues are needed.

• While it is recognized that prejudice, emotion, and politics will never be removed from these debates, even so the scientific community has a special responsibility for improving communication between scientists and nonscientists for the purpose of assisting rational decision making.

---

*Rather than presenting a paper at the conference, the author conducted a discussion on this topic with the audience. This paper describes the process of the discussion and presents the author's perspective of both the discussion and the conference as a whole.

0077-8923/81/0365-0334 $01.75/2 © 1981, NYAS

• The traditional conference lecture format, with questions from the floor, is effective for sharing information among professional colleagues or for teaching students, but it can be a barrier to better understanding or an exchange of views about complex technical issues that are controversial. Before they will become constructive, participants in decision making must also participate in designing the process for discussion, in validating the facts and assumptions that will be used, and in reaching an understanding of the essential scientific factors that will influence the impacts of the decision.

The above premises were presented to the audience as a springboard for an impressionistic analysis of the format of the conference on the Three Mile Island (TMI) accident as a process for decreasing controversy and moving towards consensus. It was recognized that the main purpose of the conference was to review the experience of TMI and to extract from the experience lessons for the future. For such an exchange among members of the scientific community and the electric power industry, this format—and the design and execution of the meetings—admirably suited this purpose. But where there was controversy, the format provided little opportunity for reducing it.

On a subject as polarized as atomic energy, to aim for total agreement is unrealistic. Under favorable conditions, however, it can be expected that progress can be made towards the following less ambitious goals:

• A narrowing of extreme positions.
• Better understanding of the underlying reasons for different values and opinions.
• An understanding of the consequences of the "bad" and "worst" outcomes of various solutions.
• A consensus on basic facts upon which predictions can be based; barring such consensus, on understanding the specific factual issues in dispute and some narrowing of the differences.

With this background, attention was then turned to the recent experience of the conference. For those papers that were primarily descriptive (e.g., John R. Lamarsh on the design and operation of light water reactors), there appeared to be a good exchange of information and the question-and-answer periods were largely noncontroversial and served to clarify difficult or misunderstood points.

But in the more controversial sessions, the podium-to-floor dialogues, restricted to one (or at most two) exchanges between speaker and questioner, usually produced more heated emotions than enlightened minds. Several examples will serve to illustrate this effect.

Most of the talks were full of facts—many quite difficult to comprehend under the best of circumstances, and especially so when presented only once and orally. Arthur Upton, in his talk on public health implications, cited some statistics from a survey of infant mortality in two periods: six months before and six months after the accident. During his presentation, some of the audience became concerned that Harrisburg seemed included in one set of data and excluded from the other. In the brief exchange on this comparatively insignificant discrepancy in a long and authoritative report, the confusion was never remedied. But because health implications are at the heart of the controversy, the exchange distorted the importance of this single comparison, failed to resolve a factual dispute that was certainly susceptible to a clear-cut resolution, and served primarily to escalate a feeling of mutual distrust.

On another occasion, during the discussion on plant and site decontamination, one questioner asked about evidence—reported in the newspaper only a few days before the conference—of leakage of radioactive substances from the plant that had been found in wells dug for this purpose. The answer given was that the detected radiation levels were only 10% of those allowed by the Environmental Protection Agency. "Maybe so," was the reply—one of the few follow-up questions permitted by the format, "but this morning another speaker said there has been no release of radioactivity since the accident. This is an example of why the public is confused and the credibility of both the industry and the government is low." Again, this is a point that could and certainly should have been clarified—at least the actual facts, if not the implications. But there was no opportunity to pursue this further, with the result that the polarization of the pro- and antinuclear participants was frozen tighter rather than thawed.

There was another incident. The venting of the remaining krypton in the plant was of course very much in the minds of all the participants. Representatives from both the government agencies and the utility were convinced that the only safe and economical alternative was to vent. Each of the other alternatives—liquefaction by freezing, passing the krypton through charcoal, or storing it in containers under pressure—had, in their unified view, elements of danger that surpassed a controlled release. On the other hand, some of the environmental representatives (and apparently some citizens of the surrounding area) were skeptical and unconvinced. This was, and remained, a highly charged issue. The representative of the Nuclear Regulatory Commisson, while admitting the political reality of "citizen participation," wondered out loud if it was responsible and moral to "permit the public to endanger itself." The vocal environmentalists wondered if the economic considerations were not predominant in the alleged safety of venting. There was no room in the conference format to pursue this matter; but if not under the auspices of the New York Academy of Sciences, then where?

One final example is that on the first morning of the conference, the mid-Atlantic representative of the Friends of the Earth, Lorna Salzman, circulated a pamphlet headlined: "The New York Academy of Sciences Three Mile Island Conference: A Vehicle for Promoting Nuclear Energy and Suppressing Dissent?"

The text of this pamphlet alleged that the conference participants were pronuclear, that the public was excluded, and that minority or dissenting views would not be presented. Then, in four pages of closely printed text, the Friends of the Earth version of the TMI accident was presented. This incident led to some brief, but heated, dialogue between several of the conference sponsors and Ms. Salzman on the first morning of the conference, and eventually to the addition of Judith Johnsrud of the Environmental Coalition on Nuclear Power on the Wednesday panel on institutional reactions. Her presence on the program, and the generous amount of time permitted for her presentation and response to questions, diffused some of the pent-up feelings of exclusion held by the environmentalists. But this is an incident that need not have happened had there been more advance consultation between the conference sponsors and environmentalist organizations. The invitation to Johnsrud was wise and helpful, but too late to avoid a significant and unnecessary psychological barrier to communication.

These examples are typical but far from exhaustive. There were any number of disputed issues raised during the conference, but few opportunities for attempted resolution. Mutual perceptions of the intransigence of the other side, of the impossibility of getting a fair hearing of one's own deep convictions, and of mirror-image perceptions of each other's ulterior motives remained unchanged.

After having pointed out some shortcomings of the traditional conference format for resolution of disputes, an alternate format was advanced with the help of an outline cast on the screen with an overhead projector. It was explained that this method of presentation, together with the recording of the ensuing discussion also on the screen (as described below), was a modification of a meeting technique called "facilitation." By keeping participants' attention focused on what was being said rather than on each other, it would be easier to keep the discussion focused on the agenda rather than on personalities or rather than diverting into digressions. Of course, such participant self-discipline could only be accomplished with the voluntary agreement of the participants, and the method about to be described was designed to obtain and retain such agreement.

This was the first slide displayed:

PLANNING FOR PARTICIPATION
1. Identify participants and interest groups
2. Select a planning committee
3. Planning committee to:
   - Define goals
   - Define issues
   - Decide upon format of meeting or negotiation
   - Select the active participants

It was explained, while this slide was on view, that once a problem area is perceived and a decision for or against action is required, the concerned participants must be identified. These must include the industries involved, the pertinent federal and local government agencies, and citizen interest groups. From this group a planning committee should be selected, recognizing that while large committees are difficult to manage, this difficulty is more than worth contending with to prevent the greater danger of omitting a legitimate participant who later could interfere with the process or impede implementation of a decision.

The planning committee first should seek to define the goals of the project: Is it merely to exchange information, to resolve controversy or misunderstandings, or to enter into negotiations to develop an agreed-upon course of action?

The committee should also seek consensus on a definition of the issues in dispute and the process or processes (there are many available) to be used for the conduct of meetings or negotiation (see Bibliography).

A final task of the planning committee is to decide who should be invited as active participants from the groups represented. Have some groups been omitted? Are some present who really are not sufficiently concerned to continue with the project? How many individuals should be involved, and when? Should there be a plenary group of the whole, with subcommittees to address specific parts of the problem?

Once the problem has been defined, the participants identified, and a process sequence selected, the next step is to prepare as thoroughly as possible for the conference itself (or, if so decided, for the negotiations). A small staff should be appointed for this task, preferably led by an individual with skills in process management. Among the tasks to be performed are these:

• Seek understanding of and consensus on the basic facts and data to be used in the discussions. Drawing from the experience of the TMI conference, this would be the time to resolve all remaining issues of who the speakers should be and to resolve data disputes, such as the infant mortality statistics or whether there was more or less radiation fallout beyond the 10-mile limit. The extent and nature of the radiation leakage around the plant should be examined and, where possible, resolved. This is a process now known as data mediation. Where total agreement cannot be reached, at least a full understanding of the precise nature of the disagreement can be defined and understood. It is also useful to determine, and so label, disagreements over judgments about unknown or unproven assumptions. An "agreement" that a certain fact or assumption is still subject to reasonable scientific doubt is better than to have opposite and confusing statements asserted as proven truths on the basis of different, but respectable, sources.

• It is also useful to prepare, at this time, written statements describing some of the issues. These should not try to "make an agreement" or resolve a dispute, but rather should be a single text upon which reasonable discussion could begin. Where a single text is not possible, separate texts can be prepared expressing the different versions of the issue (again without argumentation). These texts should be circulated in advance to the participants. A desirable goal is to have as few papers as possible, and those as short as possible, delivered orally at the conference.

After a data bank has been prepared and issue papers have been written, these should be circulated to the participants for comment and criticism. At this time, the formation of subcommittees may be advisable in order to modify, revise, and correct the conference materials. This is the ideal time for data mediation. These are some of the prearrangements that should be accomplished before a conference begins.

To most people, this elaborate preparation may seem unnecessary and expensive. But experience with this process indicates that in the long run, time spent at the front end of a conference, a negotiation, or other formats for citizen participation is both effective and cost efficient if full understanding of the issues and some movement towards consensus are the goals. In fact, those who have participated in this process often remark that more was accomplished in the preparatory stages than in the event itself. The process is the prophylaxis.

But even after all this preparation, there are some changes in the traditional conference format that should be made. The conference leader should always seek a logical progression towards decision making. This progression can be described in different ways and in varying detail. But in general, it goes through these stages:

Definition of the problem;
Definition of the issues in controversy;
Review and acceptance of data to be used;
Analysis of problems and issues;
Development of different solutions;
Examination of alternatives;
Decison.

There is always a natural tendency to skip all of the prior steps and begin with advocacy for a preferred decision or solution. When pointed out, it is obvious that this is not an orderly process and will surely produce more argumentation than rational discussion. But to keep a large group from skipping all or most of the above steps requires a far more structured meeting format than is usually the case. Just how to do this is now an active field of research (see Bibliography), and the process manager's skill must be used to fit the proper format to the occasion. Nor need one format be used throughout the conference.

The one format demonstrated in this discussion employed the use of a "group memory." Group memory is written by a recorder and sets down in brief notes what has been said by the participants: facts, opinions, value judgments, disagreements, and agreements. The aim is to get the group to understand from the start the goals of the meeting, the process to be used, and the agenda to be followed. The group memory will help keep the meeting on track, will assure everyone that interventions are being recorded accurately (they can make changes in the recorder's notes if what is written does not reflect their intent), and will help the group "police" itself. In small sessions, the group memory is kept on large sheets of paper taped to the walls. In this discussion, because the group was so large, the recorder kept the group memory on acetate sheets displayed with an overhead projector.

Since time was limited, only two items were discussed during the demonstration:

1. Would the process just described be worth trying?
2. If so, what might be the goals of another TMI conference?

In leading the demonstration, the facilitator stepped down from the podium so that he was on a level with the audience. A dialogue on each intervention was maintained until the intervenor was comfortable that the point being made was clearly understood and accurately recorded. Discussion on the first issue resulted in a consensus that it would be interesting to try the proposed format in another conference, but first the following "caveats" and objections were recorded:

- Could this process handle political disputes?
- Before using the process, there must be some attempt made to assure the good faith participation of those involved.
- Perhaps the size of the group would be a limiting factor.
- One intervenor objected to any collaborative process while atomic plants were operating.

In "real life," each of the above objections would be discussed at greater length than in this session. Presumably the first three could be answered satisfactorily. The process has been used to handle "value" and "political" questions, and these are techniques for handling large groups. "Good faith" requires more definition, and under most conditions, this important concern can be sufficiently resolved to let the process begin. With success, "good faith" increases; and even without full success, greater understanding usually results. The last question is more difficult. If an "unconditional demand" is made that plants be shut down before "good faith" participation will be given, that individual or organization may simply choose not to participate. Depending upon the strength and importance of the organization, the conference organizing committee has two options: (1) continue efforts to develop a format that will satisfy the objections of the dissident organization, or (2) proceed without it.

Discussion on the second topic, goals, resulted in a list of suggestions. These were:

1. Plot energy use for 1970;
2. Provide an opportunity for all to participate;
3. Exchange information;
4. Develop a list of concerned citizen groups and their positions;
5. Assess the risks of nuclear energy from various points of view in face-to-face discussions;

6. Assess the impact of various proposals on various societal groups;
7. Try to separate facts from values;
8. Determine the credibility of participants;
9. Develop a method of implementing recommendations;
10. Seek a definition of "safe";
11. Compile a data basis and critical comments about it before the conference starts.

After this list was developed and recorded, the time for this session ended. Again, under "real" conditions, the facilitator would next have used various techniques to get the group to narrow the list and agree upon a set of practical goals. As tedious and troublesome as this may sound, the successful completion of this task will greatly expedite the remaining tasks of conference organization.

The discussion began by reviewing some shortcomings in the traditional conference format for seeking understanding of complex technical issues. Some alternative formats and techniques were then described, and one was demonstrated.

The principal lessons are these:

1. New formats are needed;
2. There are a number now available, with others still being developed;
3. More innovation in conferencing and negotiating techniques should be attempted.

BIBLIOGRAPHY

1. DOYLE, M. & D. STRAUS. 1976. How to Make Meetings Work: The New Interaction Method. Playboy Press. New York, N.Y.
2. COOVER, V., E. DEACON, C. ESSEX & C. MOORE. 1977. Resource Manual for a Living Revolution. New Society Press. New York, N.Y.
3. MILLER, G. A. 1967. The Psychology of Communication. Basic Books, Inc. Scranton, Penn.
4. Interaction Associates. 1972. Strategy Notebook. San Francisco, Calif.
5. STRAUS, D. B. 1978. Mediating environmental disputes. Arbitration J. (December): 5.
6. STRAUS, D. B. 1979. Managing complexity—a new look at environmental mediation. Environ. Sci. Technol. **18**(6).
7. RESOLVE. Environmental Consensus. Palo Alto, Calif.
8. BALDWIN, P. 1978. Environmental Mediation—An Effective Alternative? RESOLVE. Palo Alto, Calif.
9. Laboratory of Architecture and Planning. 1978. EIA Review. Massachusetts Institute of Technology. Cambridge, Mass. (The October issue contains articles on environmental mediation.)
10. Urban Land Institute. 1977. Environmental Comment. Washington, D.C. (The May issue contains articles on environmental mediation.)
11. STRAUS, D. B. & P. B. CLARK. 1980. Bigger problems need better tools. Environmental Professional **2**(1).
12. BUSTERUD, J. 1980. Mediation: state of the art. Environmental Professional **2**(1).

Index of Contributors

(Italicized page numbers refer to comments made in discussion.)